The Performance of Photovoltaic (PV) Systems

Related titles

Concentrating Solar Power Technology
(ISBN 978-1-84569-769-3)

Functional Materials for Sustainable Energy Applications
(ISBN 978-0-85709-059-1)

Sørensen and Sørensen, *Wind Energy Systems: Optimising Design and Construction for Safe and Reliable Operation*
(ISBN 978-1-84569-580-4)

Woodhead Publishing Series in Energy:
Number 105

The Performance of Photovoltaic (PV) Systems

Modelling, Measurement and Assessment

Edited by

Nicola Pearsall

ELSEVIER

AMSTERDAM • BOSTON • CAMBRIDGE • HEIDELBERG
LONDON • NEW YORK • OXFORD • PARIS • SAN DIEGO
SAN FRANCISCO • SINGAPORE • SYDNEY • TOKYO
Woodhead Publishing is an imprint of Elsevier

WP
WOODHEAD
PUBLISHING

Woodhead Publishing is an imprint of Elsevier
The Officers' Mess Business Centre, Royston Road, Duxford, CB22 4QH, United Kingdom
50 Hampshire Street, 5th Floor, Cambridge, MA 02139, United States
The Boulevard, Langford Lane, Kidlington, OX5 1GB, United Kingdom

Notices
Knowledge and best practice in this field are constantly changing. As new research and experience broaden our understanding, changes in research methods, professional practices, or medical treatment may become necessary.

Practitioners and researchers must always rely on their own experience and knowledge in evaluating and using any information, methods, compounds, or experiments described herein. In using such information or methods they should be mindful of their own safety and the safety of others, including parties for whom they have a professional responsibility.

To the fullest extent of the law, neither the Publisher nor the authors, contributors, or editors, assume any liability for any injury and/or damage to persons or property as a matter of products liability, negligence or otherwise, or from any use or operation of any methods, products, instructions, or ideas contained in the material herein.

Library of Congress Cataloging-in-Publication Data
A catalog record for this book is available from the Library of Congress

British Library Cataloguing-in-Publication Data
A catalogue record for this book is available from the British Library

ISBN: 978-1-78242-336-2 (print)
ISBN: 978-1-78242-354-6 (online)

Publisher: Joe Hayton
Acquisition Editor: Cari Owen
Editorial Project Manager: Alex White
Production Project Manager: Poulouse Joseph
Cover Designer: Victoria Pearson

Typeset by SPi Global, India

Contents

Part Three Maximising photovoltaic (PV) system performance 233

Contributors

F. Baumgartner Zurich University of Applied Sciences ZHAW, Winterthur, Switzerland

A.W. Bett Fraunhofer Institute for Solar Energy Systems ISE, Freiburg, Germany

T.R. Betts Loughborough University, Loughborough, United Kingdom

I.R. Cole Loughborough University, Loughborough, United Kingdom

D. Dirnberger Fraunhofer Institute for Solar Energy Systems, Freiburg, Germany

C. Ferrara Fraunhofer Institute for Solar Energy Systems, Freiburg, Germany

V. Fthenakis Columbia University, New York; Photovoltaic Environmental Research Center, Upton, NY, United States

T. Gerstmaier Soitec Solar GmbH, Freiburg, Germany

B. Goss Loughborough University, Loughborough, United Kingdom

R. Gottschalg Loughborough University, Loughborough, United Kingdom

S. Goy 3E, Brussels, Belgium

D. Jordan National Renewable Energy Laboratory (NREL), Golden, CO, United States

E. Koubli Loughborough University, Loughborough, United Kingdom

S. Kurtz National Renewable Energy Laboratory (NREL), Golden, CO, United States

D. Palmer Loughborough University, Loughborough, United Kingdom

N.M. Pearsall Northumbria University, Newcastle upon Tyne, United Kingdom

M. Raugei Oxford Brookes University, Oxford, United Kingdom

V. Salas Charles III University of Madrid, Madrid, Spain

W. Sprenger Fraunhofer Institute for Solar Energy Systems, Freiburg, Germany

M. Steiner Fraunhofer Institute for Solar Energy Systems ISE, Freiburg, Germany

H.R. Wilson Fraunhofer Institute for Solar Energy Systems, Freiburg, Germany

A. Woyte 3E, Brussels, Belgium

Preface

Photovoltaic (PV) systems, converting incident solar energy directly into electricity, are an elegant and versatile way of harnessing renewable energy and have a major part to play in the transition in our energy systems required by the issues of climate change, reducing fossil fuel resources and energy security. Over the last decade, the installed capacity has grown rapidly and, at the time of writing, stands at over 300 GW worldwide. This is still a small percentage of the overall installed capacity of electricity generation, but PV is becoming a mainstream technology as the costs continue to decline. However, in order to ensure the maximum contribution from PV, we must not only install the systems but operate them in an efficient manner across their lifetime. This requires a consistent approach to determining the performance and quality of the components and the systems, a clear understanding of the loss mechanisms and how to minimize them and the long-term measurement and analysis of field performance to ensure maximum output.

Most previous books on photovoltaic technology have concentrated on the theory and design of the photovoltaic cells or systems. In contrast, this book focusses on how to measure and assess the performance of the components of the PV systems and on how to define and address the performance of the PV system in the field. The book is split into three main parts, following an introductory chapter that aims to provide an overview of PV systems and their performance as both a primer for the inexperienced reader and a reminder for those already familiar with the technology. Part One deals with the performance of the PV module, which performs the actual conversion process. Topics covered by the chapters in this part include measurement of the module rating, prediction of energy yield and changes to module performance in operation. Part Two considers the performance of the PV system, now including chapters on balance of system components, prediction and monitoring of PV system output and the environmental assessment of the systems. Finally, Part Three considers individual system types, including grid-connected and stand-alone system configurations, concentrator systems and specific performance aspects relating to building integration. This part addresses the performance issues relating to these different applications from the general viewpoint of maximizing performance in each case.

The individual chapters have been authored by experts in their field, representing several of the major research organizations in photovoltaics in Europe and the United States of America. In each case, the authors have drawn on their research expertise to provide a state-of-the-art treatment of their topic. Nevertheless, the field of photovoltaic technology is advancing rapidly, so each chapter also includes some consideration of the likely future developments in that subject area.

Because of the wide range of applications of photovoltaics and the rapid progress being made, it is impossible to cover all the aspects of PV performance in a single volume of reasonable size. However, the rationale behind this book is to focus clearly on component and system performance so as to provide a detailed introduction to the issues and to encourage the assessment of performance at all stages of the development, design and operation of PV systems.

Nicola Pearsall
July 2016

Woodhead Publishing Series in Energy

1. **Generating power at high efficiency: Combined cycle technology for sustainable energy production**
   Eric Jeffs
2. **Advanced separation techniques for nuclear fuel reprocessing and radioactive waste treatment**
   Edited by Kenneth L. Nash and Gregg J. Lumetta
3. **Bioalcohol production: Biochemical conversion of lignocellulosic biomass**
   Edited by Keith W. Waldron
4. **Understanding and mitigating ageing in nuclear power plants: Materials and operational aspects of plant life management (PLiM)**
   Edited by Philip G. Tipping
5. **Advanced power plant materials, design and technology**
   Edited by Dermot Roddy
6. **Stand-alone and hybrid wind energy systems: Technology, energy storage and applications**
   Edited by John K. Kaldellis
7. **Biodiesel science and technology: From soil to oil**
   Jan C. J. Bart, Natale Palmeri and Stefano Cavallaro
8. **Developments and innovation in carbon dioxide (CO_2) capture and storage technology Volume 1: Carbon dioxide (CO_2) capture, transport and industrial applications**
   Edited by M. Mercedes Maroto-Valer
9. **Geological repository systems for safe disposal of spent nuclear fuels and radioactive waste**
   Edited by Joonhong Ahn and Michael J. Apted
10. **Wind energy systems: Optimising design and construction for safe and reliable operation**
    Edited by John D. Sørensen and Jens N. Sørensen
11. **Solid oxide fuel cell technology: Principles, performance and operations**
    Kevin Huang and John Bannister Goodenough
12. **Handbook of advanced radioactive waste conditioning technologies**
    Edited by Michael I. Ojovan
13. **Membranes for clean and renewable power applications**
    Edited by Annarosa Gugliuzza and Angelo Basile
14. **Materials for energy efficiency and thermal comfort in buildings**
    Edited by Matthew R. Hall
15. **Handbook of biofuels production: Processes and technologies**
    Edited by Rafael Luque, Juan Campelo and James Clark

16 **Developments and innovation in carbon dioxide (CO_2) capture and storage technology Volume 2: Carbon dioxide (CO_2) storage and utilisation**
Edited by M. Mercedes Maroto-Valer
17 **Oxy-fuel combustion for power generation and carbon dioxide (CO_2) capture**
Edited by Ligang Zheng
18 **Small and micro combined heat and power (CHP) systems: Advanced design, performance, materials and applications**
Edited by Robert Beith
19 **Advances in clean hydrocarbon fuel processing: Science and technology**
Edited by M. Rashid Khan
20 **Modern gas turbine systems: High efficiency, low emission, fuel flexible power generation**
Edited by Peter Jansohn
21 **Concentrating solar power technology: Principles, developments and applications**
Edited by Keith Lovegrove and Wes Stein
22 **Nuclear corrosion science and engineering**
Edited by Damien Féron
23 **Power plant life management and performance improvement**
Edited by John E. Oakey
24 **Electrical drives for direct drive renewable energy systems**
Edited by Markus Mueller and Henk Polinder
25 **Advanced membrane science and technology for sustainable energy and environmental applications**
Edited by Angelo Basile and Suzana Pereira Nunes
26 **Irradiation embrittlement of reactor pressure vessels (RPVs) in nuclear power plants**
Edited by Naoki Soneda
27 **High temperature superconductors (HTS) for energy applications**
Edited by Ziad Melhem
28 **Infrastructure and methodologies for the justification of nuclear power programmes**
Edited by Agustín Alonso
29 **Waste to energy conversion technology**
Edited by Naomi B. Klinghoffer and Marco J. Castaldi
30 **Polymer electrolyte membrane and direct methanol fuel cell technology Volume 1: Fundamentals and performance of low temperature fuel cells**
Edited by Christoph Hartnig and Christina Roth
31 **Polymer electrolyte membrane and direct methanol fuel cell technology Volume 2: *In situ* characterization techniques for low temperature fuel cells**
Edited by Christoph Hartnig and Christina Roth
32 **Combined cycle systems for near-zero emission power generation**
Edited by Ashok D. Rao
33 **Modern earth buildings: Materials, engineering, construction and applications**
Edited by Matthew R. Hall, Rick Lindsay and Meror Krayenhoff
34 **Metropolitan sustainability: Understanding and improving the urban environment**
Edited by Frank Zeman
35 **Functional materials for sustainable energy applications**
Edited by John A. Kilner, Stephen J. Skinner, Stuart J. C. Irvine and Peter P. Edwards
36 **Nuclear decommissioning: Planning, execution and international experience**
Edited by Michele Laraia
37 **Nuclear fuel cycle science and engineering**
Edited by Ian Crossland
38 **Electricity transmission, distribution and storage systems**
Edited by Ziad Melhem

39 **Advances in biodiesel production: Processes and technologies**
Edited by Rafael Luque and Juan A. Melero
40 **Biomass combustion science, technology and engineering**
Edited by Lasse Rosendahl
41 **Ultra-supercritical coal power plants: Materials, technologies and optimisation**
Edited by Dongke Zhang
42 **Radionuclide behaviour in the natural environment: Science, implications and lessons for the nuclear industry**
Edited by Christophe Poinssot and Horst Geckeis
43 **Calcium and chemical looping technology for power generation and carbon dioxide (CO_2) capture: Solid oxygen- and CO_2-carriers**
Paul Fennell and E. J. Anthony
44 **Materials' ageing and degradation in light water reactors: Mechanisms, and management**
Edited by K. L. Murty
45 **Structural alloys for power plants: Operational challenges and high-temperature materials**
Edited by Amir Shirzadi and Susan Jackson
46 **Biolubricants: Science and technology**
Jan C. J. Bart, Emanuele Gucciardi and Stefano Cavallaro
47 **Advances in wind turbine blade design and materials**
Edited by Povl Brøndsted and Rogier P. L. Nijssen
48 **Radioactive waste management and contaminated site clean-up: Processes, technologies and international experience**
Edited by William E. Lee, Michael I. Ojovan, Carol M. Jantzen
49 **Probabilistic safety assessment for optimum nuclear power plant life management (PLiM): Theory and application of reliability analysis methods for major power plant components**
Gennadij V. Arkadov, Alexander F. Getman and Andrei N. Rodionov
50 **The coal handbook: Towards cleaner production Volume 1: Coal production**
Edited by Dave Osborne
51 **The coal handbook: Towards cleaner production Volume 2: Coal utilisation**
Edited by Dave Osborne
52 **The biogas handbook: Science, production and applications**
Edited by Arthur Wellinger, Jerry Murphy and David Baxter
53 **Advances in biorefineries: Biomass and waste supply chain exploitation**
Edited by Keith Waldron
54 **Geological storage of carbon dioxide (CO_2): Geoscience, technologies, environmental aspects and legal frameworks**
Edited by Jon Gluyas and Simon Mathias
55 **Handbook of membrane reactors Volume 1: Fundamental materials science, design and optimisation**
Edited by Angelo Basile
56 **Handbook of membrane reactors Volume 2: Reactor types and industrial applications**
Edited by Angelo Basile
57 **Alternative fuels and advanced vehicle technologies for improved environmental performance: Towards zero carbon transportation**
Edited by Richard Folkson
58 **Handbook of microalgal bioprocess engineering**
Christopher Lan and Bei Wang

59 **Fluidized bed technologies for near-zero emission combustion and gasification**
Edited by Fabrizio Scala
60 **Managing nuclear projects: A comprehensive management resource**
Edited by Jas Devgun
61 **Handbook of Process Integration (PI): Minimisation of energy and water use, waste and emissions**
Edited by Jiří J. Klemeš
62 **Coal power plant materials and life assessment**
Edited by Ahmed Shibli
63 **Advances in hydrogen production, storage and distribution**
Edited by Ahmed Basile and Adolfo Iulianelli
64 **Handbook of small modular nuclear reactors**
Edited by Mario D. Carelli and Dan T. Ingersoll
65 **Superconductors in the power grid: Materials and applications**
Edited by Christopher Rey
66 **Advances in thermal energy storage systems: Methods and applications**
Edited by Luisa F. Cabeza
67 **Advances in batteries for medium and large-scale energy storage**
Edited by Chris Menictas, Maria Skyllas-Kazacos and Tuti Mariana Lim
68 **Palladium membrane technology for hydrogen production, carbon capture and other applications**
Edited by Aggelos Doukelis, Kyriakos Panopoulos, Antonios Koumanakos and Emmanouil Kakaras
69 **Gasification for synthetic fuel production: Fundamentals, processes and applications**
Edited by Rafael Luque and James G. Speight
70 **Renewable heating and cooling: Technologies and applications**
Edited by Gerhard Stryi-Hipp
71 **Environmental remediation and restoration of contaminated nuclear and NORM sites**
Edited by Leo van Velzen
72 **Eco-friendly innovation in electricity networks**
Edited by Jean-Luc Bessede
73 **The 2011 Fukushima nuclear power plant accident: How and why it happened**
Yotaro Hatamura, Seiji Abe, Masao Fuchigami and Naoto Kasahara. Translated by Kenji Iino
74 **Lignocellulose biorefinery engineering: Principles and applications**
Hongzhang Chen
75 **Advances in membrane technologies for water treatment: Materials, processes and applications**
Edited by Angelo Basile, Alfredo Cassano and Navin Rastogi
76 **Membrane reactors for energy applications and basic chemical production**
Edited by Angelo Basile, Luisa Di Paola, Faisal Hai and Vincenzo Piemonte
77 **Pervaporation, vapour permeation and membrane distillation: Principles and applications**
Edited by Angelo Basile, Alberto Figoli and Mohamed Khayet
78 **Safe and secure transport and storage of radioactive materials**
Edited by Ken Sorenson
79 **Reprocessing and recycling of spent nuclear fuel**
Edited by Robin Taylor
80 **Advances in battery technologies for electric vehicles**
Edited by Bruno Scrosati, Jürgen Garche and Werner Tillmetz
81 **Rechargeable lithium batteries: From fundamentals to applications**
Edited by Alejandro A. Franco

82 **Calcium and chemical looping technology for power generation and carbon dioxide (CO_2) capture**
Edited by Paul Fennell and Ben Anthony
83 **Compendium of Hydrogen Energy Volume 1: Hydrogen Production and Purificiation**
Edited by Velu Subramani, Angelo Basile and T. Nejat Veziroglu
84 **Compendium of Hydrogen Energy Volume 2: Hydrogen Storage, Transmission, Transportation and Infrastructure**
Edited by Ram Gupta, Angelo Basile and T. Nejat Veziroglu
85 **Compendium of Hydrogen Energy Volume 3: Hydrogen Energy Conversion**
Edited by Frano Barbir, Angelo Basile and T. Nejat Veziroglu
86 **Compendium of Hydrogen Energy Volume 4: Hydrogen Use, Safety and the Hydrogen Economy**
Edited by Michael Ball, Angelo Basile and T. Nejat Veziroglu
87 **Advanced district heating and cooling (DHC) systems**
Edited by Robin Wiltshire
88 **Microbial Electrochemical and Fuel Cells: Fundamentals and Applications**
Edited by Keith Scott and Eileen Hao Yu
89 **Renewable Heating and Cooling: Technologies and Applications**
Edited by Gerhard Stryi-Hipp
90 **Small Modular Reactors: Nuclear Power Fad or Future?**
Edited by Daniel T. Ingersoll
91 **Fuel Flexible Energy Generation: Solid, Liquid and Gaseous Fuels**
Edited by John Oakey
92 **Offshore Wind Farms: Technologies, Design and Operation**
Edited by Chong Ng & Li Ran
93 **Uranium for Nuclear Power: Resources, Mining and Transformation to Fuel**
Edited by Ian Hore-Lacy
94 **Biomass Supply Chains for Bioenergy and Biorefining**
Edited by Jens Bo Holm-Nielsen and Ehiaze Augustine Ehimen
95 **Sustainable Energy from Salinity Gradients**
Edited by Andrea Cipollina and Giorgio Micale
96 **Membrane Technologies for Biorefining**
Edited by Alberto Figoli, Alfredo Cassano and Angelo Basile
97 **Geothermal Power Generation: Developments and Innovation**
Edited by Ronald DiPippo
98 **Handbook of Biofuels' Production: Processes and Technologies (Second Edition)**
Edited by Rafael Luque, Carol Sze Ki Lin, Karen Wilson and James Clark
99 **Magnetic Fusion Energy: From Experiments to Power Plants**
Edited by George H. Neilson
100 **Advances in Ground-Source Heat Pump Systems**
Edited by Simon Rees
101 **Absorption-Based Post-Combustion Capture of Carbon Dioxide**
Edited by Paul Feron
102 **Advances in Solar Heating and Cooling Systems**
Edited by Ruzhu Wang and Tianshu Ge
103 **Handbook of Generation IV Nuclear Power Reactors**
Edited by Igor Pioro
104 **Materials for Ultra-Supercritical and Advanced Ultra-Supercritical Power Plants**
Edited by Augusto Di Gianfrancesco
105 **The Performance of Photovoltaic (PV) Systems: Modelling, Measurement and Assessment**
Edited by Nicola Pearsall

106 **Structural Materials for Generation IV Nuclear Reactors**
Edited by Pascal Yvon

107 **Organic Rankine Cycle (ORC) Power Systems: Technologies and Applications**
Edited by Ennio Macchi and Marco Astolfi

Introduction to photovoltaic system performance

1

N.M. Pearsall
Northumbria University, Newcastle upon Tyne, United Kingdom

1.1 Introduction

Photovoltaic (PV) systems, which convert solar irradiation directly to electricity, can be used for a wide range of applications, from small systems powering loads such as sensors or domestic lighting through to large systems feeding electrical power directly into the electricity grid. Despite this variety of uses, the basic system configuration is rather simple. The heart of the system is the PV array itself, a collection of PV modules totalling the required power capacity of the system. The balance of system (BOS) components then include equipment for power conditioning and energy storage, mounting and support of the array, measurement of system performance and safety assurance.

The performance of the system, which mainly concerns the energy output, both in quantity and in timing, depends on the operating conditions and the detailed configuration of the system. In turn, the operating conditions depend on the location of the system, which governs the solar irradiation received, the ambient temperature and other climate-related aspects that influence the system performance. How the system performs determines its technical and economic feasibility and whether it is the best solution in terms of a source of electricity for any given application.

This chapter provides an introduction to the configuration of PV systems, what influences their performance and how that performance is expressed. It is intended for readers who are unfamiliar with PV systems, but can also be used as a summary chapter for those who want to refresh their knowledge about the systems. It provides a foundation for the later chapters in this book, which consider various aspects of system performance in greater detail.

1.2 The PV system

1.2.1 System categories

PV systems can be classified into two main categories, depending on whether or not they are connected to the local electricity network (or grid). Grid-connected systems are generally designed to produce their maximum energy output at all times and to contribute to meeting electrical loads, either close to the system (generally termed *local loads*) or via the electricity network. Stand-alone PV systems operate

The Performance of Photovoltaic (PV) Systems. http://dx.doi.org/10.1016/B978-1-78242-336-2.00001-X

independently from the grid supply and are often located where there is no electricity grid or the grid is difficult to access. In this case, the system is designed to supply the required amount of electricity for the specified load, but does not always necessarily provide the maximum output from the system. This is discussed further in Section 1.3, in terms of expressing these differences in performance mode.

Grid-connected PV systems accounted for around 99% of the cumulative capacity installed worldwide at the end of 2014, according to the International Energy Agency Photovoltaic Power Systems Programme (IEA PVPS) [1], and the market in 2015 is expected to have been similarly dominated by this system category. The concept of the grid-connected PV system is simple and is shown schematically in Fig. 1.1. The PV array is connected to an inverter, which converts the DC output of the array into AC output matching the voltage and frequency of the local grid supply. The system can be connected in parallel with the grid and used to meet local loads. In this case, the output from the PV system will be fed to the load and any shortfall will be supplied from the grid. If there is an excess of generation from the PV system, this excess will be exported to the grid. Alternatively, if there are no significant local loads to be met, the entire output of the PV system may be fed into the grid. It should be noted that the system shown in the figure does not include any facility for the storage of electricity and this is the usual configuration of most current grid-connected systems. However, as the penetration of PV systems on the grid increases and, in some cases, for economic reasons related to the value of local use of electricity, there is a growing interest in storage, either local or central, for grid-connected systems and it can be expected that this will become more common in the future.

The grid-connected system is often classified into distributed and centralized systems, but there are different definitions in regard to how this classification is made. The definition can be made in regard to size (eg, small systems less than, say, 100 kW in capacity would be classed as distributed, whilst larger systems would be considered as centralized), configuration (eg, those meeting local loads are distributed, those only feeding into the grid are centralized) or grid connection point (eg, systems connected to the low voltage distribution feeder are classed as distributed, those connected at a higher voltage are centralized). The IEA PVPS defines distributed (or decentralized) systems in terms of their function being to meet the needs of a specified customer or group of customers, whereas centralized systems feed into the general grid supply. Their data show that distributed systems represented around 56% of the cumulative installed capacity at the end of 2014 [1]. Building-related systems, where the PV array

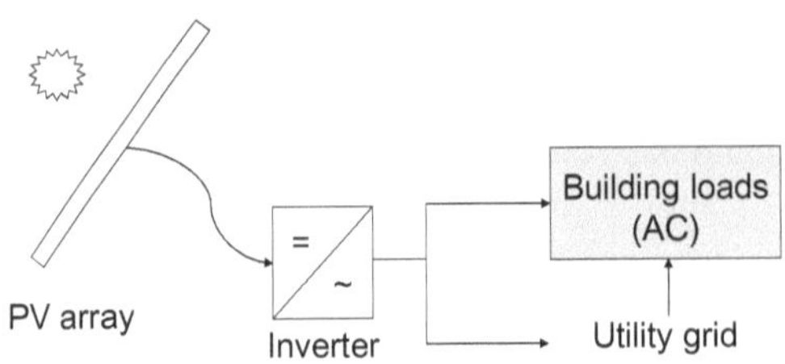

Fig. 1.1 Schematic diagram of a grid-connected PV system providing electricity to local loads.

is integrated into or attached to the roof or façade of a building to meet loads within that building, are typical examples of a distributed system.

The stand-alone system operates independently from the grid and provides the only power source to meet a particular specified load, for example, telecommunications mast, water pumping, medical refrigeration and many more. The most common configuration of the system comprises the PV array, battery storage, a charge controller (to control the charging and discharging of the batteries and to provide maximum power point (MPP) tracking) and the specified load. This is shown schematically in Fig. 1.2. The load must form part of the design of the system, so as to ensure the PV array and battery storage are sized to meet the load under all relevant weather conditions. In some cases, particularly when there is a large variation of the load profile and/or a large seasonal variation in sunlight levels, a hybrid system is used, where this includes one or more additional electricity generation technologies (eg, wind turbine, diesel generator).

PV systems can be classified as either flat plate or concentrator systems. In the latter, optical components, namely lenses or mirrors, are used to increase the intensity of the sunlight falling on the PV devices. Concentrator systems can be further classified into low concentration (up to 30 ×) and high concentration (> 300 ×) and require the modules to be moved to track the sun, since it is only possible to optically concentrate direct sunlight. Flat plate PV arrays can also be operated in sun-tracking mode, if desired and where the application allows.

When considering the performance of the system, it is clear that the operating mode, in terms of connection options, concentration level and static/tracking options, must be considered in determining the expected values.

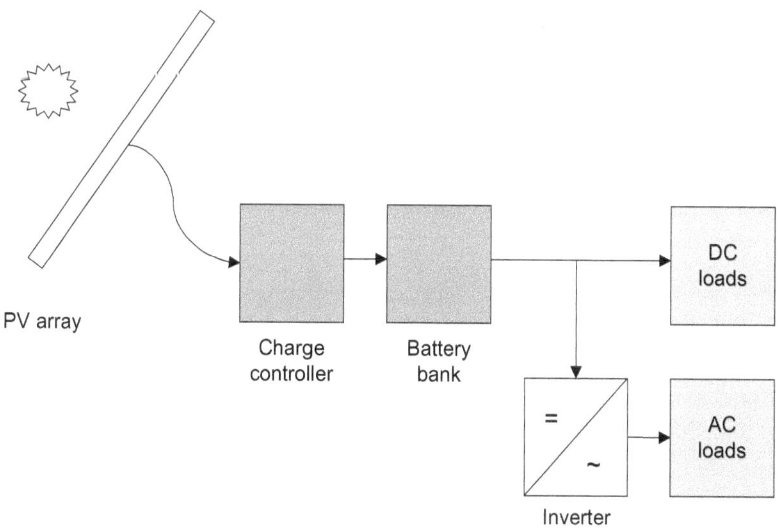

Fig. 1.2 Schematic of a stand-alone PV system providing electricity for both DC and AC loads.

1.2.2 PV module technologies

A range of semiconductor materials could be used in PV modules, but to achieve commercial viability, the modules have to have a suitably high performance coupled with stability in operation, have good manufacturability, have low cost production and be able to achieve a long lifetime. Manufacturability includes aspects such as ease of manufacture, availability of materials, health and safety, environmental impacts, reproducibility and yield. The current flat plate market is dominated by crystalline silicon, both monocrystalline and multicrystalline, with the latter having the highest market share. In 2014, crystalline silicon modules are estimated to have accounted for almost 91% of the PV market, with multicrystalline silicon contributing 55% and monocrystalline silicon 36% [2]. Other commercial flat plate technologies include cadmium telluride (CdTe), copper indium gallium diselenide (CIGS), amorphous silicon (a-Si) and several hybrid designs incorporating crystalline, nanocrystalline and a-Si (eg, heterojunction with intrinsic thin layer cells, micromorph cells). Organic and polymer based cells have seen rapid developments over recent years, although are not yet utilized in the power market. In the concentrator market, low concentration systems use silicon technology whereas the high concentration systems are based on high efficiency multijunction cells utilizing various III–V compound materials (eg, GaInAs/GaInP/Ge triple junction cells). For a more in-depth discussion of the status of current and emerging PV cell technologies, see, for example, the recently published study by MIT on the future of solar energy [3].

In terms of performance, the most obvious difference between the module types is the rated conversion efficiency, that is, the ratio between the electrical output of the module and the solar irradiation received under specified operating conditions (see Section 1.3.3 for a full definition). However, there is also variation in the voltage and current parameters, with both parameters being a function of the energy bandgap of the semiconductor material used for the absorber layer of the PV device. Thus, for example, the voltage per cell of a CdTe device is higher than for a silicon device, due to the higher bandgap of the CdTe material, but the current density is lower (under the same operating conditions for both devices).

There is continual development of module designs and production methods, leading to an increase in efficiency over time. Also, whilst it is straightforward to find the highest reported laboratory cell or small module efficiencies at any given time, the average efficiency of commercial modules is more difficult to establish. According to the status report by the Fraunhofer Institute for Solar Energy Systems published in 2015 [2], the efficiency ranges of commercial PV modules are as shown in Table 1.1.

Other module parameters that lead to a difference in module performance include spectral response (the dependence of the output on the spectral content of the incident light), temperature coefficients for voltage, current and power, cell stability and variations in module design (eg, cell spacing, module materials). Whilst all of these can be determined for the module, care must be taken in both the measurement of these parameters and their use in determining module performance under conditions which differ from those used during measurement. For example, because the basic cell shape

Table 1.1 Efficiency ranges for commercial PV modules—reported values by Nov. 2015

Module technology	Efficiency range under standard test conditions[a] (%)	Highest reported laboratory efficiency (module) (%)
Monocrystalline silicon	16–19	22.9
Multicrystalline silicon	15–17	18.9
Thin-film CdTe and CIGS	14–15	17.5
Advanced silicon designs (heterojunction, interdigitated back contact, etc.)	18–21	Not available (best cell efficiency of 25.6%)
Concentrator (>400 ×)[b]	27–33	38.9

[a] See Section 1.3.2 for an explanation of standard test conditions (STC).
[b] Note that the STC for concentrator modules differ from those for flat plate modules, as discussed in Section 1.3.2.
Adapted from data presented in Fraunhofer Institute for Solar Energy Systems, Photovoltaics report, http://www.ise.fraunhofer.de, November 2015.

varies with technology (square or pseudo-square for crystalline silicon, long narrow strips for thin film), the electrical effect of shading may differ for the same shading pattern since a different number of cells will be affected (see Section 1.3.1 for more information on the effect of shading).

In general, the information in the rest of this chapter applies to all current commercial cell technologies, unless stated, although some special conditions apply to concentrator systems and readers are referred to Chapter 10 of this volume for a more in-depth discussion of their performance.

1.2.3 BOS equipment

The BOS equipment is generally defined as all the components of the system other than the PV array itself. This can include power conditioning (eg, inverter, DC–DC converter), electricity storage (eg, rechargeable battery), wiring, fuses and switches, monitoring equipment, etc., depending on the system details. It can be seen from the system schematic diagrams in Figs 1.1 and 1.2 that the BOS components depend on whether the system is grid connected or stand-alone, whether the loads to be met are AC or DC and whether the system includes any storage facility.

The main aim of the system design in terms of choice and specification of BOS components is to achieve the highest system efficiency possible for the choice of module. This may include both selecting the BOS components that are suitable for the system and modifying the PV array specifications to best match to the rest of the system (see Section 1.3.1 for further explanation of this aspect). Wiring cross-sections and cable lengths must be chosen for minimum energy loss, especially for low voltage systems.

Component efficiency can be expressed by the usual ratio of energy out to energy in, measured at the appropriate point in the system. The lifetime of the BOS components and any degradation characteristics must also be taken into account in determining the effect of the BOS equipment on system performance. The performance parameters of the main items of BOS equipment are discussed further in Section 1.3.4 and the effects of various system losses are considered in Section 1.3.3.

1.3 Expressing PV system performance

In this section, we will consider the main methods of expressing the performance of PV systems and their constituent parts of PV modules and BOS components. The performance parameters described are then used in later chapters in this book to provide a more detailed consideration of predicting and measuring the performance of a PV system.

1.3.1 Current–voltage and power–voltage characteristics

Most current commercial solar cells have the structure of a diode, a junction between p- and n-type semiconductor materials. In some cases, an insulating layer is introduced to create a p–i–n junction or a series of junctions are created. However, from the point of view of the electrical characteristics of the device, the behaviour is similar. The relationship between the current, I, and the voltage, V, can be expressed, in the ideal case, by the relationship given in Eq. (1.1). This is the standard diode equation, with the addition of the light generated current, I_L, and a diode factor, n, which is related to the recombination mechanisms in the cell.

$$I = I_0\left[\exp\left(\frac{qV}{nkT}\right) - 1\right] - I_L \tag{1.1}$$

Here, I_0 is the reverse saturation current of the diode, q is the charge on the electron (1.602×10^{-19} C), k is Boltzmann's constant (1.38×10^{-23} m^2 kg/s^2 K) and T is the temperature in kelvin. The parameters I_0, I_L and n are dependent on the cell materials and design.

For a real device, it is also necessary to include the resistances relating to carrier transport in the semiconductor material, the ohmic contacts at the front and rear of the cell and the bulk resistance of the semiconductor layers. These are often represented by two lumped resistances, known as the series resistance, R_s, and the shunt resistance, R_{sh}. Including these in the preceding equation results in the I–V characteristic of a real solar cell being represented by Eq. (1.2):

$$I = I_L - I_0\left[\exp\left(\frac{q(V + IR_s)}{nkT}\right) - 1\right] - \frac{V + IR_s}{R_{sh}} \tag{1.2}$$

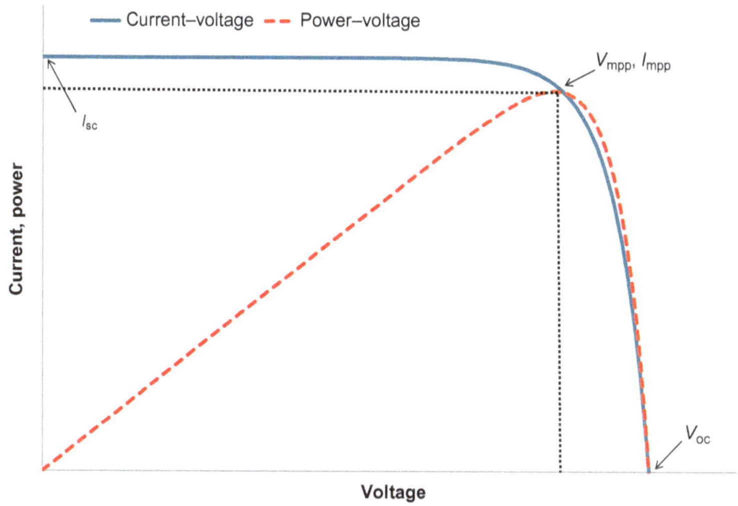

Fig. 1.3 The current–voltage and power–voltage curves for a PV cell, module or array, showing the current and voltage parameters usually considered. Note that the scales of the two curves are chosen so that the curves coincide at the maximum power point.

The resulting *I–V* characteristic is shown in Fig. 1.3, along with the power–voltage characteristic, where the power output, *P*, of the cell is determined from the usual relationship, $P = I \times V$. The characteristic is shown for an arbitrary operating condition (see Section 1.3.2 for a discussion of the effect of operating conditions).

Eq. (1.2) applies to a variety of cell types and gives a good representation of the *I–V* relationship. In some cases, the cell is more closely represented by two diode expressions in the equation, each having different I_0 and n values, but this consideration is beyond the scope of this chapter. Interested readers are referred to volumes dealing with the detailed analysis of solar cell structures [4]. Nevertheless, the two-diode model also leads to an *I–V* characteristic of the same general shape.

The *I–V* characteristic shown in Fig. 1.3 applies not only to the individual solar cells, but also to PV modules (a number of electrically connected PV cells, encapsulated into a single item) and to PV arrays (a collection of electrically connected PV modules). Clearly, there is some modification of the resistance values as we move from cells to modules to arrays, but the general shape of the characteristic remains. Thus, the following discussion applies equally to cells, modules and arrays.

The following parameters are generally used to describe the performance of the PV device, as illustrated in Fig. 1.3:

- Short circuit current, I_{sc}—this is the current when there is zero resistance between the device terminals. It represents the maximum current for the specific operating conditions of the device.
- Open circuit voltage, V_{oc}—this is the voltage across the device terminals in open circuit. It represents the maximum voltage for the specific operating conditions.

- MPP—the power level varies along the I–V characteristic and the MPP represents the point at which the power is the maximum value.
- Current at MPP, I_{mpp}—the current value at the position on the characteristic for which the $I \times V$ value is maximum.
- Voltage at MPP, V_{mpp}—the voltage value at the position on the characteristic for which the $I \times V$ value is maximum.
- Fill factor, FF—the ratio of the product of I_{mpp} and V_{mpp} to the product of I_{sc} and V_{oc}, as represented by Eq. (1.3).

$$\mathrm{FF} = \frac{I_{mpp} V_{mpp}}{I_{sc} V_{oc}} \tag{1.3}$$

The MPP is also used to define the efficiency of the device, which is given by the standard equation,

$$\text{Efficiency}, \eta = \frac{\text{Power out}}{\text{Power in}} = \frac{P_{max}}{G_i} = \frac{\mathrm{FF}\, I_{sc} V_{oc}}{G_i} \tag{1.4}$$

where $P_{max} = I_{mpp} \times V_{mpp}$ and G_i is the incident global irradiance.

The PV device will operate at the point of the I–V characteristic where the relationship given by Ohm's Law is satisfied for the resistance of the load across the terminals, that is, at the point where $R = V/I$. In the ideal case, we would wish to be operating at the MPP, such that $R_{mpp} = V_{mmp}/I_{mpp}$. However, the values of the parameters that define the I–V characteristic are dependent on the operating conditions, particularly irradiance and operating temperature, and so the value of R_{mpp} will change as the conditions change. Therefore, most PV systems will include a maximum power point tracker (MPPT), which modifies the operating point on the I–V curve in order to operate for as much of the time as possible at the MPP.

The PV array is a collection of PV modules that are electrically connected, either in series or parallel connection. The choice of connections determines the voltage and current output of the array. Series connection (the positive terminal of one module connected to the negative terminal of the next) increases voltage, leaving the current equivalent to that of the single module. Parallel connection (positive terminals connected together, as are negative terminals) increases the current, whilst leaving the voltage the same as for a single module. By means of a combination of series and parallel connections, the required current and voltage outputs for the array can be achieved.

For best performance, it is important that the performance of all modules that are electrically connected is as uniform as possible. If the modules do not have uniform performance, then the voltage at the MPP will be different for different modules. Whilst it is possible to determine the MPP for the combined set of modules, each individual module will not be operating at its MPP and therefore this results in a loss of potential output. In practice, this means that electrically connected modules should be all of the same type and model, that their individual output characteristics should be as similar as possible and that they should be operating under the same conditions of irradiance and module temperature. This implies similar mounting systems and the same orientation (tilt angle and azimuth angle).

One of the most common sources of nonuniformity is shading of the array, although the system design should attempt to keep this as low as possible. If one module is shaded whilst the other modules to which it is connected are in sunlight, the shaded module can act as a load and may be damaged by localized heating due to high current levels. To protect the module, it is usual to incorporate one or more bypass diodes, through which the current is diverted to prevent the shaded cells drawing current. However, this results in a modification of the power–voltage curve and a loss of output. The loss can be minimized by separating shaded modules from unshaded modules in the connection design, so ensuring that the MPP tracking is performed on modules with similar performance levels, but this requires careful design in both the connections and the number of MPPTs used. There is a range of options for the configuration of modules and inverters, ranging from a single inverter serving the whole array (with one or more MPPTs) to inverters attached to single modules (so-called microinverters). The system design can also mix module level MPPTs with a single central inverter. The best system design depends on the extent of the nonuniformity on the array and the economic evaluation of each option.

1.3.2 Operating conditions

As has already been discussed, the output of a PV device is dependent on its operating conditions, in particular the solar irradiance and the operating temperature as can be seen from Eqs (1.1), (1.2). The output is also dependent on the solar spectrum, since the spectral response of the device varies with the absorber material used and the device structure. Therefore, in order to be able to compare performance, a standard set of operating conditions has been defined by the PV community for the measurement of PV cells and modules.

The standard test conditions (STC) are [5]

- Irradiance of 1000 W/m^2
- AM1.5 standard global spectrum
- Operating temperature of 25°C
- Normal incidence irradiance

Note that these are the conditions for flat plate PV systems. A discussion of the testing conditions for concentrator PV modules can be found in Chapter 10.

The standard spectrum is defined for a specific path length through the atmosphere that is represented by the air mass (AM) number. The AM number can be calculated as $1/\sin(\alpha)$, where α is the solar altitude angle (the angle between the direct line to the sun and its projection on the horizontal, measured in the vertical plane). AM1 is the condition for a location at sea level with the sun directly overhead. In order to determine the spectrum, the atmospheric conditions must also be specified in order to determine the absorption and scattering of the sunlight. The standard global spectrum is given in the IEC standard 60904-3 [6].

These test conditions are suitable for the measurement of PV modules in the factory, since they do not require specific heating or cooling of the modules to achieve the required operating temperature. However, they do not represent the conditions that the module is likely to experience in practice, apart from under very specific circumstances. At the STC irradiance level of 1000 W/m^2 and an ambient temperature consistent with a low altitude site (say, in the range of 10–40°C), then the module operating temperature will be considerably higher than the STC value of 25°C. The temperature of the module in an operating PV system depends on

- the irradiance level, since this is the energy input to the system,
- the construction of the module, especially in regard to heat transfer through the module and from the module surfaces,
- the module efficiency, since any solar energy not converted to electricity will be converted to heat,
- the mounting method, particularly in regard to the air flow around the module and
- the ambient temperature, wind speed and wind direction, which impact on the heat removal from the module.

In order to consider the difference in operating conditions in the field from those at STC, it is usual to also measure the nominal operating cell temperature (NOCT), which is determined from field measurements. Nevertheless, this is still only for one specific set of operating conditions, namely:

- 800°W/m^2 irradiance
- Ambient temperature of 20°C
- Module is mounted on an open rack at an angle of 45 degrees
- Wind speed is 1 m/s
- Module at open circuit (no load)

The procedure is to measure the module temperature at several times during the test period when the conditions are close to those required and then extrapolate to the NOCT condition [5]. The NOCT value is usually provided in the module datasheet and is typically between 45°C and 50°C, although this will depend on module construction.

The NOCT value can be used directly to predict the module temperature for any combination of irradiance and ambient temperature, provided that the other system parameters are closely matched to the NOCT measurement conditions (ie, open rack mounting with good ventilation, low wind speed). The translation formula is given in Eq. (1.5):

$$T_{\text{mod}} = T_{\text{amb}} + \frac{G_{\text{mod}}}{800}(\text{NOCT} - 20) \tag{1.5}$$

where T_{mod} is the module temperature, T_{amb} is the ambient temperature and G_{mod} is the global irradiance incident on the module. Nevertheless, the NOCT only represents module temperature for one specific set of conditions, even if these are more likely to represent operational conditions than do STC parameters. Therefore, it is useful to have a model for the prediction of module temperature under a range of conditions.

Eq. (1.5) can be rearranged to state that the difference between the module temperature and ambient temperature is proportional to irradiance and this has also been shown empirically, to at least a good approximation. Most module temperature models include this assumption, although there may also be other terms. The simplest concept is that originally proposed by Ross [7], as shown in Eq. (1.6):

$$T_{\mathrm{mod}} = T_{\mathrm{amb}} + kG_{\mathrm{mod}} \tag{1.6}$$

where k is a constant which depends on the cell technology, the module construction, the mounting method and the environmental conditions. It should be noted that the preceding two relationships do not include thermal lag, which would have some influence on module temperature. Thus, more representative values are obtained for hourly average irradiance values than for instantaneous irradiance values. Other models include explicit terms for wind speed or are based on thermal transfer equations. An in-depth review of thermal models for the estimation of PV module temperatures has been published by Segado et al. [8].

The variation of PV module (or array) output with changes in operating conditions can be summarized as follows:

- The current has a linear relationship with irradiance level, whereas the voltage varies logarithmically due to the exponential term in the *I–V* equation.
- Current increases slightly with temperature, but the voltage reduces more strongly, leading to a reduction in power output with increasing temperature—the temperature coefficients for both current and voltage are dependent on the PV cell material and cell design.
- Power increases proportionally with irradiance and reduces with increasing temperature.
- The current is dependent on the spectral content of the incident sunlight and the properties of the absorber material.

For a practical PV system, the operating conditions will change consistently through the day, due to the change in sun position, temperature and weather conditions.

1.3.3 Efficiency, yield and performance ratio

The performance of a PV system is generally expressed by one of three parameters: the system efficiency, the energy yield or the performance ratio (PR). These parameters are linked but express different aspects of the overall performance.

The efficiency of the PV device, as represented by Eq. (1.4), has already been discussed and expresses how well the device converts the incoming solar energy to electricity. For a PV module, the efficiency is usually quoted under STC, although it will vary with operating conditions as described in the previous section. At the system level, we must include the efficiency of all other components in the system as well, to determine the overall system efficiency. Examples of this would be the inverter efficiency for the conversion of DC to AC electricity, any losses in the wiring, MPPT losses, etc. Clearly, the higher the system efficiency, the greater the output of the system for a given solar input and we would want to achieve the highest system efficiency that is reasonably practical for that system. However, the efficiency value depends directly on the efficiency of the PV array and therefore on the cell technology chosen, as well as some other design features of the module. Thus, it is difficult to compare absolute values of efficiency between systems of different technologies in order to determine information on system losses.

The energy yield of a system expresses the energy output and so is of particular interest to the system user. Three energy yield values have been defined, with, in each case, the energy output normalized to the PV array rating. The most useful of these is termed the final yield and represents the energy delivered to the load per unit capacity of the system. Note that, for a grid-connected system feeding power directly to the grid rather than to local loads, the final yield would be calculated using the energy to the grid.

Thus,

$$\text{Final yield, } Y_\text{f} = \frac{\text{Energy to lond}}{\text{Nominal system rating}} \tag{1.7}$$

where the energy to load is often expressed in kilowatt hours and the system rating in kilowatts and the yield is calculated over a specified period, for example, annual. Since the output of the system is dependent on the solar energy input, the yield value will clearly be location dependent and it is possible to determine typical yields for different regions or countries. Yield values are also dependent on system details, such as the orientation of the array and whether or not tracking is used.

The second yield is known as the array yield, Y_A and is defined in a similar way to the final yield, but now using the energy output from the array only (ie, before conversion to AC in the inverter). This gives a measure of the performance of the array and can help to distinguish between operational issues relating to the array and to the BOS components. However, it is only possible to determine array yield if there is a measurement of the array output at the relevant point in the system.

The third yield value is known as the reference yield. This expresses the output of the system if there were no losses, that is, in the ideal case. Thus, it considers the case when the array operates continually under STC, there is perfect MPP tracking, the DC to AC conversion efficiency is 100%, etc. However, it does take into account the solar energy received by the system. Because the system rating, which appears in the denominator for all the yield calculations, is defined as the output under STC, we can determine reference yield from the following equation:

$$\text{Reference yield, } Y_\text{r} = \frac{G_\text{A}}{G_\text{STC}} \tag{1.8}$$

where G_A is the irradiation received by the array over the period considered, expressed in kWh/m^2, and G_STC is the STC irradiation, that is, 1 kWh/m^2.

We have seen that the system efficiency is dependent on module choice and the final yield is dependent on the solar irradiation received, so making it difficult to use these values to determine performance quality without some benchmark of efficiency or yield for each system. The PR takes into account both system design and solar irradiation by comparing the final yield of the system with its reference yield. Thus,

$$\text{Performance ratio, PR} = \frac{Y_f}{Y_r} \tag{1.9}$$

As can be seen, the PR compares the actual yield achieved by the system to the yield in the ideal case and so provides a measure of the overall losses, and hence the quality, of the system. The PR value is widely used to allow comparison between systems and in the monitoring of performance over time, since changes in the PR value indicate a change in the system losses. In practice, the determination of PR requires only the measurement of energy output from the system and the solar irradiation received by the PV array over the defined period, together with knowledge of the nominal rating of the system.

We can define the following typical loss factors for a grid-connected PV system, starting with the losses at the array level and moving through the system components to the grid connection (note that the order does not imply the importance of the loss):

- Module operating temperature—expresses the difference in performance due to the module not operating at 25°C, usually a loss because the module temperature is higher but can be a gain in low temperature locations.
- Module rating—because the final yield is normalized to the array rating, it assumes that all modules achieve that rating. In practice, modules may have a higher or lower rating within the manufacturer specification and this will affect the value of PR calculated.
- Angle of incidence effects—STC requires irradiance at normal incidence, whereas in practice the irradiance is received at a range of incidence angles depending on module orientation and solar position. At high incidence angles, more light is reflected from the module surface and the absorption path of the light in the cell is changed, so leading to reduced output.
- Low light level effects—for some modules, the response reduces at low light levels (below 100 W/m^2) and the inverter will also have reduced efficiency for low power levels.
- Module mismatch—if modules with different operating characteristics under STC are electrically connected, there will be a loss since it will not be possible to operate all the modules at their MPP for a given operating voltage. The magnitude of the loss depends on the level of mismatch and it is common to select modules so as to minimize the mismatch when designing the array.
- Shading—the shading of parts of the array results in both a loss of output due to the reduced output of the shaded modules and a loss due to the increase in the mismatch between shaded and unshaded modules. The extent of the loss depends on the degree of shading and the connection of modules in the array.
- Dirt or ice/snow accumulation—in both cases, the light transmitted into the module will be reduced and this will reduce the system output. The extent of the loss depends on the soiling (or snow) at the location and any maintenance activities implemented (eg, cleaning, clearing the snow).
- MPP tracking losses—as the operating conditions change, the MPP also changes and the tracker must then reacquire the MPP. The losses relate to the period spent away from the MPP whilst the tracker searches for the correct operating point and depend on the detailed algorithm used for the tracker and the rate of variation of conditions.

- Inverter efficiency—the efficiency of the inverter varies with input power, input voltage and the inverter design (eg, single or multistage, with or without transformer). Included here would be the efficiency of any power conditioning equipment (eg, DC to DC conversion) that is required for the particular system.
- Inverter threshold—inverters consume small amounts of power in the conversion and control processes and so there is a set threshold of output from the PV array below which the inverter will be shut down so that it does not draw more power from the grid than is being generated. The appropriate threshold depends on the inverter, with modern insulated-gate bipolar transistor-based inverters having a much lower required threshold than the older thyristor-based inverters.
- Cabling losses—all electrical systems have losses in the cables linking parts of the system and these are dependent on the voltage levels, cable cross-sectional area and cable length.
- Reliability of system components—any down time of system components that affects the magnitude of the system output will influence the PR value calculated.
- Reliability of grid connection—in most cases, the grid-connected PV system will be required to shut down if the local grid supply to which it is connected is out of specification in terms of voltage or frequency. The details of the conditions for shut down are specific to the connection regulations for the grid in question, but it is clear that any lack of reliability on the grid, even if unrelated to the PV system, may cause periods of no output and this will be reflected in the calculated PR value.

It can be seen from the preceding list that there are many factors that influence the PR value, but nevertheless it is the most useful of the three parameters to determine long-term performance. With knowledge of the system and, where necessary, the use of other measured parameters to allow quantification of some of the losses detailed here, it becomes possible to identify the system losses that can be addressed and so improve the system performance. When the PR value is used over time to indicate changes in system performance, it is important to make sure that the measurement equipment is reliable and the sensors are suitably calibrated (and cleaned in the case of the irradiance sensor). If measurement of the irradiation is not made on-site, it is possible to use satellite-derived values or measurements at a nearby site provided that it is sufficiently close to provide a meaningful value. In all cases, the accuracy of the values should be taken into account in determining the accuracy with which the PR value is known.

Since the module operating temperature is one of the strongest influences on module performance, it is generally possible to observe a seasonal variation of PR due to the change in ambient temperature across the year. Therefore, if the assessment of other losses is the main focus of the analysis, a temperature corrected PR value is sometimes used. This modifies the reference yield value to account for the variation of temperature from STC and thus allows the seasonal temperature variation to be removed from consideration. This allows the variation of other loss mechanisms to be seen more clearly in the analysis.

The discussion so far has considered grid-connected systems, but the same parameters of efficiency, yield and PR can also be calculated for stand-alone systems. However, since the stand-alone system is designed to meet a specific load, an inherent property of that design is that the system would be able to provide excess energy at some times of the year since it is required to meet load under the worst-case conditions. In practice, this means that the array will sometimes be operated away from

MPP, so as to only produce the amount of energy required for the load. As a result, the yield will be limited to the load requirement and both efficiency and PR will be substantially lower than for the grid-connected system, which is designed for continuous MPPT operation. Indeed, for a stand-alone system that is designed to meet a critical load (such as health care or communications), the additional capacity that is needed to provide the required safety margin would lead to a reduction in the calculated PR value, so the PR cannot be taken as an indication of quality. The performance parameters that can be considered for a stand-alone system are discussed in the next section.

1.3.4 Other performance parameters

The three performance parameters discussed in the previous section express the overall performance of the system but it is sometimes useful to consider the specific performance of certain parts of the system in order to ensure the correct design and operational choices.

It is possible to determine the inverter efficiency if measurements of both DC input and AC output are provided. In general, the efficiency of a PV inverter is a function of the input power and input voltage, with a typical set of efficiency curves being shown in Fig. 1.4. At medium to high light levels and therefore input power from the array, the inverter has a high efficiency, generally well in excess of 90%. At low irradiance levels, the efficiency drops off sharply. This means that we can determine an optimum inverter capacity in comparison with the array capacity, such that the balance between energy loss at the low irradiance end due to reducing efficiency is balanced against energy loss at the high irradiance end due to limiting because of the maximum inverter capacity. This is illustrated in Fig. 1.5, assuming that the basic shape of the inverter efficiency curve does not depend on inverter capacity. Clearly, the balance between energy generation at low and high irradiance values is dependent on the climate and, therefore, so is the optimum inverter/array ratio, with the general approach of a reduction in this ratio as the latitude increases.

Because of the variation of efficiency with input power, and therefore irradiance on the array, the average operating efficiency of the inverter will vary with climate. In general, the technical information for a PV inverter will include both the peak efficiency (usually between 95% and 98% depending on the inverter technology) and a weighted efficiency to account for the operation at different irradiance levels. In Europe, this weighted efficiency is termed the Euro efficiency [9] and can be represented by Eq. (1.10), as follows:

$$\begin{aligned}\text{Euro efficiency} = {} & 0.03\text{Eff@}5\% + 0.06\text{Eff@}10\% + 0.13\text{Eff@}20\% \\ & + 0.1\text{Eff@}30\% + 0.48\text{Eff@}50\% + 0.2\text{Eff@}100\%\end{aligned} \tag{1.10}$$

Here, Eff@5% is the DC to AC conversion efficiency at an input power of 5% of the inverter capacity, with similar definitions for the other terms. The weighting reflects the amount of energy predicted to be gained at each energy level. An alternative

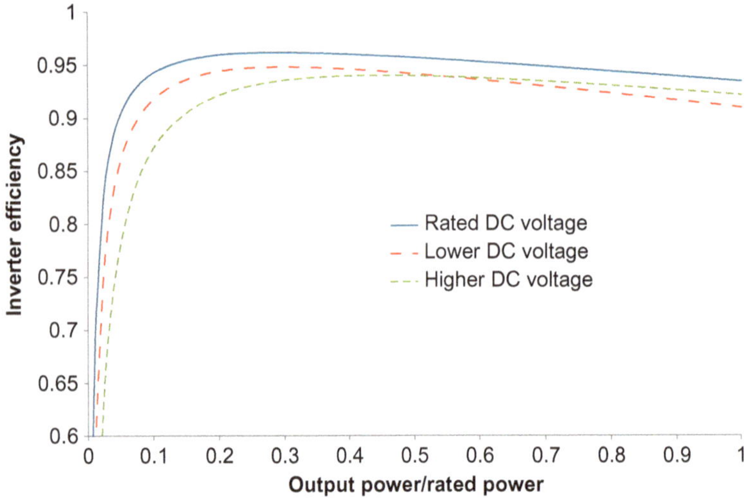

Fig. 1.4 Typical inverter efficiency curve as a function of DC input voltage. The rated input voltage will be a value around the middle of the allowed voltage range. Curves are also shown for higher and lower voltages, closer to the limits of the voltage range.

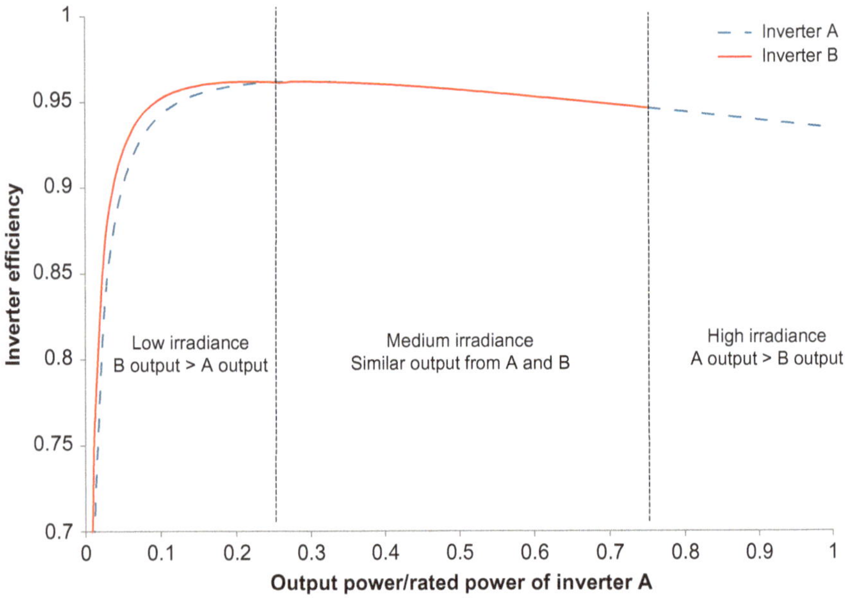

Fig. 1.5 Comparison of inverter efficiency for two inverters, A and B, where the rated capacity of B is 75% of that of A. The general shape of the efficiency curve is assumed to be unaffected by the capacity and the capacity of the PV array is assumed to be equal to A. It can be seen that inverter B provides more output than A at low irradiance values but limits at PV array output values higher than 75% of rated capacity. The optimum capacity of the inverter will depend on the ratio between the energy received at low irradiance and at high irradiance for that particular system.

weighting, using the same rationale and approach, is the California Energy Commission or CEC efficiency, calculated as follows [10]:

$$\begin{aligned}\text{CEC efficiency} = {} & 0.04\text{Eff}@10\% + 0.05\text{Eff}@20\% + 0.12\text{Eff}@30\% \\ & + 0.21\text{Eff}@50\% + 0.53\text{Eff}@75\% + 0.05\text{Eff}@100\%\end{aligned} \tag{1.11}$$

Here the weighting factors are more heavily biased towards the higher irradiance levels, to reflect the difference between the climates in the southwestern USA and central Europe. The use of either the Euro or CEC efficiency gives a lower but more representative value than the peak efficiency in terms of considering the overall inverter efficiency across a period of operation. Clearly, other weightings could be developed for other climate options, although there needs to be consensus within the community for widespread use.

For a stand-alone PV system, the important parameter is not the total energy generated but whether the load is met for the required time, that is, the service provided by the system. This is sometimes assessed by direct reference to the load, for example, amount of water pumped, amount of product manufactured using PV electricity. However, it is also possible to define parameters to express the system performance, such as the total amount of time for which the load is not met (to be compared with the loss of load probability defined in the system design) and the battery index, which is the percentage of days in a given period when full charge of the batteries in the system is achieved. In general, values over 30% are considered as good, although very high values may indicate an oversized array.

1.4 Lifetime and quality

From both a financial and practical viewpoint, the viability of the system in terms of providing the required service at an appropriate cost is based on an assumption of system lifetime and maintenance of quality of output to a specified level throughout that lifetime. It is typical to consider PV system lifetimes of 20–30 years, based on PV module performance warranties providing for less than 20% reduction in output over a period of, for example, 20 years. However, it is recognized that other system components, such as inverters, batteries, etc., would be expected to require replacement during this system lifetime. Therefore, prediction of the lifetime performance of the system and periodic assessment of the performance should take account of any expected degradation of system components.

In general, electrical components can be considered to follow the well-known bathtub curve in terms of failures, that is:

- A relatively high rate of failure in the initial phase of operation due to manufacturing faults, poor design or incompatibility with operating conditions.
- A relatively low rate of failure across much of the component lifetime.
- An increase in failure rate towards the end of the component lifetime.

These failure rates will also be affected by maintenance and repair activities undertaken during the system operation.

The increasing availability of performance data from a wide range of systems allows operational issues to be identified and addressed. Coupled with the development of international standards for all PV system components, this has led to an increase in the average performance of PV systems in recent years. Further discussions of component performance and lifetime can be found in several other chapters in this book, most notably Chapters 3 and 5.

1.5 Summary

In this chapter, the main aspects of PV system performance have been summarized, concentrating on the design aspects that influence the system output. The main influences on the system performance are the irradiation received in the plane of the PV array and the array temperature, with notable contributions from the inverter in terms of efficiency and MPP tracking. The location of the system is also important, with this governing the amount of shading or soiling experienced. Whilst these two loss factors are very dependent on the system details, they can lead to major losses of output.

The purpose of this chapter is to introduce the reader new to PV to the basic principles of PV systems and to remind the experienced reader of those principles. This information should underpin the more in-depth discussions of performance in the remainder of this book.

References

[1] IEA Photovoltaic Power Systems Programme, Trends 2015 in photovoltaic applications, Report IEA-PVPS T1-27:2015, ISBN 978-3-906042-37-4, 2015.

[2] Fraunhofer Institute for Solar Energy Systems, Photovoltaics report, Annual status report, *https://www.ise.fraunhofer.de/en/renewable-energy-data/data-and-facts-about-pv*, 2015.

[3] MIT, Photovoltaic technology, Section II, Chapter 2, in: The Future of Solar Energy, Massachusetts Institute of Technology, Cambridge, MA, 2015.

[4] See, for example, Chapters 2–6 of M.D. Archer, M.A. Green (Eds.), Clean Electricity From Photovoltaics, second ed., Imperial College Press, London, 2015.

[5] IEC 61215:2005, Crystalline Silicon Terrestrial Photovoltaic (PV) Modules—Design Qualification and Type Approval, International Electrotechnical Commission, Geneva, 2005.

[6] IEC 60904-3:2008, Photovoltaic Devices—Part 3: Measurement Principles for Terrestrial Photovoltaic (PV) Solar Devices With Reference Spectral Irradiance Data, International Electrotechnical Commission, Geneva, 2008.

[7] R.G. Ross, Interface design considerations for terrestrial solar cell modules, in: Proceedings of the 12th IEEE Photovoltaic Specialists Conference, Baton Rouge, LA, USA, 1976, , pp. 801–806.

[8] P.M. Segado, J. Carretero, M. Sidrach-de-Cardona, Models to predict the operating temperature of different photovoltaic modules in outdoor conditions, Prog. Photovoltaics Res. Appl. 23 (2015) 1267–1282.

[9] B. Bletterie, R. Bründlinger, G. Lauss, On the characterisation of PV inverters' efficiency—introduction to the concept of achievable efficiency, Prog. Photovoltaics Res. Appl. 19 (2011) 423–435.
[10] California Energy Commission (CEC), New Solar Homes Partnership Guidebook, California Energy Commission (CEC), California, 2008. CEC-300-2008-006.

Part One

Photovoltaic module performance

Photovoltaic module measurement and characterization in the laboratory

2

D. Dirnberger
Fraunhofer Institute for Solar Energy Systems, Freiburg, Germany

2.1 Introduction

The output of a photovoltaic (PV) system depends essentially on the electrical performance of the PV modules in use, as outlined in Chapter 1. The 'electrical performance' of PV modules is described by different module characteristics that can be determined in the laboratory or outdoors. These module characteristics include electrical parameters at STC (standard testing conditions) as well as temperature coefficients (TCs), low light behaviour, spectral response (SR), angular response, thermal behaviour and module stability. STC are conditions that are widely used as reference conditions, and are represented by a temperature of 25°C and a broadband irradiance of 1000 W/m^2 with a spectral distribution according to IEC 60904-3 [1].[1]

The 'electrical parameters' are specific points of the so-called *I–V* curve of the PV module. The electrical current and power created by a PV module[2] under prevailing conditions depend on the voltage that is applied to the module terminals. The resulting current–voltage characteristic is shown exemplarily in Fig. 2.1, and is referred to as the *I–V* curve. Measurement of the *I–V* curve at defined conditions is the basis for determination of most of the previously mentioned module characteristics. The power–voltage characteristic, which is also shown in Fig. 2.1, is calculated by multiplying the current and the voltage values ($P = VI$). As it is inconvenient to report the whole *I–V* and its corresponding *P–V* curve, typically the following electrical parameters are presented as results: current at short circuit I_{SC}, voltage at open circuit V_{OC}, the maximum power P_{MPP}, current I_{MPP} and voltage V_{MPP} in the point where power is maximal (maximum power point, MPP) and the fill factor (FF). These parameters are derived from the curves and, if necessary, used for further analysis (eg, TC, Section 2.3). The relevant equations for deriving the electrical parameters from an *I–V* curve and its corresponding *P–V* curve are given in the following:

[1] Note that there is no specific standard defining the standard testing conditions, but they are frequently used in the series of standards for PV characterization IEC 60904, and the standard IEC 60891 for translation of *I–V* curves to given conditions.

[2] The semiconductor structure in the PV module converts irradiance (electromagnetic power) to electrical power, which can be collected at the terminals of the PV module using a load.

The Performance of Photovoltaic (PV) Systems. http://dx.doi.org/10.1016/B978-1-78242-336-2.00002-1

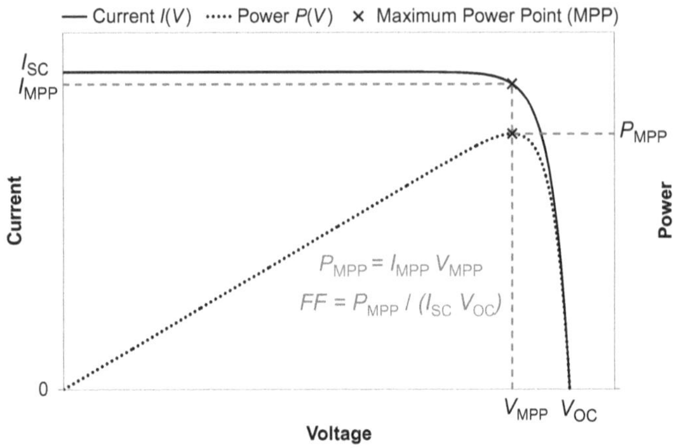

Fig. 2.1 Current–voltage characteristic (*I–V* curve) and power–voltage characteristic (*P–V* curve) of an ideal PV device with indication of the most important electrical parameters. Figure from D. Dirnberger, Uncertainties in energy rating for thin-film PV modules (Doctoral Thesis), Carl von Ossietzky Universität Oldenburg/Fraunhofer Institute for Solar Energy Systems ISE, Freiburg, 2015, with added equation for fill factor (*FF*).

$$I_{SC} = I(V = 0) \tag{2.1}$$

where I_{SC} is the short circuit current, which is the current when the voltage at the device terminals is zero.

$$V_{OC} = V|_{I(V)=0} \tag{2.2}$$

where V_{OC} is the open circuit voltage, which is the voltage when no current is produced.

$$P_{MPP} = \max(P(V)) = P(V)|_{\frac{dP(V)}{dV}=0} \tag{2.3}$$

where P_{MPP} is the power in MPP. It is the maximum power the module can produce under prevailing conditions, and is usually determined by applying a fifth-order polynomial fit to the measured *P–V* curve around the expected P_{MPP}.

$$V_{MPP} = V(P = P_{MPP}) = V|_{\frac{dP(V)}{dV}=0} \tag{2.4}$$

where V_{MPP} is the voltage in the MPP, which is the voltage that occurs where the power is maximal, that is, where the derivation of power with respect to voltage is zero.

$$I_{MPP} = P_{MPP}/V_{MPP} \tag{2.5}$$

where I_{MPP} is the current in MPP.

$$FF = \frac{P_{\mathrm{MPP}}}{I_{\mathrm{SC}} V_{\mathrm{OC}}} \tag{2.6}$$

where *FF* is the fill factor. *FF* basically indicates the magnitude of losses in the semiconductor structure [2].

The power conversion in a PV module and thus the values of the electrical parameters depend on module temperature and available irradiance (see also Chapter 5). Thus, the electrical parameters determined by measurement are influenced by the conditions during the *I–V* curve measurement. In order to achieve comparability of reported PV module parameters, the community agreed on the previously mentioned STC. As a consequence, PV module measurements involve not only measurements of electrical current and voltage, but also of module temperature and irradiance. For precise measurements of STC parameters, controlling the conditions during measurement (temperature, broadband irradiance and spectral distribution) and achieving only small deviations from STC is very important. Today, solar simulators are optimized for measurements at STC as a consequence of the key role of these conditions in the PV industry.[3]

Measurements at STC are also referred to as calibration or STC power rating of PV modules. The term *module calibration* can be used if the measurement is performed by an accredited calibration laboratory. The calibrated module is then used as a reference device for further measurements by testing or inhouse laboratories, or in a module production line for calibration of the simulator. The term *STC power rating* in this context designates the process of assigning an STC power value to a newly produced module and binning modules to STC power classes.

While modules can be characterized both by indoor and outdoor measurements, this section focuses on indoor (laboratory) measurements. The advantage compared to outdoor measurements is that the measurement conditions are much more reproducible and independent of current weather conditions or local climates. The role of PV outdoor performance assessments is explained in Chapter 3. Laboratory measurements are applied for several different purposes in practice. It is helpful for further understanding to be aware of the following purposes:

- Module development: Measurements of different module characteristics are performed in order to track the outcome of optimizations of module design and/or to document new record values. These measurements are carried out by inhouse or service laboratories and, mainly for external verification, by accredited laboratories.
- Module production: At the end of the production, each module is measured in order to assign the correct rated power (or STC power bin) to the module on its label. Calibrated reference modules are required for accurately setting the measurement parameters.
- Quality assurance: Out of a batch of purchased modules, a sample is measured in an independent laboratory to check data sheet specifications regarding STC power and other module characteristics. The acquired data can be used for performance prediction.

[3] Note that this is rather a consequence of the historical development, not of the significance of STC for PV system operation. STC power is an important reference value, but not sufficient to indicate the overall performance of a PV module in the field.

The following sections give more detailed explanation regarding the topics that were hinted at in this introduction. *I–V* curve measurements at STC and measurement uncertainty evaluation are discussed in detail in Section 2.2. This serves as a basis for the explanation of measurement of other module characteristics in Section 2.3. Section 2.4 briefly discusses aspects regarding variability of module characteristics and sampling of modules from batches for the purpose of quality assurance. Section 2.5 gives a short outlook on future perspectives.

2.2 PV module STC measurements using solar simulators

2.2.1 Measurement set-up and procedure

The typical set-up for laboratory *I–V* curve measurements is shown in Fig. 2.2. The set-up consists of:

- a light source
- a reference device for determination of (broadband) irradiance during the measurement
- a structure to mount the device under test (DUT)[4]
- temperature sensors (Pt100, Pt1000, thermocouples or sometimes infrared sensors)

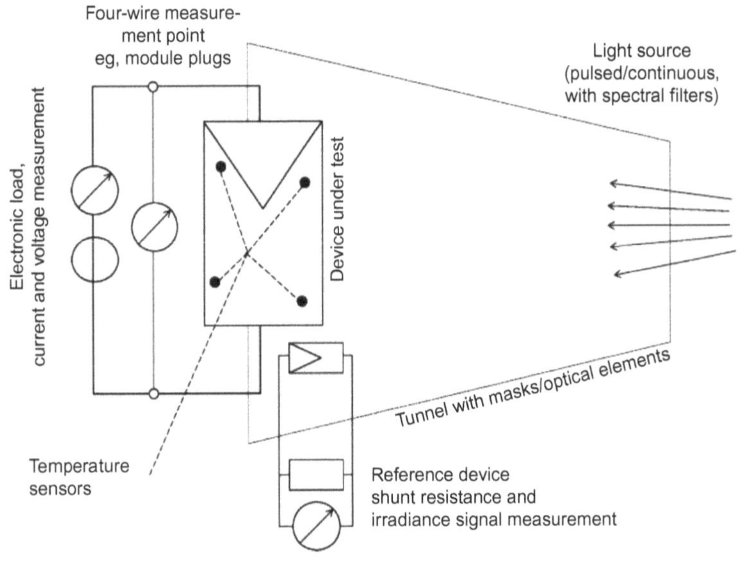

Fig. 2.2 Schematic set-up for *I–V* curve measurements. Device under test and reference device must be installed at the same distance from the light source.

[4] The DUT can be a full-size PV module, a mini module or an encapsulated cell with electrical connectors. The focus herein is on full-size PV modules.

- an electric load to apply a changing voltage (voltage ramp) to the DUT while it is illuminated
- measurement instruments to record
  - the *I–V* curve (ie, the current the module produces at the applied voltage levels under illumination)
  - the temperature
  - the irradiance signal (ie, voltage or current of reference device)

Modern measurement systems also include software solutions for treatment of the measured data, most of all the correction to STC and determination of *I–V* curve parameters.

In the following, the measurement procedure will be described with a focus on secondary module calibration. Secondary calibration is calibration using a primary-calibrated reference device and creating a secondary reference device (Section 2.2.10). Details can be different for other measurement procedures depending on their purpose. The state-of-the-art rules for performing *I–V* curve measurements are also given in the IEC 60904-1 [3].

Prior to the measurement, the DUT is mounted so that a perpendicular line can be set-up between light source and module plane. DUT and reference device must have exactly the same distance from the light source. The temperature sensors are placed in a way that allows for correct measurement results (ie, Pt100 sensors are attached to the back of the DUT, infrared sensors must be properly placed and calibrated). The cables of the PV device (positive and negative terminal) are connected to a four-wire measurement set-up and electrical load. When the measurement is started, the electrical load applies a defined voltage sweep to the module while it is illuminated. The voltage at the device terminals and the resulting device current are measured. If a pulsed solar simulator (Section 2.2.2) is used, the electronic load must be triggered by the light pulse. After the measurement, the measured data must be treated, for example, smoothed, interpolated to standard voltage steps, and corrected to STC. Finally, the previously mentioned electrical parameters are derived and reported. The final measurement result should be the average of several measurements.

2.2.2 Characteristics and characterization of solar simulators

Different types of solar simulators are available. The main differentiation is between pulsed and continuous solar simulators. Pulsed solar simulators work with electric arc lamps (eg, Xenon arc lamps) that create a light pulse with a duration of much less than 1 s, while continuous solar simulators use lamps that provide constant light for many hours.

Typically, the manufacturers of PV measurement systems offer complete systems including the light source (solar simulator), a control unit for the light source and the measurement electronics (electrical load and measurement equipment). The characteristics and performance of the measurement system are essential for the achievable measurement uncertainty. While the quality of the measurement electronics affects the quality of the *I–V* curve measurement (Section 2.2.3), the quality of the light source affects the uncertainty of the irradiance measurement (Section 2.2.4). The most important characteristics of the light source are:

- Spectral distribution: the closer the match to the reference spectral distribution, the smaller are spectral mismatch errors even for different SRs of reference device and DUT (compare Fig. 2.5). The spectral distribution depends first on the type of lamp and second on filters that can be installed in between lamp and DUT.
- Nonuniformity: the better the uniformity, the smaller are errors due to misplacement of the reference cell, which should be placed in a spot with irradiance equal to the average irradiance available over the area of the module. Furthermore, good nonuniformity minimizes influences on the *I–V* curve due to cell currents that are mismatched because different cells receive more or less irradiance depending on their position.
- Stability of the light characteristics: stability of irradiance and spectrum during the measurement (ie, typically in an order of magnitude of 10–1000 ms) is essential, but also longer term stability is desirable as this facilitates measurement procedures. For example, it was observed at Fraunhofer ISE that the spectral distribution red-shifts with increased number of flashes, and also nonuniformity can change during the lifetime of the lamps used.
- Directional distribution of the light: The PV module is mounted so that a perpendicular line can be set-up between light source and module plane. In the ideal case, all the light is incident on the module plane perpendicularly. In reality, there is usually a specific aperture angle of the light source and optics, which causes widening of the light beam and leads to a specific distribution of angles of incidence on the module plane. Furthermore, there is a certain amount of stray light due to reflections of the surroundings, even though in high-quality simulators this is minimized as far as possible (eg, by using aperture masks).
- For pulsed solar simulators, an important feature is the pulse length. The pulse length determines the maximum measurement time for one *I–V* curve sweep, unless special methods such as section measurements are applied (see Section 2.2.3). Available pulse lengths are 10 ms to around 300 ms; long pulse simulators offer pulse lengths of up to around 800 ms.

The first three of the mentioned characteristics are used in the standard IEC 60904-9 for classification of solar simulators [4]. The directional distribution of the light is mentioned here for the sake of completeness. Looking at a solar simulator as a point source of light is a common approach, but only valid for very basic considerations. It is very difficult to accurately determine the directional distribution of a solar simulator, but fortunately this has not presented severe consequences for standard measurements in the past. However, measurement results can be influenced if DUT and reference device have significantly different reflection characteristics (Section 2.3.6).

Following the IEC 60904-9, simulators are attributed to a class for each of the first three of the mentioned characteristics, leading to classifications like 'AAA' or 'ABA'. The criteria for the classification are summarized in Table 2.1. Following is a brief discussion of how the characteristics can be determined.

Measurement of the spectral distribution of a solar simulator can be quite complicated, especially in the case of a pulsed solar simulator. In this case, spectroradiometers with very short integration time and sophisticated equipment for correctly timing the measurement must be used. More details on the challenges can be found in Refs. [5,6]. The question of whether the spectrum changes during the pulse can be

Table 2.1 **Solar simulator classification according to IEC 60904-9**

Class	Spectral match	Spatial nonuniformity (%)	Temporal instability (STI[a]/LTI[b]) (%)
A	0.75–1.25	2	0.5/2
B	0.6–1.4	5	2/5
C	0.4–2.0	10	5/10

[a] STI—short-term instability, the observed variation of irradiance from one measurement point to the next during one *I–V* curve measurement.
[b] LTI—long-term instability, the observed variation (max/min) of irradiance during one *I–V* curve measurement.
From IEC 60904-9 Ed. 2.0, Photovoltaic Devices—Part 9: Solar Simulator Performance Requirements, International Electrotechnical Commission (IEC), Geneva, Switzerland, 2007.

answered by repeating the measurement several times within one flash, which further increases the integration speed requirement. For accurate measurement results, the spectroradiometer must be well-calibrated, even though relative calibration is sufficient. Relative calibration means that the spectral distribution is measured in counts, which gives information only about the relative irradiance at a specific wavelength (eg, 60% compared to the wavelength with maximum irradiance) but not about the absolute irradiance in W/nm or W/m^2 nm. Further information on spectroradiometer calibration and spectroradiometry in general can be found in Ref. [7].

The 'spectral match' according to IEC 60904-9 is determined with respect to the wavelength range from 400 to 1100 nm. For six defined wavelength intervals (100 nm step width except for 900–1100 nm), the percentage of irradiance within the interval compared to the total irradiance is calculated and compared to the corresponding percentage for the reference spectrum. The spectral match is the ratio of the percentage for the simulator and the reference spectrum:

$$\text{Spectral match}_{ab} = \frac{\left[\dfrac{\int_a^b E_{\text{simulator}}(\lambda)d\lambda}{\int_{400\,\text{nm}}^{1100\,\text{nm}} E_{\text{simulator}}(\lambda)d\lambda}\right]}{\dfrac{\int_a^b E_{\text{reference}}(\lambda)d\lambda}{\int_{400\,\text{nm}}^{1100\,\text{nm}} E_{\text{reference}}(\lambda)d\lambda}} \tag{2.7}$$

where a and b are the limits of the wavelength interval under consideration, $E(\lambda)$ is the spectral distribution as measured for the simulator and for the reference, respectively.

The classification of the simulator is equal to the worst classification of all intervals. As the spectral distribution of solar simulators has improved considerably in the past years, additional classifications like 'A+' (indicating a spectral match between

0.88 and 1.12) have appeared. While this is useful to better differentiate simulators with class A spectral classification in a general way, it has to be kept in mind that the significance of the classification in terms of measurement uncertainty is limited: the wavelength range under consideration is smaller than the SR of most PV modules, and the calculation procedure allows that deviations in opposite directions within one wavelength interval cancel out. Such deviations do not necessarily cancel out in the spectral mismatch calculation (Section 2.2.4), so that high uncertainties in spectral mismatch calculation can remain despite A+ classification.

The measurement of spatial nonuniformity can be performed in two ways. In the first option, the area under consideration is scanned by one measurement device (eg, a calibrated industrial PV 15.6×15.6 cm^2 cell or a reference cell with area of 2×2 cm^2) that is moved from one position to the next. This method requires a rigid structure for exact positioning of the reference device and an additional reference device that is positioned in one constant spot throughout all measurements to check on the irradiance level at the time of each measurement. The second option is to build special measurement instruments that are comprised of several cells enabling the simultaneous measurement of irradiance in several spots (eg, the nonuniformity module used at Fraunhofer ISE as described in Ref. [8]). For both methods, the result is a matrix of irradiance values and the nonuniformity equals the difference between minimum and maximum irradiance, divided by the sum of these values. Fig. 2.3 shows example results for two different solar simulators.

The determined nonuniformity value depends on the size of the device used for the determination, the coverage of the whole area of interest with measurement spots (ie, how large the space is between measurement points), and the magnitude of the area of interest (module area). Therefore, comparability of nonuniformity values is limited. At least, the considered area should be stated along with the nonuniformity value—while 1% for 2 m × 1 m is a rather typical value, it would be a very good result for 3 m × 3 m. The measurement uncertainty of the determined nonuniformity

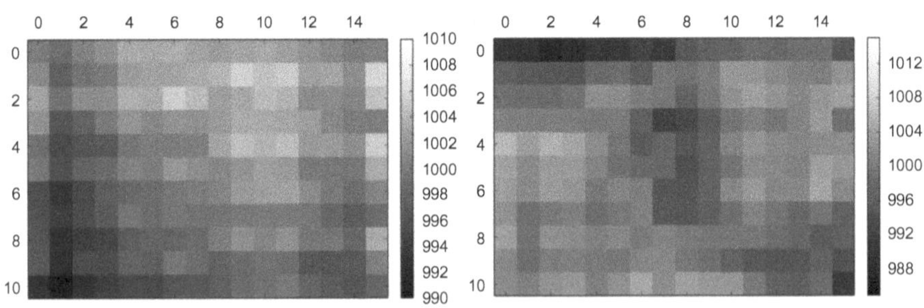

Fig. 2.3 Matrix resulting from a nonuniformity measurement of two different solar simulators. The area covered by the measurement is c.2 m × 1.4 m. The nonuniformity depends on the considered area [eg, *left* figure excluding the last and second-last row (c.2 m × 1.1 m): c.0.7%; *right* figure excluding the first row (c.2 m × 1.3 m): c.0.9%; note different scaling of the colourmap].
Figures provided by Fraunhofer ISE.

depends a lot on the measurement procedure and, especially for the second method, on the calibration of the nonuniformity device.

The temporal instability can be measured with basically any PV device that is stable during the measurement. The determined value can be influenced by the sampling rate. Compared to the spectrum and the spatial nonuniformity, temporal instability is typically not an issue. Furthermore, the influence of temporal instability can be corrected after the measurement (point-by-point correction of the *I–V* curve; see Section 2.2.6).

2.2.3 I–V *curve measurement*

The conditions during an *I–V* curve measurement usually differ a little from STC, even though they are controlled. Therefore, the result of an *I–V* curve measurement always depends on the prevailing conditions, but this can be corrected before reporting the final STC result, as long as the *I–V* curve itself was measured correctly. This section on *I–V* curve measurement refers to aspects related to the measurement of the *I–V* curve itself, disregarding the conditions (but assuming small deviations from STC).

As mentioned previously, the positive and negative terminals of the DUT are connected to a four-wire measurement set-up and an electrical load. Four-wire measurement is essential for correctly recording both current and voltage—additional resistance in the connection before the four-wire measurement point creates a voltage drop which introduces a current-dependent error in the measured voltage:

$$V_{\text{applied}} = V_{\text{PV}} + RI(V_{\text{PV}}) \tag{2.8}$$

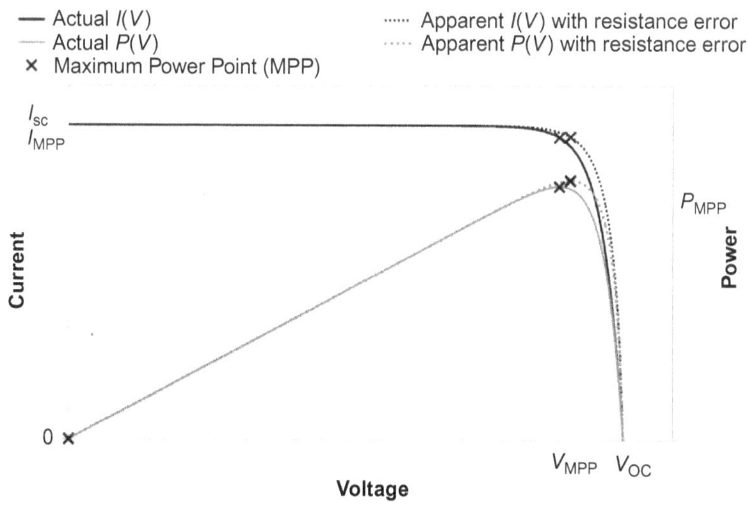

Fig. 2.4 Comparison of the actual *I–V* curve and the apparent *I–V* curve that is measured in presence of an additional ohmic resistance before the four-wire measurement point. The power and fill factor appear larger than they are.

where $V_{applied}$ is the voltage as applied to the module (and measured), V_{PV} is the actual voltage at the PV device terminals, and R is the series resistance causing the error (Fig. 2.4).

Such errors affect above all P_{MPP} and FF. The four-wire measurement point for a PV module is typically the module plugs. For other PV devices, for example, with just busbars coming out of the mini module or encapsulated cell, one should report in detail how the DUT was electrically connected to avoid differences in measurements that cannot be explained.

After plugging DUT and measurement equipment, the *I–V* curve measurement can be started. The *I–V* curve measurement is essentially applying a voltage sweep to the DUT and in parallel measuring voltage and current of the module under illumination. For continuous solar simulators, the measurement can be started manually. When a pulsed solar simulator is used, it must be triggered by the light pulse. The electronics must be capable of measuring with a sufficiently high sampling rate, as the pulse length is often in the order of magnitude of only 10 ms. Prior to starting the measurement, the sweep profile of the electronic load must be defined considering the module characteristics and the available features of the measurement equipment. Finding the optimal parameters is usually an iterative process.

If a three-quadrant load[5] is available, typically the voltage ramp starts a little below 0 V in order to overcome ohmic resistances in the cables, and to obtain a clear intersection with the *y*-axis (0 V; this is important for accurate determination of I_{SC}). Similarly, it ends at a value somewhat larger than open circuit voltage to obtain a clear intersection with the *x*-axis (0 A). The sweep speed and direction play an important role. The sweep speed is the total voltage difference divided by the time, for example, 48 V/8 ms = 6 V/ms. Each solar cell and thus each PV module has a specific capacitance [9, p. 67], which causes a specific dependence of the current on the applied voltage and the direction of the voltage sweep. For 'forward measurements' (load voltage is increased from 0 V to open circuit), one can think of a capacitor within the module which is charged with the current, so that the power in MPP can appear smaller than the actual power. For 'backward measurements' (load voltage is decreased from open circuit to 0 V), this capacitor is discharged and adds current, so that the power in MPP can appear larger. The actual amount of over- or underestimation of P_{MPP} depends on the combination of module characteristics and sweep speed.

To prevent measurement errors due to the described hysteresis effect, different methods can be used. Hysteresis measurements consist of one forward and one backward measurement. The difference between both measurements indicates the severity of the capacity effect. If the detected hysteresis $\left(\text{hysteresis} = \frac{P_{backward} - P_{forward}}{P_{backward} + P_{forward}} \times 100\%\right)$ is too large, it can be decreased by splitting up the *I–V* curve in several sections (ie, measuring parts of the *I–V* curve with several light pulses). This decreases the sweep speed and thus the hysteresis. A reasonable range is 0.5–1%. Alternatively, important points of the *I–V* curve

[5] Electrical load that is capable of operating the module not only at positive currents and voltage (first quadrant, illuminated *I–V* curve) but also at negative currents (dark *I–V* curve) and negative voltage.

can be measured using a 'multiflash' procedure, where the voltage is kept constant during the light pulse. The desired voltage level is set per pulse. Further special methods to deal with capacitive effects exist, such as the Pasan-developed Dragon Back method [10] or an algorithm developed by Halm for temperature dependence or power rating measurements [11] (compare also Section 2.3).

2.2.4 Measurement of irradiance

The irradiance is measured using a reference device. Generally, the reference device can be smaller or larger than the DUT, or can have the same size. For secondary calibrations, the reference device is typically a primary-calibrated reference cell in WPVS (World Photovoltaic Scale) design [12]. For measurements in production, the reference device is typically a PV module of the same module type, ie, same kind and same number of cells compared to the produced modules.

In general, the irradiance indicated by the reference device is calculated using the calibration value and the measured signal:

$$G_{\text{ref}} = \frac{X_{\text{signal}}}{X_{\text{calibration}}} 1000 \frac{W}{m^2} \tag{2.9}$$

where G_{ref} is the irradiance as measured by the reference device, X_{signal} the signal (short circuit current, or the corresponding voltage measured via a resistance) of the reference device, and $X_{\text{calibration}}$ the calibration values, that is, the signal of the reference device at STC.

The calibration value of the reference device is one of the most important parameters affecting the measurement result. Therefore, it is essential that a reference device calibrated by an experienced and accredited institution is used.

The irradiance measured by the reference device can differ from the irradiance that effectively contributes to current generation in the DUT ('effective irradiance', [13]) mainly for the following reasons (compare Section 2.2.2):

- Spectral effects, quantified using the spectral mismatch factor (MM)
- Spatial effects, depending on nonuniformity
- Reflection/stray light effects

Theoretically, the effective irradiance can be calculated using correction factors as follows:

$$G = G_{\text{ref}} MM \, F_{\text{nonuniformity}} F_{\text{reflection}} \tag{2.10}$$

where G is the effective irradiance available to the DUT, G_{ref} the irradiance as indicated by the reference device, MM the spectral MM for correction of spectral mismatch, $F_{\text{nonuniformity}}$ a correction factor for nonuniformity errors in case the reference cell cannot be positioned correctly (see the following), and $F_{\text{reflection}}$ a correction factor for reflection-related errors.

The correction factors are strongly linked to the quality of the simulator light in terms of spectral irradiance, spatial nonuniformity and the directional distribution of the light as was discussed in Section 2.2.2. Including the *MM* quantitatively in measurements can be considered as state-of-the-art. The spatial nonuniformity factor is usually taken into account by positioning the reference cell correctly, that is, in a spot with irradiance equal to the average irradiance over the module area. The reflection factor is rather a wildcard character to illustrate here that such effects exist, but is typically not included quantitatively in measurements.

The spectral mismatch refers to the fact that PV modules have individual SRs, which means the conversion efficiency of light to current depends on the wavelength of the light. This causes the irradiance effectively contributing to current generation to be different for the reference device and the DUT—even at a spectral distribution equal to the reference spectral distribution. Fig. 2.5 depicts the spectral distribution of a c-Si and a CdTe module and a c-Si reference cell and demonstrates that differences in the SRs can be considerable. Furthermore, the spectral distribution of simulators differs from the reference spectral distribution (Fig. 2.5), which causes additional differences. The overall difference in the effective irradiance (ie, the irradiance that contributes to current generation in the DUT) is quantified using the spectral MM, which is calculated as follows:

$$MM = \frac{\int_a^b SR_{\text{DUT}}(\lambda) E_{\text{meas}}(\lambda) d\lambda}{\int_a^b SR_{\text{DUT}}(\lambda) E_{\text{ref}}(\lambda) d\lambda} \cdot \frac{\int_a^b SR_{\text{ref}}(\lambda) E_{\text{ref}}(\lambda) d\lambda}{\int_a^b SR_{\text{ref}}(\lambda) E_{\text{meas}}(\lambda) d\lambda}$$

$$= \frac{I_{SC}{}_{\text{DUT}}^{\text{simulator}}}{I_{SC}{}_{\text{reference}}^{\text{simulator}}} \Bigg/ \frac{I_{SC}{}_{\text{DUT}}^{\text{reference}}}{I_{SC}{}_{\text{reference}}^{\text{reference}}} \tag{2.11}$$

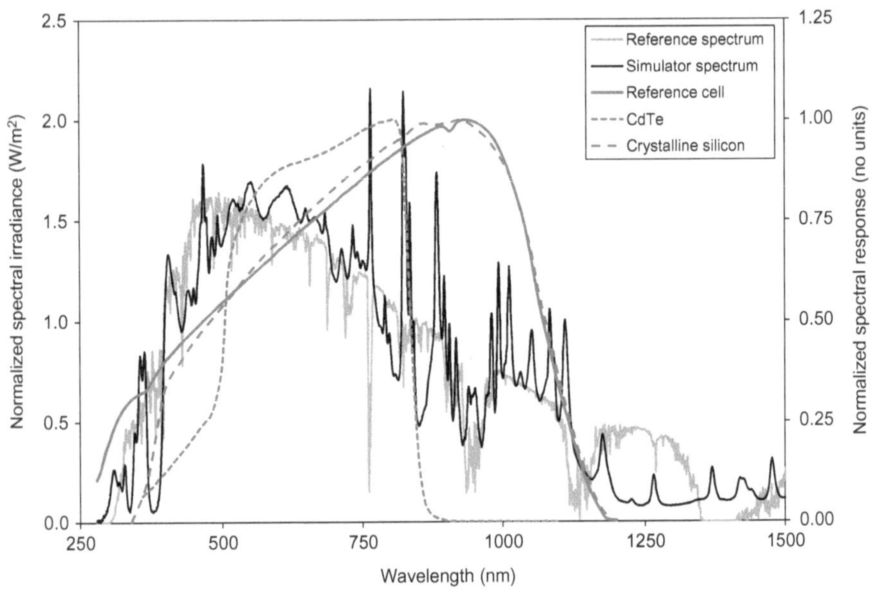

Fig. 2.5 Reference spectral distribution, realistic spectral distribution of a solar simulator (normalized so that the integrated irradiance in the range of 280–1700 nm is equal to that of the reference spectrum) and different spectral responses of PV devices.

where $SR_{DUT}(\lambda)$ and $SR_{ref}(\lambda)$ are the (relative) SR of the PV DUT and the reference device, respectively, $E_{meas}(\lambda)$ is the measured (relative) spectral irradiance under actual conditions and $E_{ref}(\lambda)$ is the reference spectral irradiance according to IEC 60904-3 [1]. The wavelength limits a and b should span the full range of spectral sensitivity of the DUT and the reference device. The transformation of Eq. (2.11) shows that *MM* can be written as a ratio of short circuit current values. This indicates that *MM* is the ratio of the effective irradiance difference between DUT and reference device at simulator spectrum, and the corresponding difference at reference spectrum.

The spectral MM can be used either to set the correct irradiance during the measurement or to correct the results afterwards. In the first case, the calibration value of the reference device is divided by *MM*, so that the indication of irradiance by the reference device during measurement already represents the effective irradiance for the DUT. A *MM* value larger than unity basically indicates that, effectively, the DUT can use a larger part of the incident irradiance than the reference device, that is, 1000 W/m^2 effective irradiance is reached at a slightly smaller power level of the lamp. If the MM correction is performed after the measurement, this is technically synonymous to corrections to STC for cases where the irradiance level during the measurement was set too low or high (Section 2.2.6). A useful approximation is to divide I_{SC} or P_{MPP} by *MM*.

Spatial nonuniformity can cause offsets in measured irradiance, if the reference cell is placed in a spot that is particularly dark or bright (compare Fig. 2.3). The offset should be minimized by placing the reference cell in a spot with irradiance equal to the average irradiance in the module plane (see IEC 60904-1, Sections 6 and 7 [3]), but could also be corrected by a nonuniformity factor after the measurement. It can be difficult to find a generic location for the reference cell for arbitrary module sizes.

Apart from the discussed influences, the characteristics of the reference device are essential for correct measurement of irradiance. Most important is long-term stability, meaning that the calibration value (but also other characteristics such as SR or reflection behaviour) must be constant over a period of several years. Furthermore, linearity is important (compare IEC 60904-10 [14]), and the reference device should have a high FF in order to limit influences if a shunt resistance is used.[6]

2.2.5 Measurement of module temperature

The module temperature is usually measured by attaching temperature sensors (Pt100, Pt1000 or thermocouples) to the back of the module. Sometimes, infrared sensors are used. The 'module temperature' is the average of the indication of all sensors. Typically, laboratories set-up limits for deviation from the target temperature and limits for

[6] When operated with a shunt resistance, the operating point of the reference device is not exactly short circuit while the calibration value typically is short circuit current. However, the current does not change significantly for operating points close to short circuit for reference devices with high fill factor.

the deviation of individual sensors from the average. To ensure module temperatures close to 25°C, the laboratory and the area where modules are stored prior to the measurement have to be carefully temperature controlled. This helps to limit uncertainties due to two frequently discussed issues with module temperature determination:

- Is the temperature of the backside equal to the temperature of the cell and the *p–n* junction?
- Is the spatial nonuniformity of temperature within the module sufficiently small?

For measurements with pulsed solar simulators, this question is not crucial if good air conditioning is available, but is harder to handle for STC measurements using continuous solar simulators or natural sunlight. It is a really difficult question for continuous outdoor monitoring. In a production line, where the air temperature (and thus module temperature) cannot be controlled as tightly as in a laboratory, these two issues lead to a larger variation of module temperature and cause larger associated measurement uncertainty.

2.2.6 The correction of I–V curves to STC

The STC parameters are obtained by combining measured *I–V* curve, measured module temperature and irradiance. Even if measurements are made at conditions very close to STC, small deviations of 5–10 W/m^2 or 1°C typically occur. These should be corrected to achieve best possible repeatability and comparability of results. Otherwise results from repeated measurements in one laboratory, or measurements from different laboratories, will differ from each other only because conditions were not identical.

Missing correction regarding irradiance mainly affects current (ie, I_{SC}, I_{MPP} and P_{MPP}). The influence is linear, that is, a deviation of 10 W/m^2 at 1000 W/m^2 leads to a difference of 1% in reported current values. Deviations of temperature from 25°C mainly affect voltage (V_{OC}, P_{MPP} and FF). The deviation propagates depending on the TC: a temperature difference of 1 K will lead to a difference in voltage and power values of around 0.3% and 0.4–0.5%, respectively (for standard crystalline silicon modules).

The correction of the *I–V* curve should be performed point-by-point and using one of the procedures described in IEC 60891 [15]. Especially for smaller corrections (around 10%), both procedure 1 and 2 deliver reliable results and do not offer significant advantages one over the other [16]. In the following, the equations for procedure 1 in IEC 60891 Ed. 2 are given for reference:

$$I_{STC} = I + \frac{G^*}{G_{SC}} I_{SC}\left(\frac{1000w/m^2}{G^*} - 1\right) + \alpha(25°C - T) \tag{2.12}$$

where I_{STC} is the current at STC, I the current as measured, T the temperature during the measurement, and α the absolute TC of short circuit current. G^* is the (effective) irradiance at time of data acquisition of individual *I–V* data points, and G_{SC} is the irradiance at the time of I_{SC} measurement. This method is used to correct for unstable irradiance during the measurement. It is mandatory that the irradiance be measured point-by-point like current and voltage.

$$V_{STC} = V + R_S(I_{STC} - I) - \kappa I_{STC}(25°C - T) + \beta(25°C - T) \quad (2.13)$$

where V_{STC} is the voltage at STC, I_{STC} is the current at STC, V and I are the voltage and current as measured, R_S is the series resistance, T is the temperature during the measurement, k is the curve correction factor, and β is the absolute TC of open circuit voltage.

The parameters k and R_S are determined from measurements at different temperatures (like TCs) and different irradiances, respectively (for details see Ref. [15]). For very small corrections (few percentage points), they can be neglected. For larger corrections, it is possible to work with generic module type or technology specific values [17]. The values strongly differ between thin-film and crystalline silicon PV modules, and the suitability of generic values for thin-film PV module types must always be evaluated carefully.

2.2.7 Module stability

The considerations made so far assumed PV modules are stable with regard to their electrical parameters. A PV module can be considered as stable 'if its electrical characteristics neither depend on its history of exposure to irradiance and temperature, nor on previous electrical bias or operating conditions, nor on combinations of these' [18]. Strictly speaking, this is never really the case for PV modules, especially not for most thin-film PV modules. Depending on the state of the module, the determined STC parameters may be different after a certain amount of dark storage or exposure to light. This can imply either power increase or decrease. For significant interpretation of measured parameters, it is necessary to characterize the stability of a module (or module type) in terms of time scale, magnitude and external causes (irradiance, temperature and electrical operation). It is useful to differentiate between irreversible and reversible (metastable) changes of the module *I–V* curve.

Known (meta)stability effects depend on module technology, and were summarized, for example, in Refs. [18,19]. It must be noted that the most relevant stability or metastability effects depend on the PV technology group. A-Si is affected by the Staebler–Wronski effect which causes, on the one hand, severe light-induced degradation (LID) in the first (hundreds) of kilowatt hours of exposure but, on the other hand, allows for annealing of defects at higher temperatures. This is visible as seasonal annealing in the performance of PV systems. Chalogenide technologies (mainly CdTe and chalcopyrite technologies (CI(G)S)) are affected by dark storage. The power change induced by dark storage can be recovered by exposure to light. Table 2.2 summarizes the most important effects per technology (table reprinted from [18]).

There are several terms used to refer to stability effects or the methods to limit their impact. An overview on these terms according to [18] is given in the following:

- Preconditioning/pretreatment procedures: conducted in order to remove metastable effects. Such procedures involve exposure of the module to irradiance and/or temperature at electrical bias or in open circuit. Relevant especially for chalcogenide PV technologies (cadmium telluride, CI(G)S) that are sensitive to dark storage.

Table 2.2 Reported impact of (meta)stability effects on module STC power

	c-Si	Amorphous silicon (a-Si, a-Si/a-Si, a-Si/μ-Si, triple junction)	CdTe	CI(G)S
Initial light-induced degradation	Around 1% [20], up to 4% [19]	Often around 15–20% [21], but up to 35% [22]; takes 100 or more kWh to stabilize [23], stabilized power is dependent on exposure conditions [24,25]	4–7%, but also power increase observed [26,27], superposition of recovery from dark storage and LID is possible	Both decrease and increase in the range of 1% to more than 10% observed [23,27], superposition of recovery from dark storage and LID is possible
Seasonal annealing	Not relevant	Amplitude of 3–6%, dependent on location and operating temperature [28–31], tends to be in the lower range for a-Si/μ-Si modules [32]	Not observed in published literature	Not observed in published literature
Sensitivity to unbiased dark storage (reversible by preconditioning)	Not relevant if module is stabilized	Not relevant if module is stabilized and kept in the dark at constant temperature	Several percent, up to 10% [23,27]	Several percent, up to 15% [23,27,33,34]

Summary of observed magnitude of stability problems in the mentioned literature, given by technology and group of effect. Note that all results are specific to material and processing of the modules, so that they can be transferred to behaviour of other module types of same technology only qualitatively and approximately.
From D. Dirnberger, Uncertainty in PV module measurement—Part II: verification of rated power and stability problems, IEEE J. Photovolt. 4 (2014), 991–1007.

- Sensitivity to dark storage: modules change their STC power as a consequence of dark storage. Power loss is more common, but power gain has also been observed.
- Stabilization: Procedure conducted in order to permanently remove initial stability effects, such as LID. Modules are exposed to 1000 W/m^2 in operation at MPP or open circuit until they are stable. The stability is determined either by subsequent *I–V* curve measurement in the laboratory (eg, every 43 kWh [35]) or by in situ *I–V* curve measurements.
- LID: Permanent degradation that occurs in the first tens or hundreds of kilowatt-hour exposure. Occurs for crystalline silicon (typically between 0% and 2%) and amorphous silicon (15–25% or more), and can occur for chalcogenide PV modules. For the latter, LID may be hard to determine because the power loss due to initial degradation could be masked by the power increase due to recovery from dark storage.
- Seasonal annealing: reversible change of module STC power in the field (increase in warmer months, decrease in colder months). Mainly observed for a-Si PV modules due to the Staebler–Wronski effect.
- Long-term degradation: power loss that occurs when modules are operated continuously, for example, as part of a PV system. See Ref. [36] for a summary of observed degradation rates.

One must be aware that the mentioned effects can influence the result of laboratory measurements, while the goal of measurements sometimes is to characterize these stability effects. Such measurements need to be planned properly, and it is important to consider the impact of stability effects depending on the timescale in which they occur (ie, cause the apparent PV parameters to change):

- ms: such very short timescale effects (eg, transient processes in the semiconductor structure) influence the *I–V* curve measurement itself, and, in extreme cases, may exclude the use of pulsed solar simulators. They increase measurement uncertainty and decrease reproducibility.
- Hours/days/weeks (eg, sensitivity to dark storage): the measurement result depends on the time that passed, for example, after dismounting the modules from an outdoor test set-up, or after a specific preconditioning procedure. Measurements must therefore be very well planned and documented, but are able to characterize the stability effects.
- Month/years: laboratory measurements can be used to characterize these changes without constraints, for example, long-term degradation of c-Si modules.

Solutions for stability problems are required for significant quality assurance. They depend on module technology and the purpose of the measurement. In Ref. [18], four different practical cases are explained:

- Verification of module power of new modules: The module power must be determined with initial effects removed, as the rated STC power must be representative of field operation. For crystalline silicon modules, this is typically tested by 20 kWh exposure (20 h, 1000 W/m^2, operation in V_{OC}; lower irradiance levels are allowed for outdoor exposure). For a-Si modules, the exposure time must be longer [35]. In situ *I–V* curve measurements during the stabilization period are recommended as this can considerably decrease measurement effort for determination of stability. Often, manufacturers of thin-film PV modules give indications of how the initial stabilization has to be conducted.
- Verification of power loss during accelerated ageing tests: The module power must be measured in the most reproducible way. Metastable effects can alter module power in between two measurements and must be dealt with under consideration of the specific characteristics of the module type under scrutiny.

- Verification of module power on field-exposed modules: Depending on the module technology and potential module-type-specific characteristics, sensitivity to dark storage (during transport) can affect the result.
- Calibration: above all, reference devices have to be stable. The short circuit current is often not or only little affected by (meta)stability issues, so that one solution can be to strictly use only the current for calibration. In cases where power is also used for calibration (eg, in some production lines), the module behaviour should be known very well.

As stability or metastability issues depend strongly on the detailed semiconductor composition and details in the production process, every module type must be looked at separately. For quality assurance purposes for large projects, it is reasonable to investigate the characteristics of one module type periodically. For the evaluation of the stability behaviour, comparison of indoor and outdoor measurements is essential. Fig. 2.6 shows an exemplary procedure of how a thorough investigation could be conducted [18].

2.2.8 Summary of influences on the I–V curve parameters I_{SC}, V_{OC}, P_{MPP}, FF

For analysis of deviations between two measurement results for the same DUT, it is helpful to keep in mind which influences most affect which electrical parameters.

As the photocurrent generation depends on the incident irradiance, but not much on temperature, I_{SC} depends on all mentioned aspects regarding measurement of irradiance. These are calibration, spectral mismatch and nonuniformity. Note that the calibration affects all measurements regardless of the characteristics of the DUT, while the other two also depend on module characteristics. I_{SC} is a rather stable parameter, but can be affected by metastability issues for a-Si devices. V_{OC} is affected mainly by temperature and, in the case of thin-film modules, by stability issues. When differences in FF are observed, this is typically related to contacting issues, capacitive effects and, in the case of thin-film modules, again stability issues. Differences in all mentioned parameters accumulate in differences of P_{MPP}, as $P_{MPP} = I_{SC} \cdot V_{OC} \cdot FF$. If a positive deviation in I_{SC} is combined with a negative deviation in FF, this can lead to no deviation in P_{MPP}. To judge on the overall quality of a comparison, it is therefore important to compare all parameters.

A summary of the most relevant influences per electrical parameter is given in Table 2.3. Note this represents the typical case, but can serve only as the starting point for a specific real case. The exact reason for deviations must be carefully analysed based on the exact measurement procedures that were applied by the involved parties. Examples for the interpretation of measurement results following this precept can be found in Ref. [37].

2.2.9 Measurement uncertainty for STC measurements

Measurement uncertainty, whether for *I–V* curve measurements or any other discipline, is information about the significance of the result of a measurement. The introduction to the *Guide to the Expression of Uncertainty in Measurement* (GUM)

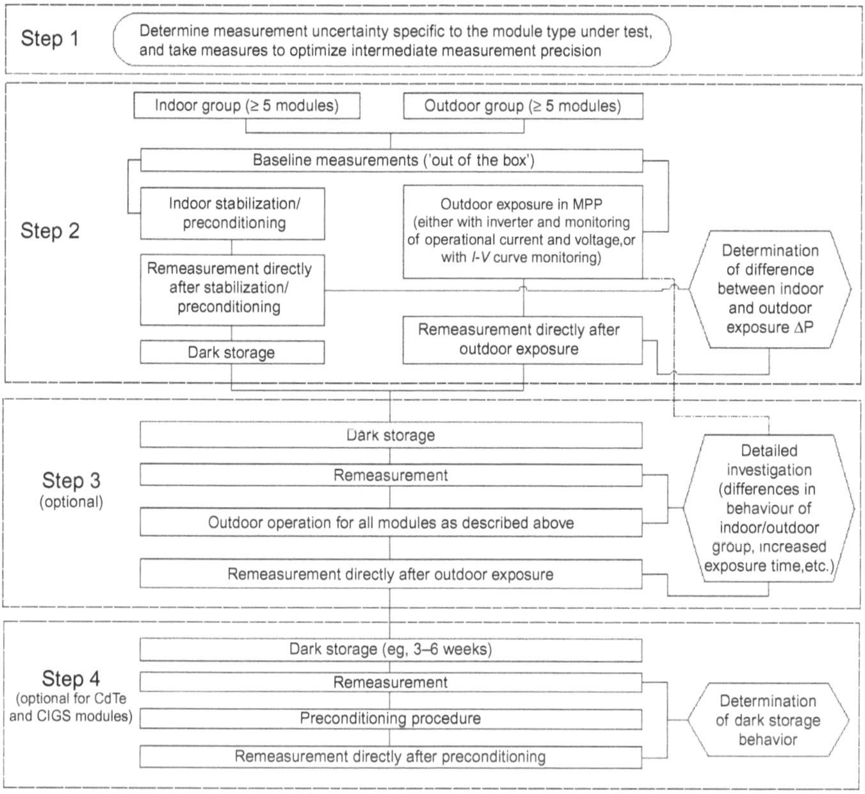

Fig. 2.6 Exemplary procedure to determine the STC power as representative for field operation and the corresponding uncertainty.
From D. Dirnberger, Uncertainty in PV module measurement—Part II: verification of rated power and stability problems, IEEE J. Photovolt. 4 (2014) 991–1007.

Table 2.3 **Summary of the main influences on measurement results for electrical parameters of PV modules**

Main influence	I_{SC}	V_{OC}	FF	P_{MPP}
Calibration	x			x
Spectral mismatch	x			x
Nonuniformity	x		(x)	x
Temperature		x	(x)	x
Contacting			x	x
Capacitive effects			x	x
Stability	(x)	x	x	x

describes measurement uncertainty as an indication of 'how well one believes one knows' [38, p. 3] the true value of a quantity by the measurement result. The 'true value' is a theoretical concept and can never actually be known. Uncertainty in more detail is probabilistic information—the probability density function that is attributed to a measurement result shows the range of values around the measurement result which has a specific probability to contain the true value. To somewhat standardize the information included in the probability density function, the coverage probability[7] is used. This is the probability that the true value lies within a specific interval, for example, the interval represented by the measurement result plus and minus the standard uncertainty associated with the measurement result. This interval is called coverage interval. Its magnitude depends on the selected coverage probability and the probability density function. Fig. 2.7 presents a simple example to demonstrate the introduced terms. It shows a measurement result, here the measured STC power of a PV module (250 W), and the attributed probability density function, here a Gaussian function. The standard deviation of the Gaussian function is equal to the determined standard uncertainty of, here, 0.8%. For a desired coverage probability of 95%, the Gaussian function requires the standard uncertainty to be multiplied with a coverage factor of approximately 2. The 'expanded uncertainty' is then 1.6%. The coverage factor depends on the probability density function and can often be determined only approximately.

It is important to understand the probabilistic character of measurement uncertainty in order to be able to correctly interpret measurement results under consideration of their uncertainty. Especially when evaluating differences between measurement results, or between measurements and data sheet specifications, one must be aware that a measurement result is not the absolute truth, but the best available estimate. Results that agree within the standard uncertainty can usually be looked at as essentially equal.

For the evaluation and quantification of measurement uncertainty, general rules are given by a framework of documents provided by the Bureau Internationale des Poids et Mesures (BIPM)[8] [38–41]. These documents are available free of charge from the BIPM website[9] and are also ISO standards [42–44]—except the *International Vocabulary of Metrology* (VIM) [41]. The process of measurement uncertainty evaluation is summarized in the *GUM* [39] in eight steps (see Section 8 of GUM; for a more basic and PV-related explanation see also [13], Section 2.4). These steps include identification, quantification, and combination of contributions to uncertainty. The key to sound uncertainty evaluation is that enough information about the measurand, the measurement procedure and equipment, and other influencing factors is gathered.

[7] For exact definition of this and other metrological terms used herein, the reader is referred to the '*International Vocabulary of Metrology*' which is also accessible as html version via http://jcgm.bipm.org/vim/en/index.html (accessed 06.11.15).

[8] The BIPM is 'the intergovernmental organization through which Member States act together on matters related to measurement science and measurement standards', http://www.bipm.org, (accessed 06.11.15).

[9] http://www.bipm.org/en/publications/ (accessed 06.11.15).

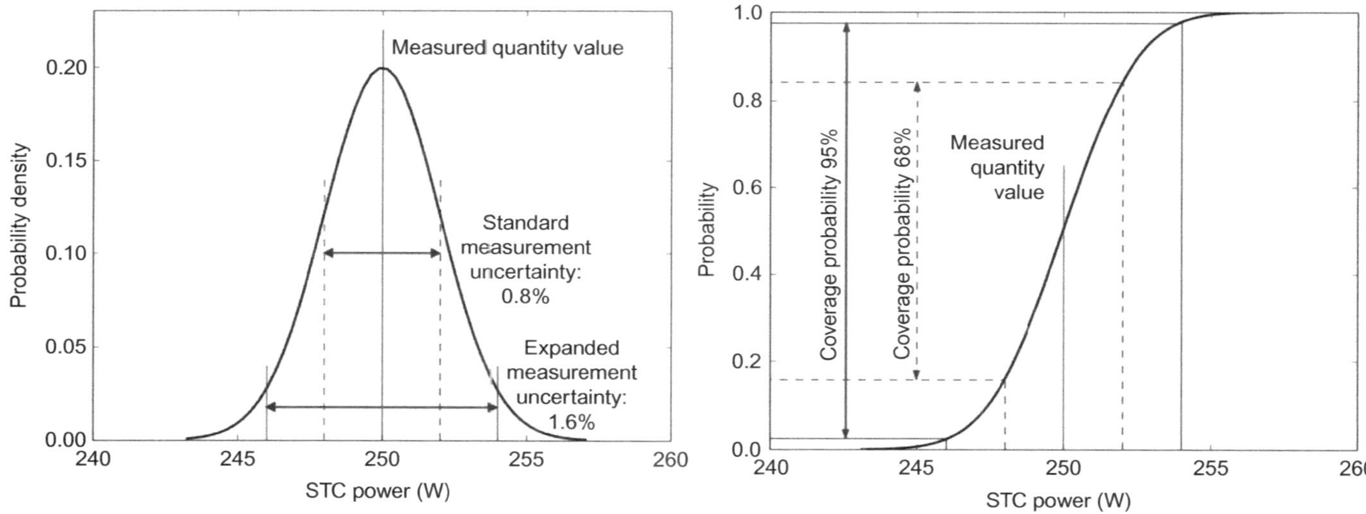

Fig. 2.7 *(Left)* Probability density function for a measurement of STC power with associated uncertainty. The measured power is 250 W and the standard measurement uncertainty 0.8%. For a coverage probability of 95%, the coverage interval is 246–254 W (68%: 248–252 W). *(Right)* (Cumulative) probability distribution function with indication of coverage probabilities and corresponding coverage intervals.
Figure from D. Dirnberger, Uncertainties in energy rating for thin-film PV modules (Doctoral thesis), Carl von Ossietzky Universität Oldenburg/ Fraunhofer Institute for Solar Energy Systems ISE, Freiburg, 2015.

In the context of PV module calibration, this concerns all influences that were discussed in the previous sections.

The first step according to GUM is to formulate a measurement equation, which expresses the relation between measurand (eg, I_{SC}, P_{MPP} and V_{OC}) and input quantities (everything that influences the measurand, eg, irradiance, reference cell, spectral mismatch, temperature, etc.). Ideally, this equation analytically represents the relation between measurand and all input quantities. In praxis, this is rarely possibly, so the GUM suggests the following simplification (paragraphs 5.1.4 and 5.1.5 in Ref. [39]):

$$Y = Y_0 + \sum_{i=1}^{N} c_i \delta_i \tag{2.14}$$

where Y is the measurand, $Y_0 = f(X_{1,0}, X_{2,0}, \ldots, X_{N,0})$ and $X_{1,0}, X_{2,0}, \ldots, X_{N,0}$ nominal values, X_i the input quantities, N the number of input quantities, $\delta_i = |X_i - X_{i,0}|$ transformations (eg, changes compared to the nominal values) of the input quantities, and c_i the sensitivity coefficients.

This equation expresses empirically how a specific change δ_i in an input quantity X_i propagates to the measurand. For example, the calibration value of the reference device is an important input quantity. According to Eqs (2.9), (2.12), it propagates directly to the STC short circuit current, that is, $c = 1$. Thus, δ of 1% in the reference device calibration would cause an error of 1% in the I_{SC}. For the temperature influence, c is equal to the TC: the temperature difference ΔT from 25°C during measurement will cause (if no correction to STC is performed) a change in the measurand with a magnitude of ΔT times the TC.

After having formulated the measurement equation, the estimated values of the input quantities must be determined (step 2). These are both quantities that result from the measurement under consideration (*I–V* curve, irradiance and temperature) as well as quantities that must be obtained from other sources (calibration value of reference cell, ambient conditions, spectrum of simulator, etc.). To visualize all input quantities, herringbone diagrams are helpful. As an example, Fig. 2.8 shows the most important input quantities as considered in the uncertainty evaluation by Fraunhofer ISE's module calibration laboratory CalLab PV Modules [8].

Then, the standard uncertainties of the input quantities must be determined (step 3). This is often an estimation based on calibration reports or characterizations of measurement equipment (eg, to determine that variability of the solar simulator spectrum) and the DUT (eg, the SR). Furthermore, possible correlations between the input quantities must be determined (step 4). A typical example for a correlation is the following: for determination of FF, both I_{SC} and I_{MPP} are required. These quantities both depend on irradiance and are thus correlated, which must be considered in the uncertainty calculation.

The result of the measurement is calculated using the formulated measurement equation (step 5). When the previous simplification (Eq. 2.14) is used, this step may be obsolete. If Y_0 is already the corrected measurement result and other corrections are not made, the only purpose of Eq. (2.14) is to formally derive Eq. (2.15) for

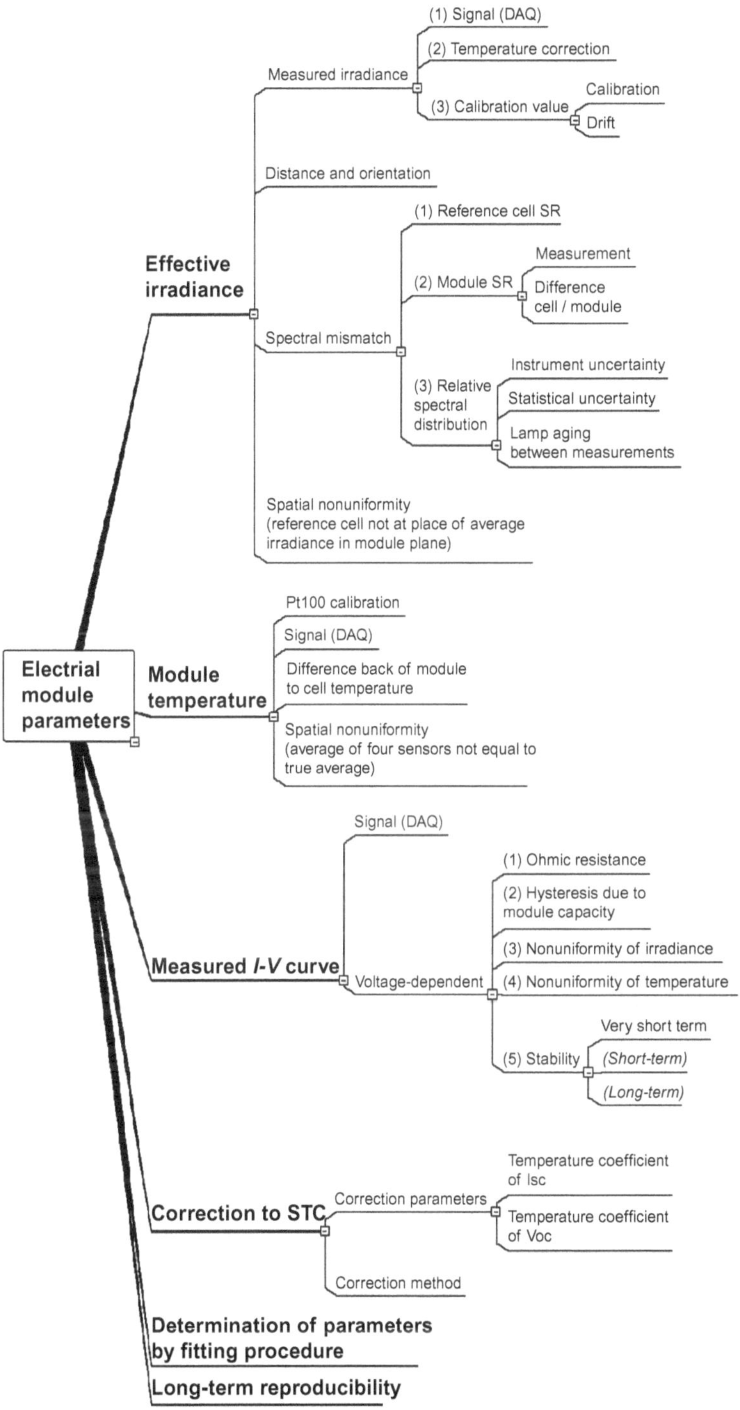

Fig. 2.8 Qualitative uncertainty budget for the determination of electrical parameter of PV devices, including all important input quantities. The combined uncertainty can be calculated subsequently for each node.
Figure reprinted from D. Dirnberger, Uncertainty in PV module measurement—Part I: calibration of crystalline and thin film modules, IEEE J. Photovolt. 3 (2013) 1016–1026.

calculation of the combined standard uncertainty. It is important to keep in mind that, even if the changes δ_i are assumed to be zero, this does not mean that δ_i is without uncertainty. The combined standard uncertainty is determined using the following equation (step 6), assuming no correlations between the input quantities:

$$u_c^2(y) = \sum_{i=1}^{N} \left(\frac{\partial f}{\partial x_i}\right)^2 u^2(x_i) = \sum_{i=1}^{N} c_i^2 u^2(\delta_i) \tag{2.15}$$

where $\frac{\partial f}{\partial x_i} = c_i$ are the partial derivatives evaluated at the expected values of X_i (sensitivity coefficients); u_c is the combined standard uncertainty; and $u(x_i)$ are the standard uncertainties of the input quantities X_i (Eq. 10 in the GUM; here: $x_i = \delta_i$).

If input quantities are correlated, an appropriate equation must be used (see GUM), or other, simplified methods to deal with the correlation must be applied. In the case of FF, for example, it is possible to determine the uncertainty for the quotients I_{MPP}/I_{SC} and V_{MPP}/V_{OC} instead of the uncertainty for the individual values, which implicitly considers correlations.

Uncertainty evaluation can consist of several subsequent steps of combining different contributions to combined uncertainties. For example, Fig. 2.8 shows that the combined uncertainty for the effective irradiance depends on contributions from the spectral MM, spatial nonuniformity and the reference cell calibration. The final combined uncertainty of the I–V curve parameters depends on the uncertainty of effective irradiance, module temperature, measured I–V curve, etc.

After having determined the combined standard uncertainty, the expanded uncertainty of the measurement result is calculated by multiplying the standard uncertainty with an appropriate coverage factor k (step 7). The coverage factor depends on the exact probability density function. This function is often not known, and is then assumed to be sufficiently well represented by a Gaussian probability density function. The coverage factor that is accordingly assumed is approximately 2, and the coverage probability approximately 95%. Finally, the measurement result is presented along with combined standard or expanded uncertainty (step 8). It should be kept in mind that an uncertainty evaluation is only valid when conditions during the measurement and characteristics of the DUT conform with the assumptions made during the uncertainty evaluation—for example, the spectral mismatch uncertainty depends on the SR of the DUT, and the nonuniformity influence is different for smaller and larger modules. Therefore, depending on the measurement equipment in use, it may be necessary to estimate some contributions to uncertainty specific to the DUT [8,37,45].

Detailed examples for measurement uncertainty budgets for PV module calibration which also include quantitative indications can be found in Refs. [8,46,47]. The lowest achieved measurement uncertainty for PV module calibration is 1.6% [8], with the largest contributions to uncertainty arising from reference cell calibration, spectral mismatch and spatial nonuniformity.

2.2.10 Importance of calibration and traceability

The role of the reference device calibration value in PV module measurement was explained previously (Eq. 2.9). It is easy to see that a change in the calibration value can shift the result of a PV module measurement to higher or lower current and thus power. For a consistent evaluation of the power of PV modules, it is thus important that the primary calibration procedures applied throughout the world achieve comparable results.

In general, national metrology institutes transfer the definition of SI and other units to accredited laboratories and to the market. In the case of PV, this is done by providing primary-calibrated reference cells. 'Primary calibration' (or, according to the VIM [41], 'primary reference measurement procedure') denotes a measurement procedure that produces a result with a specific unit without using a reference device with a calibration value with the same unit. Primary calibration procedures for PV devices transfer the measured radiated energy (in units of watts) to current generated by the PV device (in units of amperes). With a calibration value resulting from a primary calibration, the device becomes a 'primary standard'. The calibration is transferred from PV cell to PV module by *I–V* curve measurement as described in the previous sections (using a primary standard as reference cell). PV modules (or cells) with calibration values from such measurements ('secondary calibration') are 'secondary standards'. The calibration level is then further transferred by the creation of reference standards and working standards. For such measurements, a secondary standard is used to measure the irradiance instead of a primary standard. The pyramid in Fig. 2.9 depicts the multiplication of the calibration level from primary standards

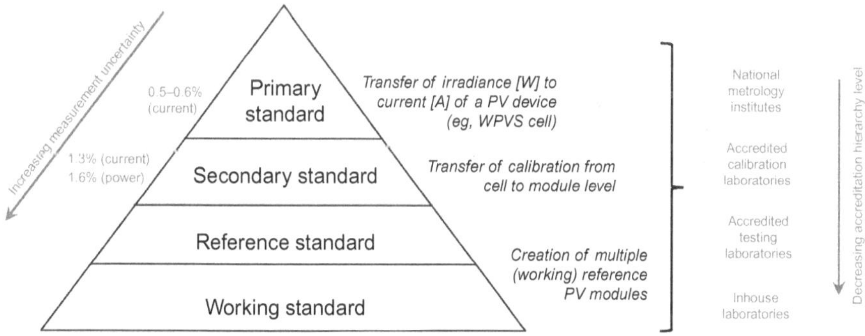

Fig. 2.9 The calibration of PV devices is based on primary standards created by national metrology institutes or other reference laboratories. The calibration is transferred to the market via calibration, testing and inhouse laboratories. With disregard of their accreditation hierarchy, laboratories can choose which standard they use. The inherent uncertainty increases with each additional transfer of calibration. The indicated uncertainties are best-case values for primary cell calibration [48] and for secondary module calibration [8].
Figure and caption reprinted from D. Dirnberger, Uncertainties in energy rating for thin-film PV modules (Doctoral Thesis), Carl von Ossietzky Universität Oldenburg/Fraunhofer Institute for Solar Energy Systems ISE, Freiburg, 2015.

to working standards. While the definition of primary and secondary standards is strict, the use of the terms *reference standard* and *working (reference) standard* can sometimes differ. Manufacturers frequently also use the terms *golden* or *mother* module for the standards they use for creation of further working standards. The established calibration and recalibration regime is specific to the manufacturer or laboratory. It is important that every standard device used be somehow traceable back to a primary standard, as each additional transfer of calibration adds uncertainty, and the mere fact that a traceable standard is used as a reference device does not guarantee low uncertainty.

At the time of writing, primary calibration of PV cells is carried out by a small number of laboratories and national metrology institutes all around the world. The most important ones are the PTB (Physikalisch-Technische Bundesanstalt, Germany's national metrology institute), NREL (National Renewable Energy Laboratories, Golden, Colorado, USA), ESTI (European Solar Test Installation, part of the European Commission's Joint Research Center in Ispra, Italy) and the AIST (National Institute of Advanced Industrial Science and Technology, part of the Research Center for Photovoltaic Technologies in Japan). These laboratories use different indoor or outdoor methods, for example, the Differential Spectral Responsivity Method at PTB [48–51], the Solar Simulator Method at AIST [52,53] and ESTI [54], the Direct Normal Sunlight Method [55,56] at NREL and ESTI [57], and the Global Sunlight Method [58] at ESTI. A comparison of several primary calibration methods is included in Ref. [54]. The procedures are traceable to SI units (PTB) or to the World Radiometric Reference [59]. The currently lowest measurement uncertainty is provided by PTB (0.5–0.6%, $k=2$ [48]). The international comparability of these procedures is traced by intercomparisons such as the WPVS [12,60,61].

Secondary calibration of PV modules is carried out by several laboratories all around the world. These are reference laboratories both for manufacturers and other service laboratories. Intercomparisons such as presented in Ref. [37] check on the comparability of secondary calibrations on module level.

While traceability is certainly most important for the reference device calibration, all measurement equipment used should be calibrated and traceable back to primary standards.

In parallel to the hierarchy in the traceability chain, the accreditation level of a laboratory is important (see Fig. 2.9). The accreditation represents the degree of third-party verification of the measurement equipment, procedures, and uncertainty estimation. The exact requirements to obtain accreditation as a calibration laboratory or testing laboratory may differ slightly from country to country, as each country has its own accreditation body (eg, DAkkS in Germany, UKAS in Great Britain, or ANSI in the United States). The national metrology institutes are usually part of the accreditation process, as they are the state organization responsible for measurement. In general, the accreditation level does not give information about whether a laboratory produces primary, secondary or working standards.

2.3 PV module characterization

2.3.1 Introduction of yield-relevant module characteristics

As mentioned previously, STC power is by far not the only important characteristic to assess PV module performance. The operating conditions span over a wide range of temperatures, irradiance levels, angles of incidence of the sunlight, and spectral distributions. The influence of these conditions on the energy produced depends on the corresponding module characteristics, which are:

- Temperature coefficients
- Irradiance dependency (or low light behaviour)
- Spectral response
- Angular response
- Thermal behaviour

In the following sections, state-of-the-art measurement methods for determination of these characteristics will be introduced. Furthermore, long-term stability of STC power and the mentioned characteristics is important. While long-term stability of STC power has been in the focus of many investigations [36], information about the long-term stability of the other characteristics is rare.

2.3.2 Measurement of temperature dependency: methods and challenges

The temperature dependency is determined by *I–V* curve measurements at a constant irradiance, typically 1000 W/m^2, and different temperature levels. Therefore, temperature dependency measurements require the same set-up as for STC measurements and in addition the possibility to heat up and cool down the DUT. Typically, temperature chambers are used. The offered temperature range is at least 25–65°C; modern temperature chambers even offer ranges from 15°C to 75°C. Data obtained in the range of 25–65°C is, however, sufficient for most practical applications.

Compared to STC measurements, the measurement of the temperature of the DUT is even more important. Using a sufficient number of temperature sensors is mandatory in order to monitor and control the temperature nonuniformity over the module during all measurements. Normative guidance for TC measurement is included in the standards IEC 60891 Ed. 2 [15] and IEC 61853-1 [62] (measurements at different temperature levels).

Using a temperature chamber, there are, in principle, two ways for performing the measurement: either the module is heated up first and *I–V* curves are measured during cooling in specific temperature increments (eg, 1 K), or the module is heated up to selected temperature levels and kept there for each *I–V* curve measurement. In both cases, keeping nonuniformity within set limits is essential. A reasonable limit when using four temperature sensors is a difference of 2 K between minimum and maximum temperature.

The first method allows for taking more data points in reasonable time, but the accuracy is limited when the pulse duration is too short for the capacity of the DUT (compare Section 2.2.3). As the cooling process goes on continuously, each *I–V* curve must be finished within one pulse as otherwise the temperature will change too much (note that most simulators require at least 10–30 s between two pulses). Customized methods for temperature correction of each *I–V* curve point might solve this issue. The second method allows keeping the module at the selected temperature level for a longer time and thus conducting multiflash or section measurements. However, the number of points is limited for time reasons—it takes a while until the temperature has stabilized. Thus, an individual data point and its statistical measurement uncertainties can influence the overall measurement result relatively strongly.

The typical results of a temperature dependency measurement are TCs for the electrical parameters I_{SC}, V_{OC}, P_{MPP} and FF. For each measured *I–V* curve, the parameters I_{SC}, V_{OC}, P_{MPP} and FF are determined and plotted versus temperature in order to calculate the respective TC (Fig. 2.10). The TC can be reported either as absolute (ie, mV/K, mA/K, mW/K, etc.) or relative value. Relative values are calculated by dividing the absolute TC by the corresponding STC value:

$$TC_{\text{rel}} = TC_{\text{abs}}/X_{\text{STC}} \tag{2.16}$$

where TC and X_{STC} are the TC and the STC value of the parameter under consideration, respectively.

The uncertainty of the absolute TC of an individual module is smaller than that of the relative TC, as the STC uncertainty does not contribute. Thus, important influences like calibration of the reference cell do not influence the uncertainty. However, the absolute TC varies depending on the module-specific voltage or power. Relative TCs eliminate this influence and can thus be used as generic values, for example, for different power classes of one module type on the datasheet or even for different module types of the same module technology (eg, standard c-Si). In the case of metastable modules (Section 2.2.7), the TC can be influenced by temperature-induced metastable changes of the module state. For example, annealing that takes place while the module is heated up to different temperature levels for subsequent *I–V* curve measurements can reduce the apparent TC: STC power increase due to annealing partly makes up for the power decrease due to rising temperature. The exact influence depends on the measurement method and the module behaviour and can thus not be quantified on a general basis.

Some module types exhibit nonlinearities, which typically show only in measurements over a larger temperature range (eg, 15–75°C; compare Fig. 2.13). Typically, these are quite small and can be neglected—especially considering that simulations are usually performed under the assumption of a linear temperature change, that is, TC is constant over temperature. If larger nonlinearities occur, it may be necessary to use different TCs

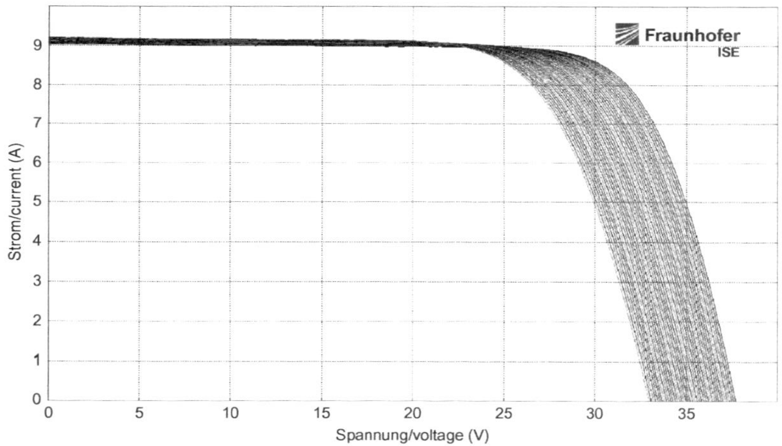

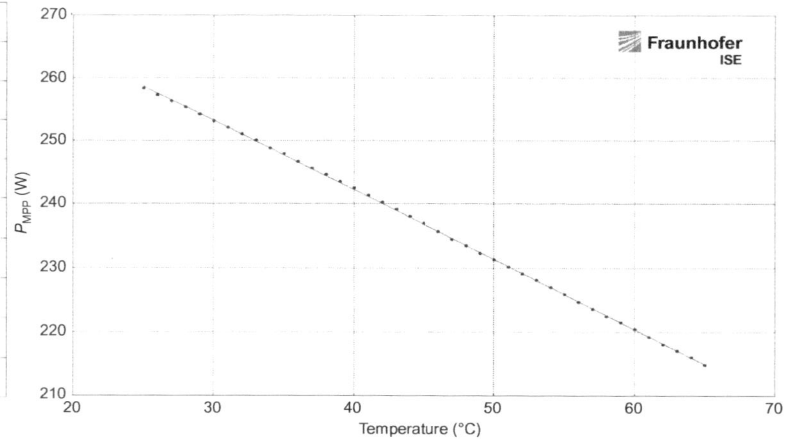

Fig. 2.10 *I–V* curves measured during a TC measurement *(left)* and the derived P_{MPP} points used for calculation of the temperature coefficient *(right)*. Figures provided by Fraunhofer ISE.

depending on the temperature range under consideration. In the case of metastable PV modules, nonlinearities can appear as a consequence of the metastability.

Uncertainty of relative (linear) TCs is currently estimated to be around 25% for I_{SC} (the influence is very small and thus difficult to measure) and around 5–10% for V_{OC} and P_{MPP}, although commonly agreed-on methods for determination of this uncertainty still have to be developed. The fact that TC is a value derived from a set of other values limits application of the GUM [39] and requires approaches such as Monte Carlo methods [40]. Prerequisites for accurate and precise results are sufficient temperature uniformity, accurate temperature measurements and high reproducibility of measurements.

2.3.3 Measurement of irradiance dependency: methods and challenges

The irradiance dependency of a PV module is often referred to as 'low light behaviour'. It is usually measured at 25°C in a range of 100–1000 W/m^2 in 100 W/m^2 increments. Some simulators allow for irradiance levels of up to 1300 W/m^2. The measurement result consists of a set of measured *I–V* curves. The 'low light behaviour' is typically reported as 'relative efficiency', which is the measured efficiency normalized to the efficiency at 1000 W/m^2. Like the relative TC, this allows for easy comparison between module types or modules with different STC power (Fig. 2.11).

In addition to the STC measurement equipment, the possibility of changing the irradiance level is required. Normative guidance can again be found in the standards IEC 60891 Ed. 2 [15] and IEC 61853-1 [62]. The technical challenge of irradiance dependency measurement is to reduce the simulator irradiance without changing other light characteristics such as nonuniformity or spectral irradiance. Otherwise, the apparent irradiance dependency is a combined effect of the actual irradiance dependency and alterations induced by the dependence on spectral irradiance or nonuniformity. State-of-the-art solar simulators are optimized for measurements at STC, so that, typically, at least one light characteristic does change. This increases measurement uncertainty. If such influences are not recognized and corrected, this has consequences for the quality of the measurement result: the relative efficiency can systematically appear better or worse (Fig. 2.12, [63]).

There are mainly two established concepts for reducing irradiance: either the power for operating the simulator lamps is reduced or the produced light is reduced by means of filters. Reducing lamp power increases the share of higher wavelength in the light, so that the spectrum changes with irradiance. Especially for large differences of SR of reference device and DUT, this can lead to large systematic errors. Spectral mismatch corrections specific to each irradiance level can solve this issue. Using grey filters usually affects nonuniformity, while keeping the spectral distribution constant on the whole (note that often a combination of grey filters with slight changes in lamp power is necessary to perform measurements in 100 W/m^2 increments). The effect on nonuniformity strongly depends on the quality of the grey filters, so that it is mandatory to measure the nonuniformity with each filter set in use (Section 2.2.2). With high-

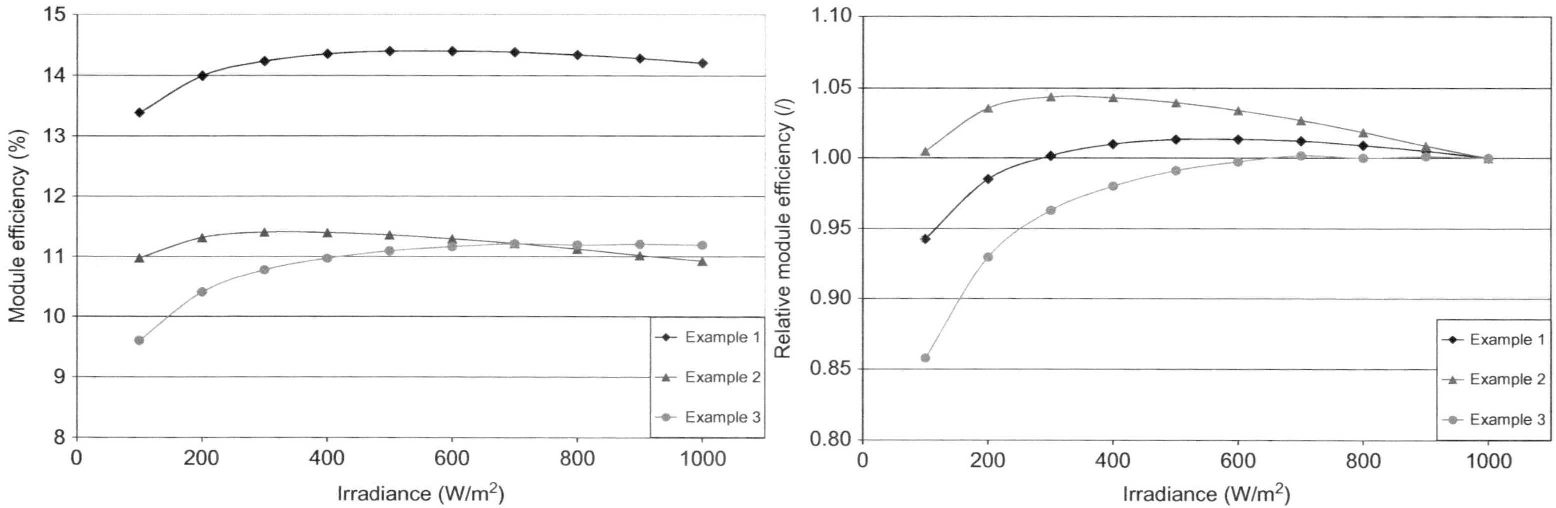

Fig. 2.11 Measured absolute irradiance dependency of the efficiency of different PV modules *(left)* and corresponding relative efficiency ('low light behaviour', *right*).
Reprinted from D. Dirnberger, B. Müller, U. Kräling, S. Dittmann, Uncertainties of laboratory measurements for energy rating, in: 27th European Photovoltaic Solar Energy Conference and Exhibition, Frankfurt, Germany, 2012, pp. 3294–3301.

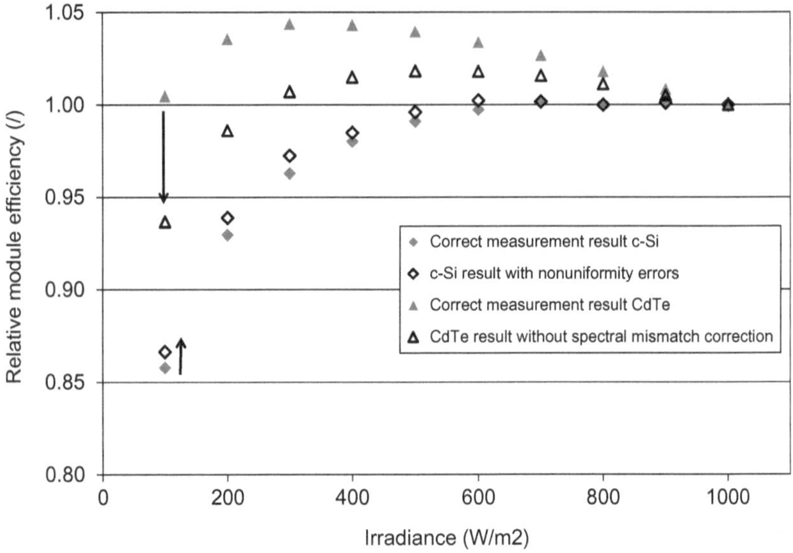

Fig. 2.12 Influence of systematic offsets on apparent low light behaviour. Lacking spectral mismatch correction can increase the error with each irradiance step (simulator spectrum changes continuously). The nonuniformity error changes when a different set of grey filters is used (at 600 and 300 W/m^2).
Figure reprinted from D. Dirnberger, B. Müller, U. Kräling, S. Dittmann, Uncertainties of laboratory measurements for energy rating, in: 27th European Photovoltaic Solar Energy Conference and Exhibition, Frankfurt, Germany, 2012, pp. 3294–3301.

quality filters, the nonuniformity can usually be kept below or around 2% for typical module areas of around 2 m × 1.1 m.

Uncertainties for 'low light efficiency' are to date reported mainly for the 200 W/m^2 efficiency (the absolute value), and are in a range of 3.5–5% depending on the laboratory. For the relative efficiency, which is used for simulations, uncertainties are smaller because systematic influences like reference cell calibration are removed due to the normalization. The uncertainty is dominated by effects that influence measurements at different irradiance levels in a different way, such as shown in Fig. 2.12. It must be kept in mind that this influence is again specific to characteristics of the DUT. Furthermore, the apparent irradiance dependency may be influenced by module metastability, so that it is recommended to perform measurements after preconditioning/stabilization procedures whenever in doubt.

2.3.4 Power rating measurements according to IEC 61853-1

Temperature and irradiance dependency are typically measured at 1000 W/m^2 and 25°C, respectively. For yield predictions and module simulations (Chapter 4), the data are also used to describe module behaviour at different combinations of irradiance and temperature levels, that is, the influences are assumed independent. In order to be able

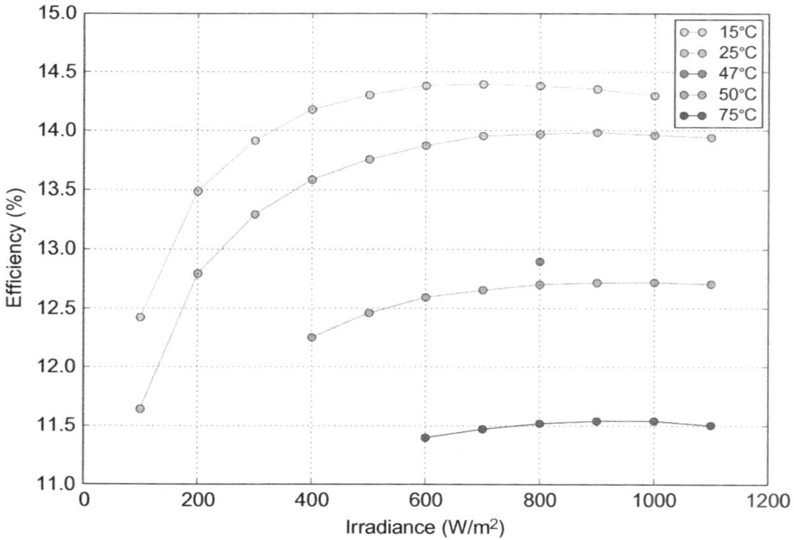

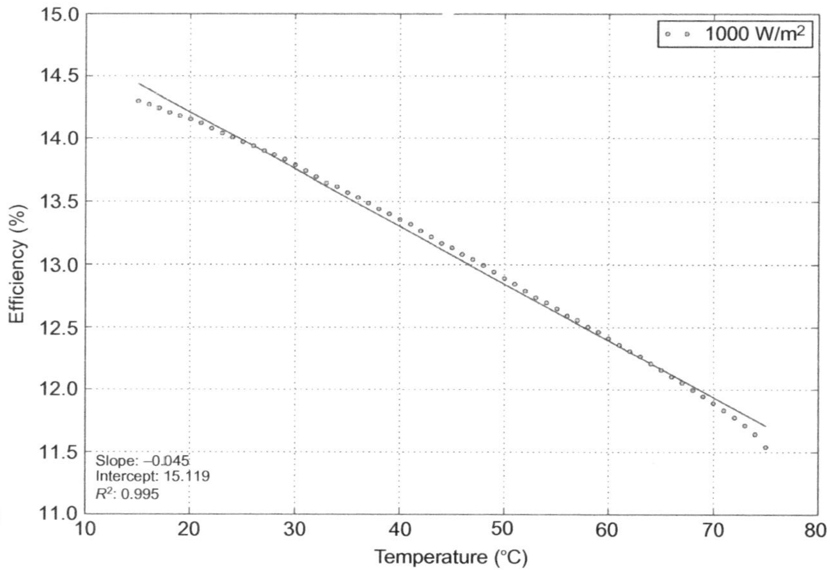

Fig. 2.13 Result of a power rating measurement (*left*: efficiency vs. irradiance for several temperature levels, *right*: efficiency vs. temperature at 1000 W/m^2) for a thin-film CI(G)S module. Slight differences in the irradiance dependency for the different temperature levels (especially 15°C and 25°C) show as well as slight nonlinearity in the temperature coefficient.

Figures provided by Fraunhofer ISE.

to verify this assumption, the power rating matrix suggested in IEC 61853-1 [62] foresees measuring *I–V* curves at different combinations of irradiance and temperature, so that, for example, the low light behaviour is available at several temperatures (Fig. 2.13).

The measurements are carried out with a simulator that is capable of reaching the required irradiance levels and has a temperature chamber. Some simulator manufacturers offer comprehensive software solutions to carry out these measurements automatically.

2.3.5 Measurement of SR: methods and challenges

The SR(λ) is the module characteristic that describes the dependence of photocurrent generation on the wavelength of the incident irradiance. It is related to the external quantum efficiency EQE(λ) as follows:

$$SR(\lambda) = \frac{q}{hc}\lambda EQE(\lambda) = 0.808\lambda\, EQE(\lambda) \quad (2.17)$$

where h is the Planck constant, c is the speed of light and EQE(λ) is the wavelength-dependent EQE ([2], p. 12).

As was mentioned before, the SR of a module plays an important role in the achievable measurement uncertainty. As this data is required for spectral mismatch correction, the measurement is very often part of module calibrations.

Measurements can be carried out on mini modules with the same cells, encapsulation, backsheet and glass as the full-size module under investigation. To exclude any effect from increased reflections, the mini module should consist of 3×3 cells with only the middle cell electrically connected. Alternatively, full-size modules can be opened up on the back side and one individual cell can be connected. SR measurements on cell level still have somewhat lower uncertainties than on module level (around 2–3% [64] compared to around 5% [65] in the range of approximately 400–1000 nm), but sufficient agreement is achieved in comparisons, as shown in Fig. 2.14.

For full-size modules, two different methods to measure the SR exist. The first one measures SR(λ) with the full module illuminated with quasi-monochromatic light (wavelength band with a width of around 50 nm or less), using a pulsed solar simulator with bandpass filters [66]. No bias light[10] is applied, which limits the application of the method for nonlinear PV modules. The nonuniformity of the quasi-monochromatic light has an important influence on the measurement uncertainty. The nonuniformity can cause different cells to limit the measured current of the module. A measurement uncertainty analysis for this method was presented in Ref. [65].

[10] Bias light is white light, that is, containing a broad range of wavelengths as opposed to the quasi-monochromatic light, which can be used to make sure a measurement is made at irradiance levels above the threshold where nonlinearities typically occur. The quasi-monochromatic irradiance can be lower than that threshold, depending on DUT and measurement set-up.

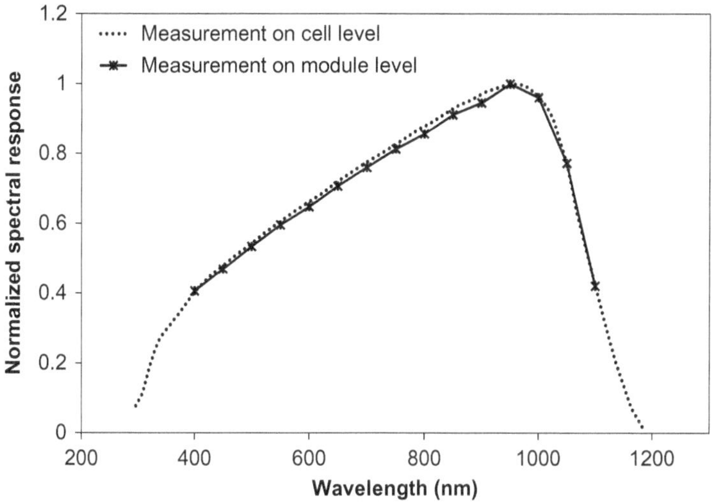

Fig. 2.14 Normalized spectral response measured on the full-size module and one individual cell of the same module. Deviations are smaller than measurement uncertainties. Data provided by Fraunhofer ISE, CalLab PV Modules.

The second method measures SR(λ) with only a small spot on the module illuminated with monochromatic light [67], and the other cells of the module flooded with bias light.

Recommendations for SR measurement on the module level are also included in the draft IEC 61853-2 [68] by reference to IEC 60904-8, Ed. 3.0 [69]. Guidance on measurement of multijunction modules can be found in Refs. [70,71].

2.3.6 Measurement of angular response: methods and challenges

The angular response of a PV module describes its relative reflection behaviour at nonperpendicular irradiance. One can speak of the 'relative transmission',[11] which is measured via the short circuit current of the PV module (see also IEC 61853-2 [68]):

$$\tau_{\text{rel}}(\text{AOI}) = \frac{I_{\text{SC}}(\text{AOI})/I_{\text{SC}}(\text{AOI} = 0°)}{\cos(\text{AOI})} \tag{2.18}$$

where τ_{rel} is the relative transmission and AOI the angle of incidence.

[11] This refers only to the change of reflection characteristics with different angles of incidence, not to reflection losses that occur at STC.

A set-up is required that allows for rotation of the DUT. As a consequence of the rotation, one part of the module will move closer to the light source than the other, and will thus receive more irradiance.[12] For typical solar simulators, this increases the nonuniformity over the whole module to an unacceptably high level. Therefore, the measurements must be performed with respect to one single cell. The axis of rotation must be the middle axis of the cell, or the part thereof, that is measured.

Four methods exist for determination of the relative transmission according to Eq. (2.18) in the laboratory:

- Measurement of the short circuit current of a mini module (eg, nine cells) with only the middle cell electrically connected, with identical design compared to the full-size PV module under investigation (with regard to cell, encapsulation, and glass, etc.)
- Measurement of the short circuit current of a single cell of the module under investigation. The cell is contacted through the backsheet. The disadvantage of this method is that the module is damaged and not suitable for further use.
- Measurement of the short circuit current of the full-size module with one shaded cell [72]. Approximately half the cell is shaded, so that the current of this cell will limit the current of the module and thus the measured short circuit current at the module terminals is the current produced by the shaded cell. The bypass diodes of the modules have to be removed.
- Measurement of *I–V* curves of the full-size module with no shading and with one shaded cell. The current value used for determination of relative transmittance is the current where both *I–V* curves have their intersection. Details on the evaluation can be found in Ref. [73].

Furthermore, the IEC 61853-2 [68] also includes an outdoor procedure.

The main contributions to uncertainty are the determination of the AOI and the effect of both stray light and the directional distribution of the light. An error of 1 degree in the determination of AOI at a nominal angle of 60 degrees causes a difference of around 3% in the relative transmission. The effect from stray light or directional distribution depends very much on the solar simulator in use.

2.3.7 Thermal behaviour

An important factor for the performance of a PV module in reality is the operating temperature that sets at specific conditions. The operating temperature depends on the thermal equilibrium that attunes, that is, on the incident irradiance, the ambient temperature, the wind speed and direction, and also the efficiency[13] and construction of the module. As it is difficult to analytically describe this dependence and to experimentally determine relevant factors, so far the 'thermal behaviour' of a module has been reported using the nominal operating cell temperature (NOCT) determined according to IEC 61215 [74].

The NOCT test is performed outdoors, the module being mounted in an open rack at a 45-degree tilt angle and operated in open circuit. Module temperature is monitored

[12] Assuming a point source of light, the irradiance at a specific point with a distance r to the light source is proportional to $1/r^2$ (inverse square law).

[13] Irradiance incident on the module that is not reflected on the glass either creates electron–hole pairs or is transformed to thermal energy in the module, which increases the module temperature.

along with irradiance, ambient temperature and wind speed. NOCT is the temperature determined according to specific rules for reference conditions of 800 W/m^2, 20°C ambient temperature and 1 m/s wind speed.

The results from this procedure were found to be location and test set-up dependent, so that the NOCT concept and the method are questioned among scientists [75–77]. For example, NOCT values indicated on the data sheet tend to differ strongly (up to 10 K) between different module types with same cell technology and similar construction, but scientific results presented in Ref. [75,76] suggest that the actual difference is negligible in such cases. Furthermore, PV modules have higher operating temperatures (several degrees) when operated in open circuit compared to MPP.

An alternative procedure to overcome these limitations is suggested by Koehl et al. [76]: a realistic module operating temperature (ROMT) shall be determined for straightforward comparison of module types, along with model parameters for description of the correlation between irradiance, ambient temperature, wind and the module operating temperature. The parameters could be used much better for simulation of module temperature under different conditions in PV performance predictions (Section 2.5), and would thus enable differentiating modules also considering their thermal behaviour in future.

While one could think of methods to determine NOCT or a comparable parameter in the laboratory, for example, using wind tunnels, controlled air temperature and suitable solar simulators, efforts to develop such methods are not known to the author at the time of writing.

2.3.8 Outlook on special requirements for outdoor calibration and characterization

In the previous sections, the calibration and characterization of PV modules was described with a focus on laboratory measurement methods. All the measurements can also be performed outdoors, and the applicable rules are the same. As outdoor measurements depend on the current weather conditions and the local climate, the specific challenges are a little bit different:

- *I–V* curve measurement: As the sunlight is constant light, there are no issues regarding sweep speed and capacity effects, as long as appropriate measurement equipment is used and as long as conditions are stable (eg, no moving clouds).
- Irradiance measurement: For outdoor measurements, pyranometers can be used as an alternative to reference cells; however, this increases uncertainties due to differences in spectral and angular response between reference device and DUT. Both for pyranometer and reference cell, care should be taken that the temperature has stabilized and that an appropriate signal temperature correction is applied. Nonuniformity of natural sunlight can be much better than in the laboratory, but this also depends on the exact location where the measurements are carried out (reflections from the surroundings can have a negative impact).
- Spectral distribution: Depending on location, time, weather and atmospheric conditions, there can be significant differences between the reference spectrum and the solar spectrum (see eg [78,79]). Therefore, the spectrum of the sun should be recorded and a spectral mismatch correction should be performed.

- Temperature: The determination of the correct module temperature is one of the largest challenges in outdoor measurement. As the irradiance will heat the DUT, the usual approach is to keep the DUT in an air-conditioned environment at a temperature somewhat below 25°C until directly before the measurement, and then make subsequent measurements during the heating up. To avoid making measurements while the temperature changes (transient processes negatively affect the temperature nonuniformity over the module and the difference between the temperature at the backside of the DUT and the actual cell temperature), one could also let the temperature stabilize and then perform corrections to STC (Section 2.2.6). In that case, uncertainties due to the correction, especially due to uncertainties in the TC, occur [17].
- Correction to STC: If conditions are stable and close to 1000 W/m^2, the situation is very similar to laboratory measurements. If this is not the case, the correction and especially the determination of the relevant parameters play a much larger role [17].

The great advantage of laboratory measurements is that conditions can be controlled and are independent of the weather and climate, which allows for very high reproducibility of measurements. Nevertheless, outdoor measurements can be as accurate and precise as laboratory measurements, if performed with the right level of expertise and at appropriate conditions. In intercomparisons, very good agreement was achieved between laboratory and outdoor calibration [37]. For new PV technologies or technologies that are sensitive to specific characteristics of solar simulators (eg, multijunction modules to simulator spectrum), outdoor measurements can offer a possibility to verify laboratory measurements.

2.4 PV module characterization in praxis for the purpose of quality assurance

2.4.1 Variability of module characteristics between modules of the same type

While the preceding sections contained descriptions of measurements of one individual module, it is important to keep in mind that characteristics can vary between modules of the same type. Thus, measurements of one individual module provide insufficient information about the quality of a whole batch of modules. In the following, the relevance of variable module characteristics for quality assurance is discussed.

First of all, the STC power varies among modules of one type due to cell power and the module production process. Typically, modules are binned to power classes with a width of 5 W, for example, 250, 255 Wp, etc. They can be 'plus-binned', that is, class 250 Wp includes only modules with STC power >250 Wp. Alternatively, symmetrical binning is common, that is, class 250 Wp contains modules with STC power of 250 ± 2.5 Wp. The STC power here is the power measured at the end of the production line, with production line solar simulators. These measurements are typically less accurate than measurements in a laboratory and are especially affected more strongly by statistical influences. The STC power of a whole batch of modules is best determined by using proper sampling procedures, which is described in Section 2.4.2.

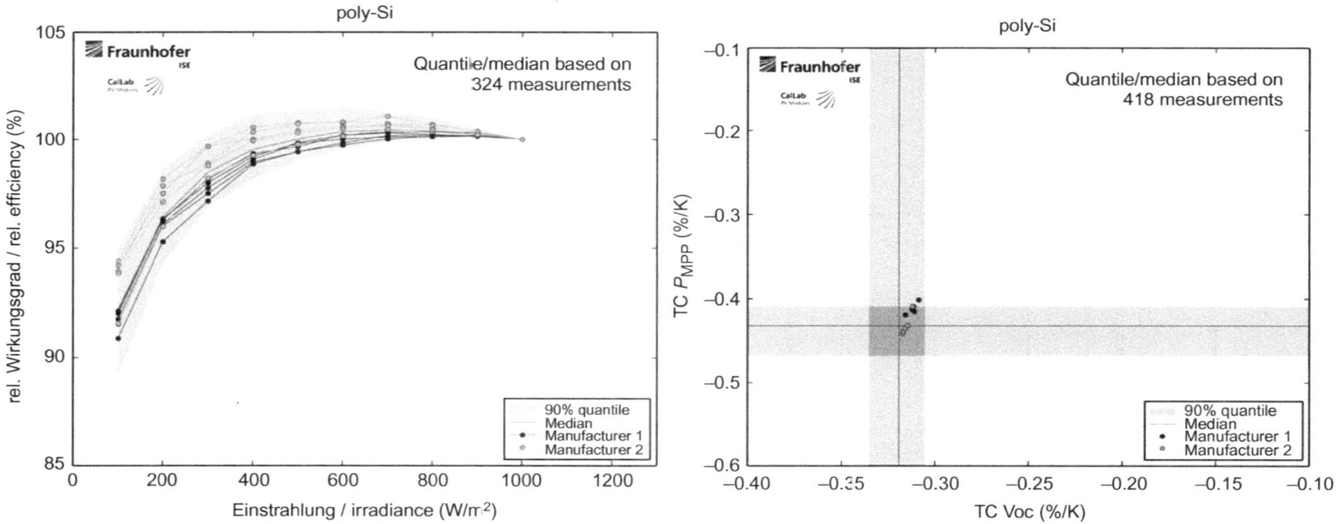

Fig. 2.15 Low light behaviour *(left)* and temperature coefficients *(right)* measured by Fraunhofer ISE's calibration laboratory CalLab PV Modules. Exemplary measurements of five modules produced by two different manufacturers are compared to the 'typical' characteristics of standard polycrystalline silicon modules. The 'typical' characteristics are represented by the 90% quantile out of 324 and 418 measurements of low light behaviour and temperature coefficient, respectively.
Figures provided by Fraunhofer ISE.

Module characteristics that are mainly determined by physical properties of the semiconductor, like SR and TC, typically do not vary strongly within one module type, and not even within one type of semiconductor. The low light behaviour additionally depends on module production-related influences, such as series resistance, so that larger variations can occur. For a high-quality module type this is, however, not expected. Typically, the variation is not larger than measurement uncertainty limits, meaning it is unclear whether variations are actual module-related differences or measurement artefacts. Fig. 2.15 shows the typical range for low light behaviour and TCs of standard c-Si PV modules according to measurements by Fraunhofer ISE. For thin-film modules, one must consider that module characteristics can vary quite a lot within the same technology group (eg, CI(G)S) [13].

The reflection behaviour is dominated by the air-glass interface and should thus not vary between modules of one type. However, extensive data to verify this do not yet exist.

Reasonable sample sizes for module characteristics are around 10–25 modules for STC measurements, 5–10 for low light behaviour and 1–3 for TC, spectral and angular response measurement. Regarding sample size, there is always a trade-off between significance of the result and costs for the measurements. The optimum depends on the specific situation. For quality assurance for thin-film modules or new PV technologies, rather higher sampling sizes are recommended.

2.4.2 Sampling schemes for STC power verification

Verification of STC power presents one of the most important steps of quality assurance when purchasing larger amounts of PV modules [80,81]. As not all modules can be measured by independent laboratories for time and cost reasons, a sample for the STC power verification must be selected.

In order to understand the underlying challenges, the following terms must be clarified:

- Flashlist: the list of STC power as determined by the module manufacturer in the production line (this measurement is the basis for binning modules to power classes). The flashlist values may include any manual offsets the manufacturer might have added, for example, a security margin in order not to overrate modules.
- Selected sample: the modules that were selected randomly or systematically for measurement in an independent laboratory. Only for this sample, measurements of STC power both by the manufacturer and the independent laboratory are available.
- Rated power: The power bin the modules were assigned. This value is constant for all modules of one type and power class, while the flashlist power is not.

The standard approach is to select a sample randomly, and to compare the STC power determined by an independent laboratory with rated power. When the average STC power of the sample is above or around rated power, the whole batch of modules is assumed to comply with the STC power rating by the manufacturer. Fig. 2.16 demonstrates why this can be misleading: The STC power according to the flashlist varies over the whole batch within the borders of the power bin width. If the sample is

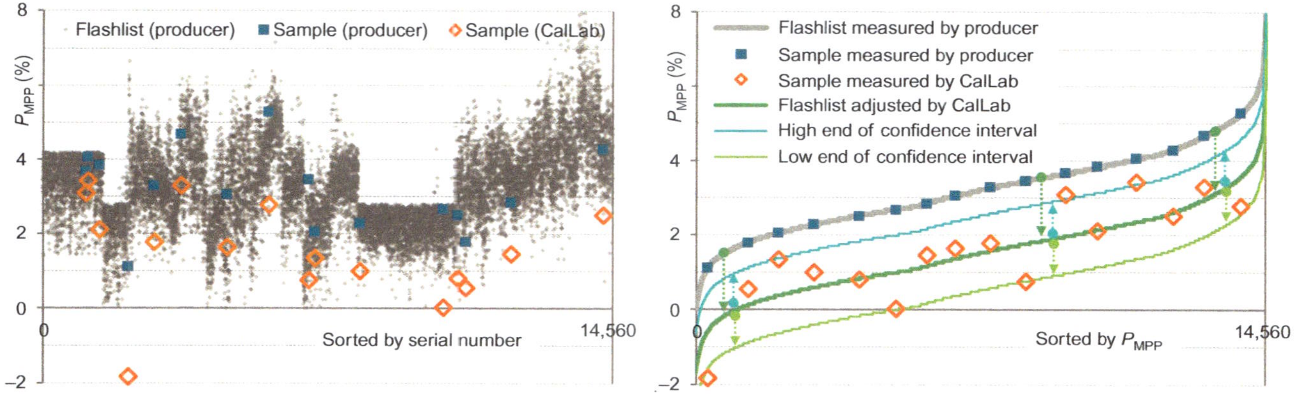

Fig. 2.16 STC power (shown as deviation from rated power in percent) of a batch of modules versus serial number *(left)* and versus number of sorting according to STC power *(right)*. The flashlist represents measurements by the module manufacturer, which are verified by measurements of a small sample of modules in an independent laboratory (here: CalLab PV Modules). High and low end of confidence intervals are indicated based on expanded measurement uncertainty of the laboratory measurements.

Figures reprinted from M. Jantsch, D. Dirnberger, F. Neuberger, Optimized procedures for representative samples from PV module batches, in: 27th European Photovoltaic Solar Energy Conference and Exhibition, Frankfurt, Germany, 2012, pp. 3214–3218.

selected randomly, the modules might by chance all be picked from the range of higher power values, instead of, as shown in the figure, equally distributed over the whole range of power values. When the measurement performed for verification in an independent laboratory reveals that the correct power is approximately equal to the rated power, the situation seems ok. What is not considered by this analysis is the difference between the actual STC power and the flashlist power, which can be several percent: while modules with rather higher power still meet the rated power, this is not the case for modules with lower power. As a consequence, the conclusion regarding whether or not a batch of modules is accepted depends also on which modules were selected.

To prevent such kind of misinterpretations, the sample is best selected systematically [82], especially if large batches of modules are purchased. The procedure presented in Ref. [82] consists of the following steps and is shown in Fig. 2.16:

- Obtain flashlist of the batch of modules
- Determine average module power according to flashlist
- Select a number of modules (at least 10) out of the flashlist that are spread over the whole power and serial number range, and whose average power is equal to the average power of the flashlist (this is done using an algorithm)
- Measure the selected modules
- Determine the difference between the average measured power and the average flashlist power of the sample
- Adjust the flashlist by the difference and compare the adjusted flashlist average with rated power

For the example shown in Fig. 2.16, the rated power (0%, power is shown as deviation from rated power) is met by most of the modules and the batch can be accepted. The statistical spread in the difference between laboratory and flashlist value demonstrates that the sample cannot be arbitrarily small.

Alternatively, statistical evaluation procedures for module selection can be applied [83]. Here, the sample size is selected with the aim to achieve a specific probability for rejecting the batch while it is ok, or accepting the batch while it is not ok. Depending on the shape of the distribution of the STC power values in the flashlist, the necessary sample size can still become quite high.

2.5 Future trends and perspectives

In the future, PV module characterization may see twofold developments.

First, measurement uncertainties will decrease. Measurement uncertainty for STC measurements can decrease to values as low as 1.2%, assuming a reduction of uncertainties in primary calibration and SR measurements [8]. Improvement regarding uncertainties of characterization measurements will be driven by more detailed analyses and subsequent optimization of measurement equipment and more sophisticated methods to evaluate measurement uncertainties.

Second, measurement equipment for module characterizations such as spectral and angular response will become more available and/or easier to use. As a consequence, more information about module characteristics apart from low light behaviour and TCs will become available, which offers new ways for energetic module optimization and for decreasing uncertainties also in simulation. Furthermore, measurement equipment will be developed for characterization of emerging PV technologies, such as bifacial PV modules.

References

[1] IEC 60904-3 Ed. 2.0, Photovoltaic Devices—Part 3: Measurement Principles for Terrestrial Photovoltaic (PV) Solar Devices with Reference Spectral Irradiance Data, International Electrotechnical Commission (IEC), Geneva, 2008.

[2] T. Markvart, L. Castaner (Eds.), Solar Cells—Materials, Manufacture and Operation, Elsevier, Oxford, UK/Amsterdam, The Netherlands, 2005.

[3] IEC 60904-1 Ed. 2.0, Photovoltaic Devices—Part 1: Measurement of Photovoltaic Current–Voltage Characteristics, International Electrotechnical Commission (IEC), Geneva, 2006.

[4] IEC 60904-9 Ed. 2.0, Photovoltaic Devices—Part 9: Solar Simulator Performance Requirements, International Electrotechnical Commission (IEC), Geneva, 2007.

[5] J. Hohl-Ebinger, G. Siefer, W. Warta, Measuring the spectral distribution of a flash simulator, in: 22nd European Photovoltaic Solar Energy Conference and Exhibition, Milan, Italy, 2007, pp. 425–428.

[6] D. Domine, G. Friesen, S. Dittmann, D. Chianese, The influence of measurement errors in spectral irradiance of flash solar simulators on the spectral mismatch factor of PV modules, in: 25th European Photovoltaic Solar Energy Conference and Exhibition/5th World Conference on Photovoltaic Energy Conversion, Valencia, Spain, 2010, pp. 3889–3895.

[7] H.J. Kostkowski, Reliable Spectroradiometry, Spectroradiometry Consulting, La Plata, MD, 1997.

[8] D. Dirnberger, Uncertainty in PV module measurement—Part I: calibration of crystalline and thin film modules, IEEE J. Photovolt. 3 (2013) 1016–1026.

[9] M.A. Green, Solar Cells: Operating Principles, Technology and System Applications, University of New South Wales, New South Wales, 1992.

[10] N. Ferretti, Y. Pelet, J. Berghold, V. Fakhfouri, P. Grunow, Performance testing of high-efficient PV modules using single 10 ms flash pulses, in: 28th European Photovoltaic Solar Energy Conference and Exhibition, Paris, France, 2013, pp. 3184–3187.

[11] K. Ramspeck, S. Schenk, L. Komp, A. Metz, M. Meixner, Accurate efficiency measurements on very high efficiency silicon solar cells using pulsed light sources, in: 28th European Photovoltaic Solar Energy Conference and Exhibition, Paris, France, 2013, pp. 1253–1256.

[12] C.R. Osterwald, S. Anevsky, K. Bücher, A.K. Barua, P. Chaudhuri, J. Dubard, K. Emery, B.R. Hansen, D.L. King, J. Metzdorf, F. Nagamine, R. Shimokawa, Y.X. Wang, T. Wittchen, W. Zaaiman, A. Zastrow, J. Zhang, The World Photovoltaic Scale: an international reference cell calibration program, Prog. Photovolt. Res. Appl. 7 (1999) 287–297.

[13] D. Dirnberger, Uncertainties in Energy Rating for Thin-Film PV Modules (Doctoral Thesis), Carl von Ossietzky Universität Oldenburg/Fraunhofer Institute for Solar Energy Systems ISE, Oldenburg, 2015.

[14] IEC 60904-10 Ed. 2.0, Photovoltaic Devices—Part 10: Methods of Linearity Measurement, International Electrotechnical Commission (IEC), Geneva, 2009.

[15] IEC 60891 Ed. 2.0, Photovoltaic Devices—Procedures for Temperature and Irradiance Corrections to Measured I–V Characteristics, International Electrotechnical Commission (IEC), Geneva, 2009.

[16] D. Dirnberger, Verbesserung des Mess- und Auswerteverfahrens bei Outdoormessungen an Solargeneratoren, Diploma, Technische Universität Dresden, Fakultät für Maschinenwesen, Dresden, Germany, 2008.

[17] D. Dirnberger, J. Bartke, A. Steinhüser, K. Kiefer, F. Neuberger, Uncertainty of field I–V-curve measurements in large scale PV-systems, in: 25th European Photovoltaic Solar Energy Conference and Exhibition/5th World Conference on Photovoltaic Energy Conversion, Valencia, Spain, 2010, pp. 4587–4594.

[18] D. Dirnberger, Uncertainty in PV module measurement—Part II: verification of rated power and stability problems, IEEE J. Photovolt. 4 (2014) 991–1007.

[19] M. Gostein, L. Dunn, Light soaking effects on photovoltaic modules: overview and literature review, in: 37th IEEE Photovoltaic Specialists Conference, Seattle, Washington, USA, 2011, pp. 003126–003131.

[20] M. Tayyib, Y.H. Rao, M. Ramanjaneyulu, T.S. Surendra, J.O. Odden, T.O. Saetre, Initial light-induced degradation study of multicrystalline modules made from silicon material processed through different manufacturing routes, in: 38th IEEE Photovoltaic Specialists Conference, 2012, pp. 2395–2399.

[21] A. Luque, S. Hegedus (Eds.), Handbook of Photovoltaic Science and Engineering, John Wiley and Sons, 2003.

[22] D.L. King, J.A. Kratochvil, W.E. Boyson, Sandia National Laboratories, Stabilization and performance characteristics of commercial amorphous-silicon PV modules, in: 28th IEEE Photovoltaic Specialists Conference, Anchorage, Alaska, USA, 2000, pp. 1446–1449.

[23] R.P. Kenny, A.I. Chatzipanagi, T. Sample, Preconditioning of thin-film PV module technologies for calibration, Prog. Photovolt. Res. Appl. 22 (2014) 166–172.

[24] J. Müller, T. Bachmann, K.-H. Stegemann, Comparison of light induced degradation of thin film a-Si:H modules under standard test conditions and long-term environmental outdoor exposure, in: 25th European Photovoltaic Solar Energy Conference and Exhibition/ 5th World Conference on Photovoltaic Energy Conversion, Valencia, Spain, 2010, pp. 3076–3077.

[25] M. Nikolaeva-Dimitrova, R.P. Kenny, E.D. Dunlop, Controlled conditioning of a-Si:H thin film modules for efficiency prediction, Thin Solid Films 516 (2008) 6902–6906.

[26] N. Strevel, L. Trippel, M. Gloeckler, Performance characterization and superior energy yield of first solar PV power plants in high-temperature conditions, Photovolt. Int. 17 (2012) 148–154.

[27] C. Deline, J. del Cueto, D.S. Albin, S. Rummel, Metastable electrical characteristics of polycrystalline thin-film photovoltaic modules upon exposure and stabilization, J. Photon. Energy. 2 (2012) 1–14.

[28] G. Makrides, B. Zinsser, A. Phinikarides, M. Schubert, G.E. Georghiou, Temperature and thermal annealing effects on different photovoltaic technologies, Renew. Energy 43 (2012) 407–417.

[29] M. Nikolaeva-Dimitrova, R.P. Kenny, E.D. Dunlop, M. Pravettoni, Seasonal variations on energy yield of a-Si, hybrid, and crystalline Si PV modules, Prog. Photovolt. Res. Appl. 18 (2010) 311–320.

[30] A. Virtuani, L. Fanni, Seasonal power fluctuations of amorphous silicon thin-film solar modules: distinguishing between different contributions, Prog. Photovolt. Res. Appl. 22 (2014) 208–217.
[31] L. Fanni, I. Pola, E. Bura, T. Friesen, D. Chianese, Investigation of annealing and degradation effects on a-si PV modules in real operating conditions, in: 24th European Photovoltaic Solar Energy Conference, Hamburg, Germany, 2009, pp. 3596–3599.
[32] R. Gottschalg, J.A. del Cueto, T.R. Betts, D.G. Infield, Seasonal performance of a-Si single- and multijunction modules in two locations, in: 31st IEEE Photovoltaic Specialists Conference, Lake Buena Vista, Florida, USA, 2005, pp. 1484–1487.
[33] C. Deline, A. Stokes, T.J. Silverman, S. Rummel, D. Jordan, S. Kurtz, Electrical bias as an alternate method for reproducible measurement of copper indium gallium diselenide (CIGS) photovoltaic modules, in: Reliability of Photovoltaic Cells, Modules, Components, and Systems V, San Diego, California, USA, 2012.
[34] L. Dunn, M. Gostein, Light soaking measurements of commercially available CIGS PV modules, in: 38th IEEE Photovoltaic Specialists Conference, Austin, Texas, USA, 2012, pp. 1260–1265.
[35] IEC 61646 Ed. 2.0, Thin-Film Terrestrial Photovoltaic (PV) Modules—Design Qualification and Type Approval, International Electrotechnical Commission (IEC), Geneva, 2008.
[36] D.C. Jordan, S.R. Kurtz, Photovoltaic degradation rates—an analytical review, Prog. Photovolt. Res. Appl. 21 (2013) 12–29.
[37] D. Dirnberger, U. Kräling, H. Müllejans, E. Salis, K. Emery, Y. Hishikawa, K. Kiefer, Progress in photovoltaic module calibration: results of a worldwide intercomparison between four reference laboratories, Meas. Sci. Technol. 25 (2014) 105005.
[38] JCGM 104:2009, Evaluation of measurement data—an introduction to the "Guide to the expression of uncertainty in measurement" and related documents, JointCommittee for Guides in Metrology (JCGM), France, 2009.
[39] JCGM 100:2008, Evaluation of Measurement Data—Guide to the Expression of Uncertainty in Measurement, Joint Committee for Guides in Metrology (JCGM), France, 2008.
[40] JCGM 101:2008, Evaluation of Measurement Data—Supplement 1 to the "Guide to the Expression of Uncertainty in measurement"—Propagation of Distributions Using a Monte Carlo Method, Joint Committee for Guides in Metrology (JCGM), France, 2008.
[41] JCGM 200:2012, International Vocabulary of Metrology—Basic and General Concepts and Associated Terms (VIM), Joint Committee for Guides in Metrology (JCGM), France, 2012.
[42] ISO/IEC Guide 98-3, Uncertainty of Measurement—Part 3: Guide to the Expression of Uncertainty in Measurement (GUM:1995), International Electrotechnical Commission (IEC), Geneva, 2008.
[43] ISO/IEC Guide 98-1, Uncertainty of Measurement—Part 1: Introduction to the Expression of Uncertainty in Measurement, International Electrotechnical Commission (IEC), Geneva, 2009.
[44] Guide 98-3/Supplement 1ISO/IEC, Propagation of Distributions Using a Monte Carlo Method, in: International Electrotechnical Commission (IEC), Geneva, 2008.
[45] D. Dirnberger, U. Kräling, B. Farnung, F. Neuberger, K. Kiefer, Mastering thin film measurements on a grand scale as a daily routine, in: 26th European Photovoltaic Solar Energy Conference and Exhibition, Hamburg, Germany, 2011, pp. 3261–3266.
[46] K. Emery, Uncertainty Analysis of Certified Photovoltaic Measurements at the National Renewable Energy Laboratory, National Renewable Energy Laboratory (NREL), Golden, CO, 2009. NREL/TP-520-45299.

[47] H. Müllejans, W. Zaaiman, R. Galleano, Analysis and mitigation of measurement uncertainties in the traceability chain for the calibration of photovoltaic devices, Meas. Sci. Technol. 20 (2009) 1–12.
[48] S. Winter, T. Wittchen, J. Metzdorf, Primary reference cell calibration at the PTB based on an improved DSR facility, in: 16th European Photovoltaic Solar Energy Conference, Glasgow, United Kingdom, 2000.
[49] S. Winter, A. Sperling, Uncertainty analysis of a photometer calibration at the DSR setup of the PTB, in: 2nd CIE Expert Symposium on Measurement Uncertainty, Braunschweig, Germany, 2006, pp. 139–142.
[50] S. Winter, T. Fey, D. Friedrich, I. Kröger, K. von Volkmann, New laser-DSR facility at PTB: concept for a next generation high accuracy primary calibration facility, in: 26th European Photovoltaic Solar Energy Conference and Exhibition, Hamburg, Germany, 2011, pp. 3466–3468.
[51] S. Winter, T. Fey, I. Kröger, D. Friedrich, K. Ladner, B. Ortel, S. Pendsa, D. Schlüssel, Laser-DSR facility at PTB: realization of a next generation high accuracy primary calibration facility, in: 27th European Photovoltaic Solar Energy Conference and Exhibition, Frankfurt, Germany, 2012, pp. 3049–3051.
[52] R. Shimokawa, F. Nagamine, Y. Miyake, K. Fujisawa, Y. Hamakawa, Japanese indoor calibration method for the reference solar-cell and comparison with the outdoor calibration, Jpn. J. Appl. Phys. 1—Regul. Papers Short Notes Rev. Pap. 26 (1987) 86–91.
[53] R. Shimokawa, H. Ikeda, Y. Miyake, S. Igari, Development of wide field-of-view cavity radiometer for solar simulator use and intercomparison between irradiance measurements based on the world radiometer reference and electrotechnical laboratory scales, Jpn. J. Appl. Phys. 1—Regul. Papers Short Notes Rev. Pap. 41 (2002) 5088–5093.
[54] H. Müllejans, W. Zaaiman, F. Merli, E.D. Dunlop, H.A. Ossenbrink, Comparison of traceable calibration methods for primary photovoltaic reference cells, Prog. Photovolt. Res. Appl. 13 (2005) 661–671.
[55] K.A. Emery, C.R. Osterwald, T.W. Cannon, D.A. Myers, J. Burdick, T. Glatfelter, W. Czubatvi, J. Yang, Methods for measuring solar cell efficiency independent of reference cell or light source, in: 18th IEEE Photovoltaic Spec. Conf., Las Vegas, Nevada, 1985, pp. 623–628.
[56] C.R. Osterwald, K.A. Emery, Spectroradiometric sun photometry, J. Atmos. Ocean. Technol. 17 (2000) 1171–1188.
[57] W. Zaaiman, H. Müllejans, E.D. Dunlop, H. Ossenbrink, SI traceability for irradiance transfer standards to silicon solar cells using direct and global irradiance, in: 21st European Photovoltaic Solar Energy Conference and Exhibition, Dresden, Germany, 2006, pp. 64–68.
[58] H. Müllejans, W. Zaaiman, E.D. Dunlop, H.A. Ossenbrink, Calibration of photovoltaic reference cells by the global sunlight method, Metrologia 42 (2005) 360–367.
[59] C. Frohlich, History of solar radiometry and the World Radiometric Reference, Metrologia 28 (1991) 111–115.
[60] K. Emery, The Results of the First World Photovoltaic Scale Recalibration, National Renewable Energy Laboratory (NREL), Golden, CO, 2000. NREL/TP-520-27942.
[61] S. Winter, J. Metzdorf, K. Emery, F. Fabero, Y. Hishikawa, B. Hund, H. Mullejans, A. Sperling, W. Warta, The results of the second World Photovoltaic Scale recalibration, in: 31st IEEE Photovoltaic Specialists Conference, Lake Buena Vista, Florida, USA, 2005, pp. 1011–1014.

[62] IEC 61853-1 Ed. 1.0, Photovoltaic (PV) Module Performance Testing and Energy Rating—Part 1: Irradiance and Temperature Performance Measurements and Power Rating, International Electrotechnical Commission (IEC), Geneva, 2011.
[63] D. Dirnberger, B. Müller, U. Kräling, S. Dittmann, Uncertainties of laboratory measurements for energy rating, in: 27th European Photovoltaic Solar Energy Conference and Exhibition, Frankfurt, Germany, 2012, pp. 3294–3301.
[64] J. Hohl-Ebinger, Untersuchungen zur Hochpräzisen Vermessung der Elektrischen Parameter von Solarzellen, Fachbereich für Physik, Universität Konstanz, Fraunhofer Institut für Solare Energiesysteme (ISE), Freiburg, 2011. Dissertation (Doctorial Thesis).
[65] J.Y. Ye, S. Guo, T.M. Walsh, Y. Hishikawa, R.A. Stangl, On the spectral response of PV modules, Meas. Sci. Technol. 25 (2014) 095007.
[66] M. Pravettoni, A. Komlan, R. Galleano, H. Müllejans, E.D. Dunlop, An alternative method for spectral response measurements of large-area thin-film photovoltaic modules, Prog. Photovolt. Res. Appl. 20 (2012) 416–422.
[67] Y. Tsuno, Y. Hishikawa, K. Kurokawa, A method for spectral response measurements of various PV modules, in: 23rd European Photovoltaic Solar Energy Conference and Exhibition, Valencia, Spain, 2008, pp. 2723–2727.
[68] IEC 61853-2 (IEC 82/606/CDV:2010), Photovoltaic (PV) Module Performance Testing and Energy Rating—Part 2: Spectral Response, Incidence Angle and Module Operating Temperature Measurements, International Electrotechnical Commission (IEC), Geneva, 2010.
[69] IEC 60904-8 Ed. 3.0, Photovoltaic Devices—Part 8: Measurement of Spectral Responsivity of a Photovoltaic (PV) Device, International Electrotechnical Commission (IEC), Geneva, 2014.
[70] M. Pravettoni, A. Virtuani, H. Müllejans, R.P. Kenny, E.D. Dunlop, K.W.J. Barnham, Spectral response measurements of large area thin film multijunction photovoltaic modules at the European Solar Test Installation and spectral mismatch calculation, in: 24th European Photovoltaic Solar Energy Conference, Hamburg, Germany, 2009.
[71] D.L. King, B.R. Hansen, J.M. Moore, D.J. Aiken, New methods for measuring performance of monolithic multi-junction solar cells, in: 28th IEEE Photovoltaic Specialists Conference, Anchorage, Alaska, 2000, pp. 1197–1201.
[72] G. Baarah, Angle of Incidence Effect on Photovoltaic Modules—Validation of IEC 61853-2 Draft Standard, Faculty of Environment and Natural Resources in cooperation with ZEE (Centre for Renewable Energy), Albert-Ludwigs-Universität, Freiburg, 2014. Master of Science Master.
[73] W. Herrmann, L. Rimmelspacher, M. Reuter, Optical characteristics of PV module front glasses—incidence angle effects of various glass types and impact on annual energy yield, in: 28th European Photovoltaic Solar Energy Conference and Exhibition, Paris, France, 2013, pp. 2882–2886.
[74] IEC 61215 Ed. 2.0, Crystalline Silicon Terrestrial Photovoltaic (PV) Modules—Design Qualification and Type Approval, in: International Electrotechnical Commission (IEC), Geneva, 2005.
[75] M. Muller, B. Marion, J. Rodriguez, Evaluating the IEC 61215 Ed.3 NMOT procedure against the existing NOCT procedure with PV modules in a side-by-side configuration, in: 38th IEEE Photovoltaic Specialists Conference, 2012.
[76] M. Koehl, M. Heck, S. Wiesmeier, J. Wirth, Modeling of the nominal operating cell temperature based on outdoor weathering, Sol. Energy Mater. Sol. Cells 95 (2011) 1638–1646.

[77] T.W. Neises, S.A. Klein, D.T. Reindl, Development of a thermal model for photovoltaic modules and analysis of NOCT guidelines, J. Sol. Energy Eng.—Trans. ASME 134 (2012) 011009–011016.
[78] D. Dirnberger, G. Blackburn, B. Müller, C. Reise, On the impact of solar spectral irradiance on the yield of different PV technologies, Sol. Energy Mater. Sol. Cells 132 (2015) 431–442.
[79] T. Ishii, K. Otani, T. Takashima, Y.Q. Xue, Solar spectral influence on the performance of photovoltaic (PV) modules under fine weather and cloudy weather conditions, Prog. Photovolt. 21 (2013) 481–489.
[80] B. Farnung, P. Bostock, J. Brückner, K. Kiefer, Complete PV power plant certification: new standards for quality assurance of large scale PV power plants, in: 40th IEEE PVSC, 2014.
[81] B. Farnung, B. Müller, P. Bostock, J. Sedgwick, K. Kiefer, All about PV power plants: challenges for technical bankability, in: 40th IEEE PVSC, 2014.
[82] M. Jantsch, D. Dirnberger, F. Neuberger, Optimized procedures for representative samples from PV module batches, in: 27th European Photovoltaic Solar Energy Conference and Exhibition, Frankfurt, Germany, 2012, pp. 3214–3218.
[83] W. Herrmann, A. Steland, W. Herff, Sampling procedures for the validation of PV module output power specification, in: 24th European Photovoltaic Solar Energy Conference, Hamburg, Germany, 2009, pp. 3540–3547.

Photovoltaic module stability and reliability

3

D. Jordan, S. Kurtz
National Renewable Energy Laboratory (NREL), Golden, CO, United States

3.1 Introduction

As the installed capacity of the photovoltaic (PV) industry is approaching 200 GW, investors seek confidence in long-term PV performance. Profitable investment in PV will benefit from decades of reliable operation. The discipline of reliability has been well established and has led to significant quality increases for many types of products, including PV systems; however, the application to PV modules is challenging for several reasons. The desired long lifetime prevents direct verification, especially because of the short product development life cycle. In addition, the size and cost of PV modules typically inhibits testing of a large number of samples. Furthermore, PV modules are complex and can display a variety of failure modes. Because of the synergistic nature of the different failure modes an understanding of the underlying physical and chemical processes is essential. Finally, the use environment can be highly variable, leading to a climate dependency of different failure modes. As the industry enhances service life prediction, improved accelerated lifetime tests need to be developed. For years the industry relied on extending the qualification testing of IEC 61215 and IEC 61646 to establish confidence in its products [1]. Arbitrary extension of accelerated tests, however, may not lead to the desired result, as failure modes observed in accelerated testing should match the ones observed in the field. Therefore, detailing and quantifying failure modes and power loss from field observation is an essential part of developing lifetime service tests. The first part of this chapter is dedicated to the lifetime and gradual power decline and the second part to the various failure modes that have been documented.

3.2 Module stability and lifetime

3.2.1 Historical survey

As modules and balance-of-system components of a PV system age, they gradually lose some performance. Most commonly, PV module manufacturers guarantee against this loss by promising 80% of nameplate power after 25 years whether in a stepped or linear fashion. The decline is expressed in relative percentage, such that a module with a hypothetical absolute efficiency of 20% today would decline to 17.5% after 25 years at an annualized (relative) rate of 0.5%/year, assuming the

The Performance of Photovoltaic (PV) Systems. http://dx.doi.org/10.1016/B978-1-78242-336-2.00003-3

decline is approximately linear. The decline or degradation rate has significant financial consequences, as this rate directly affects the yield and, therefore, the cash flow in future years. In financial projections, the value of a PV project after 25 years is often assumed to be negligible, yet our simple hypothetical example illustrates that such a module would present a considerable residual value after 25 years. At the time of this writing, more than 30 studies of systems older than 20 years have been reported, with some at 30 years and one even approaching 40 years. As such, there appears to be no end to the lifetime of certain PV systems. Yet other reports exist of premature decline due to manufacturing defects or poor installation quality.

For an investor, this presents a considerable quandary associated with financial risk. Therefore, it is important to understand, detail and quantify the various factors influencing the performance loss curves. Some of the authors first aggregated and analysed the public knowledge of PV lifetime measurements, which were later reanalysed [2,3]. As can be seen in the colour-coded map (A) and in the graph of studies exceeding 20 years (B) of Fig. 3.1, this topic has seen an increased interest in recent years. This chapter offers an updated analysis with a goal of providing better answers to some of the questions that remain open.

More than 11,000 annual degradation rates have been aggregated and analysed in this section. The first observation in the histogram of Fig. 3.2A of all data (solid blue) is that the distribution is bimodal, a peculiar property that needs further discussion.

As the degradation rates are impacted by a variety of factors, including (but not limited to) manufacturer, model, technology, climate, mounting, failure modes, measurement uncertainty, etc., it is important to delineate the factor causing the bimodality. Next to these hardware-related variables influencing the long-term performance assessments, statistical procedures including (but not limited to) methodology and sampling can also influence the determined rates and therefore the overall distribution. To accurately assess the status and the health of the industry in general, a representative sample is essential. As the present study is not based on a designed study but a correlational study, it is important to investigate the aggregated values in different

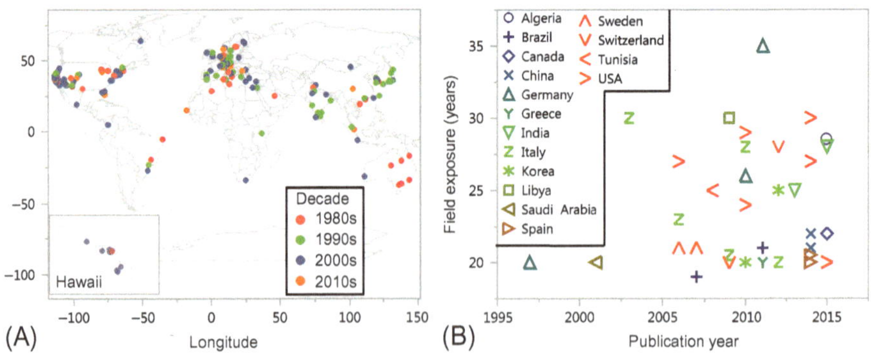

Fig. 3.1 Worldwide reported degradation rates coloured by the decade of installation (A) and system reports exceeding 20 years by publication year (B).

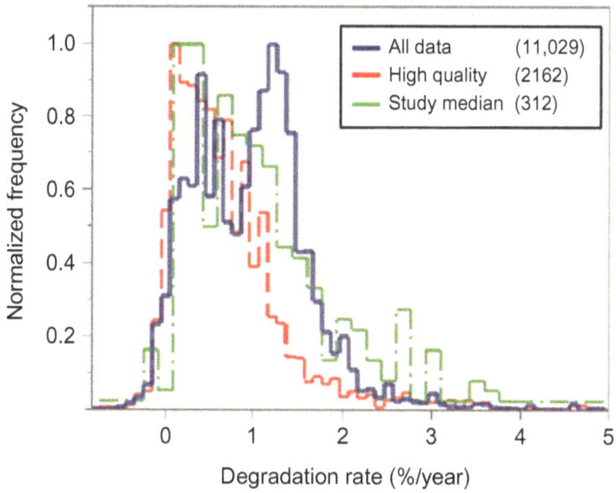

Fig. 3.2 Histograms of all data, high-quality data and the median per study and system presented as the normalized frequency. The number of data points for each respective subset is given in parentheses.

ways to ensure the absence of unintended bias. To illustrate this, we consider this hypothetical example: 10 studies investigating the long-term performance of 10 different modules, in different climates, mounting configurations, etc., analysed together with one additional study investigating a single product, in a single location but investigating 100 modules. The aggregated distribution would be dominated by the single study but might not be representative of the product population in general. Thus, to reduce sampling bias, that is, the over- or under-representation of a group, we present the data aggregated from the literature both by including all reported measurements and by using the median of each study and system. However, even this method may not be the best methodology to find a representative sample. Consider again a hypothetical example of two studies: One study examines 10 products at high accuracy and a similar study assesses an equal number of products at very low accuracy. By using the median of each study, the sampling bias is superficially reduced but this method would equally weigh each study. However, one would probably choose to give the more accurate study more weight than the less accurate study.

Therefore, we also show a degradation rate distribution of studies that we consider to be high quality based on the following criteria: multiple measurements were taken for increased confidence; the measurement methods and calibrations were clearly described and are generally similar at each measurement point; details on the installation (disregarding proprietary considerations) are provided. Each of these analytical methodologies is imperfect, but it is interesting to note that they consistently lead to a unimodal distribution centered in the 0.5–0.8%/year range. Furthermore, the aggregation of degradation rates through survey of published information may induce sampling bias because modules or systems at high decline rates may be less likely to remain exposed in the field than those with smaller decline rates. In contrast, some studies may examine prototypes and not commercially available products without specifying. Finally, every effort was made to separate light-induced degradation (LID) from long-term performance decline. LID is a short-term performance degradation upon initial light exposure because of light-induced chemical reactions that increase the effects of defects in the semiconductor material.

Nevertheless, since we are largely limited to reporting the information provided, LID artefacts cannot be completely excluded, especially when considering data collected on thin-film technologies.

3.2.2 Nameplate rating

Many studies were found to have utilized only the nameplate rating and one measurement to determine a degradation rate slope. One-measurement studies often are a result of missing beginning-of-life measurements because they were (1) never taken or (2) because of the long life of the PV system, they can no longer be recovered. To assess the durability of the module or system, the analyst has no choice but to compare the measurement to the nameplate rating. Several implicit assumptions are made in this procedure:

1. The nameplate rating was accurate at the beginning of the PV life.
2. If an ensemble of modules is measured, the ensemble of performance starts as a point source and not as a distribution with a shape defined by a convolution of the binning practices of the manufacturer and the process control. Modules are sorted according to their measured power and allocated to specific rated power products, 'bins'; therefore, as an example, modules from a nominal 250 W 'bin' may produce slightly more or less power depending on the specific manufacturer's sorting practices. Furthermore, nameplate rating constitutes a measurement by itself and several papers have discussed the uncertainties involved with the rating measurements [4–6]. The combination of these effects leads to the product tolerance, which has changed significantly over the past several decades.
3. Beginning-of-life LID is not considered consistently. For crystalline silicon (x-Si) this occurs on a short time scale but for some thin-film products the stabilization process may take from several months to years but is often included in the manufacturer's rating [7].

The difficulty is to take these effects into account to accurately assess PV durability, yet in many durability studies these uncertainties are often not discussed. To better gauge the uncertainty associated with one-measurement studies, Fig. 3.3 shows an extensive (yet most likely not comprehensive) graph on the history of module measurements with respect to their nameplate rating [8–19]. Indoor I–V measurements are indicated by red open circles and outdoor I–V data by blue open triangles. As a guide to the eye, a 10% discount line is also shown. Considerable spread exists but it appears that measurements on solar simulators tend to be closer to the nominal nameplate rating. However, the data points only go back to the mid-1990s. With the exception of a few data points in 2011, the outdoor measurements trend closer towards the nameplate rating in recent years, an effect which may be attributable to the change in manufacturers' tolerance. At the time of the writing, most PV module manufacturers warrant tolerances of 0% to positive deviations, whereas older modules

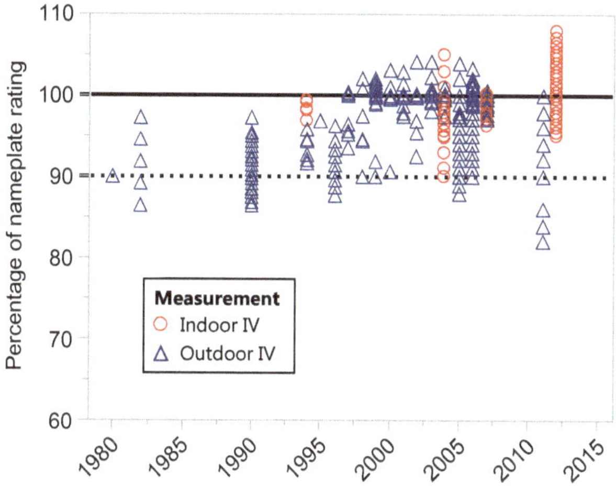

Fig. 3.3 Historical module I–V measurements with respect to their nameplate rating for new or almost new modules. Indoor measurements are given by the red circles and outdoor measurements are indicated by blue triangles. As a guide to the eye, the 90% value is indicated by a dashed line.

often carried equally symmetric positive/negative deviations. As discussed in the previous section, another reason for the larger outdoor deviation may be associated with the confounding effects of measurements and soiling.

In addition LID is often contained in these data points. However, it must be emphasized that, for long-term performance, LID must also be taken into account. The 2011 study in Fig. 3.3 includes modules that had been on sun for more than 1 year; thus these modules may not truly represent beginning-of-life measurements, a fact that may explain some of the low-lying outliers. It is evident from Fig. 3.3, even if the graph is not comprehensive, that considerable deviations at the beginning of life may be expected, depending on the specific production time and tolerances. Therefore, depending on the fielded time and the specific modules, degradation calculations may be significantly impacted. These considerations should be included in an uncertainty budget. Additionally, for these reasons, one-measurement studies, which depended upon nameplate ratings, were not included in the high quality data set category in Fig. 3.2.

3.2.3 Technology trends

In this section we examine the aggregated degradation rates by technology and the evolution over the past decades. Fig. 3.4 is colour-coded by the different technologies. Multicrystalline (multi-Si) and monocrystalline (mono-Si) have been combined into one category as a significant number of studies specify x-Si but do not disclose any other details. The type of study or measurements is symbol-coded. Degradation of 1%/year and 0.5%/year are indicated by red solid and dashed lines, respectively.

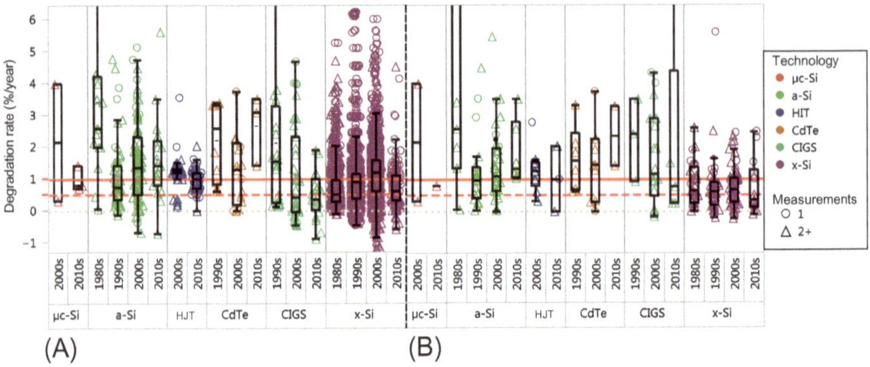

Fig. 3.4 All degradation rates (A) and the median per system and study (B) partitioned and coloured by technology and decade of installation. Studies using one measurement are indicated by an open circle and studies with two or more measurements are indicated by an open triangle. As a guide to the eye, 1%/year and 0.5%/year are indicated by solid and dashed horizontal lines, respectively.

Each technology category is partitioned into the decade of installation wherever possible. Because of considerable data variation, boxplots representing the interquartile range and the median are overlaid over the data points.

Microcrystalline (µc-Si) is a relatively new technology; therefore, long-term performance assessments are relatively scarce yet tend to aggregate around 1%/year. In contrast, amorphous silicon (a-Si) has a history almost as long as x-Si technologies. The technology initially suffered from high degradation, and then significantly reduced degradation before an increase to above 1%/year that appears to be consistent in the past 15 years. Sufficient data are available to partition the heterojunction technology (HJT) into its own category. Similarly to µc-Si, the HIT technology is not that old; only two subcategories are available with a median degradation that appears to be consistent around 1%/year. Cadmium telluride (CdTe) shows behaviour similar to a-Si as a significant decline in degradation occurs from the first to the second decade, followed by an increase in the third decade. This increased degradation rate may be influenced by one-measurement studies and longer stabilization for some products [7]. After suffering from high initial degradation followed by the first reports of low degradation, copper indium gallium selenide (CIGS) appears to have settled for a median of around 0.5%/year.

The x-Si is the largest category and shows initial degradation of around 0.5%/year in the first decade. In subsequent decades the degradation appears to increase to slightly above 1%/year before receding to 0.5%/year in the current decade. Many of the data points in the 1990–2000 decade that showed an increase in degradation were measured using the nameplate rating. In addition, as we discussed previously, sampling bias is a serious concern. Thus, Fig. 3.4B shows the same data but uses the median of each study or system. It can be seen that most categories appear unchanged and the conclusions we stated previously are the same. Considerable change, however, can be seen in the x-Si category. The median degradation for x-Si technologies now appears to be consistently around 0.5%/year, illustrating that the higher rates reported in Fig. 3.4 for x-Si may reflect a small number of studies that reported data for thousands of modules. The mean (not shown in Fig. 3.4) for x-Si dropped from 0.99%/year for the 1980s to 0.72%/year, then was 0.78%/year and 0.77%/year for the 1990s, 2000s and 2010s, respectively.

3.2.4 Climate and mounting

Recent studies describe higher degradation rates in hotter climates compared to more moderate climates [20–23], which are also reflected in the guidance provided by module manufacturers [24]. Degradation rates are influenced not only by temperature but other factors such as age, measurement uncertainties, methodologies, technology, product type and mounting. It is of considerable interest whether the aggregated data can corroborate these findings.

To avoid the confounding technology effect, Fig. 3.5 shows x-Si degradation rates (A) and the median for each study and system (B) partitioned by the number of measurements and climate zones. The climate zones are based on the Köppen–Geiger classification but aggregated into four simplified categories [25]. Statistical analysis revealed significantly higher degradation rates for studies using the nameplate rating. In addition, hotter climates show a trend towards elevated degradation rates compared to cooler climates. Mounting configuration—roof- versus rack-mounting—leading to prolonged elevated temperatures may also lead to increased degradation [23,26,27].

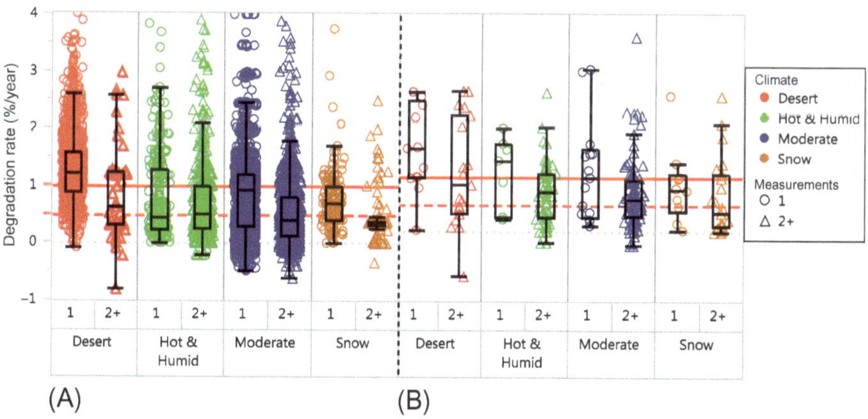

Fig. 3.5 Aggregated degradation rates for x-Si partitioned by the number of measurements (symbol coded) and climate (colour coded). Box plots of the interquartile range with the median are overlaid on the data. All degradation rates are shown (A) and the median per study and system (B) to assess sampling bias. Studies using one measurement are indicated by an open circle and studies with two or more measurements are indicated by an open triangle. As a guide to the eye, 1%/year and 0.5%/year are indicated by solid and dashed horizontal lines, respectively.

3.2.5 Module versus system degradation

Very few studies exist where module and system performance data are investigated at the same time and thus it is not clear how module long-term performance relates to system performance. In Fig. 3.6 we partition crystalline silicon (x-Si) high-quality data into system and module data. Cumulative distribution functions for all data (solid lines) and median values per study and system (dashed lines) are shown. The number of data points is again given in brackets. The cumulative probability median is indicated by a dashed horizontal line, whereas the 0.5%/year and 1%/year degradation are indicated as vertical dashed and dash-dotted lines, respectively. At the cumulative probability median, the system curves are very similar around 0.6%/year. The degradation rates for the module data are below the systems data and show some difference depending on whether all data or median per study and system are shown, which may enable us to infer some uncertainty associated with this comparison. It follows that one would expect system degradation to be similar or slightly larger than module degradation at the median. Module and system curves start to deviate more substantially for worse-performing products. This may be consistent with a recent more detailed examination of a 20-year-old x-Si system, in which it was found that the worst-performing module was limiting the string, and, in turn, the worst-performing string was limiting the system [8]. If the modules in a given system degrade similarly, the system will degrade close to the median module; conversely, if the modules show substantial spread in their degradation behaviour, the system degradation may be significantly different from the median module behaviour.

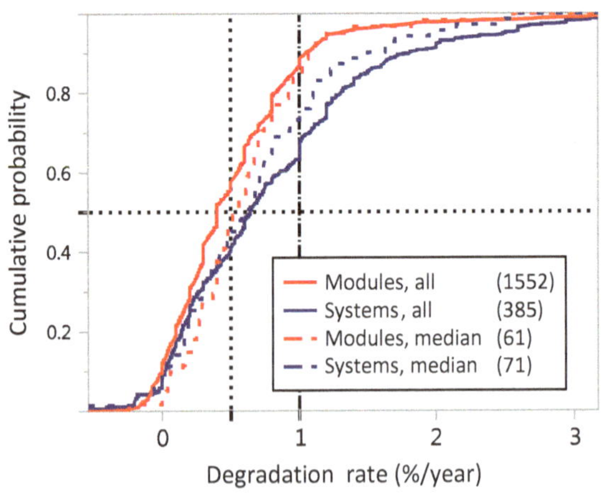

Fig. 3.6 Cumulative distribution functions for high-quality data of x-Si systems and modules. The median is indicated by a dashed horizontal line, 0.5%/year and 1%/year degradation are indicated as a dashed and dash-dotted vertical line, respectively. The number of data points for the respective subsets is given in parentheses.

3.2.6 Nonlinearities

The first challenge in long-term performance prediction is the question of linearity. Many PV technologies, especially thin-film technologies, exhibit nonlinearities at the beginning of their useful life. The initial rapid decline for amorphous silicon (a-Si) has been well documented [28]. The initial rapid decline occurs during several months before the onset of the long-term trend, as shown in Fig. 3.7. Some CdTe products show a similar behaviour [7]. The CIGS system shown in Fig. 3.7 shows a distinctly different behaviour; an initial increase during the first several months of light exposure is followed by the onset of the long-term degradation. It is evident from these examples that the initial trend is different from the long-term behaviour. In addition, if the initial phase is included in the evaluation, the long-term prediction would include a substantial, yet unintended bias.

In addition to the beginning-of-life phase, the wear-out phase in the lifecycle of PV modules can include nonlinearities. Fig. 3.8 displays four studies examining an ensemble of modules several times during the life of a PV system, each system having been fielded for at least 20 years. The sample from the Colorado, USA study is too small to confidentially establish linearity [8]. The California, USA study shows a fairly linear behaviour for the central tendency of the modules [29]. However, the worse-performing modules display distinct signs of nonlinearity. Similar trends can be observed in the studies from Switzerland [18] and Italy [30] of modules that were exposed for 30 years.

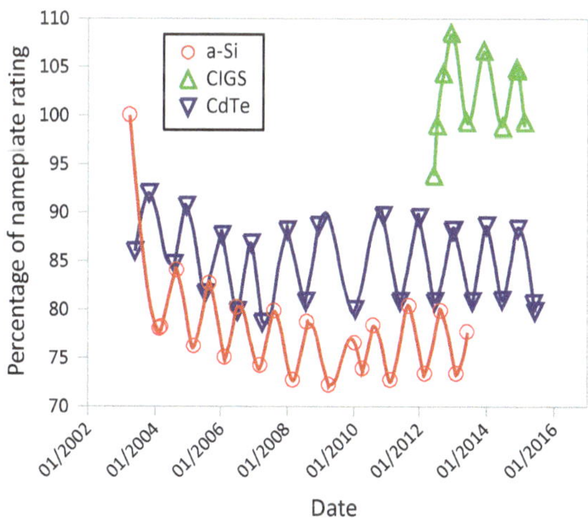

Fig. 3.7 Monthly percentage of nameplate rating not temperature corrected outdoor for thin-film technologies a-Si, CdTe and CIGS. Nonlinear tendencies can be readily observed over seasonal variations.

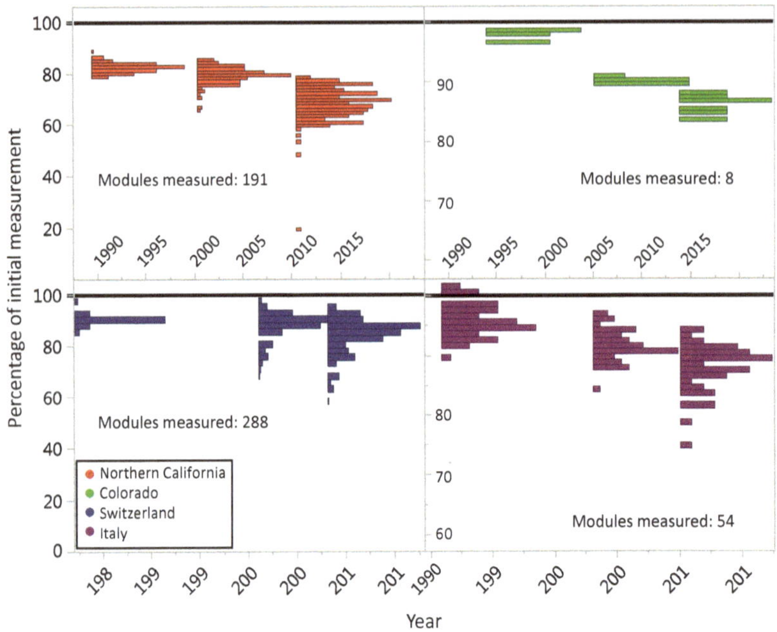

Fig. 3.8 More-than-20-year-old systems that measured an ensemble of modules multiple times during the field exposure. The modules were measured at the beginning of life, with the exception of modules for the Italian study, which were first measured after 11 years of field exposure.

3.2.7 Financial impact

On the systems level, several studies have emerged recently that partition continuous data into shorter time intervals allowing the determination of several subsequent degradation rates instead of one overall degradation rate [31,32]. This is a promising trend, as it allows delineating the properties of the degradation curve and provides more information than one overall degradation rate. Quantifying nonlinearity of degradation curves, in contrast to rates that imply linearity, can have significant impact on financial aspects of a PV project, such as the levelized cost of energy (LCOE) [33]. In summary, we may conclude that nonlinear behaviour may depend on a variety of factors, such as technology, product, climate, system load, etc., as some products will be more susceptible than others.

3.2.8 Other factors

Finally, the load under which a PV module or system is exposed may also influence the overall degradation. Crystalline Si modules were shown to have a significantly higher degradation after 20 years of field exposure when they were grid-connected through an inverter compared to open-circuit condition [34]. In contrast, for a-Si technology open-circuit conditions led to significantly higher degradation compared to under continuous load [35].

3.3 Reliability and failure modes

3.3.1 Historical survey

Service lifetime prediction, as discussed in Section 3.1, depends on developing accelerated tests that can duplicate failure modes from field observation. Information on field failure modes can be obtained empirically through site visits [36], warranty returns [37] and/or maintenance records [38]. Recently the International Energy Agency PV Power Systems Program—Task 13 published a detailed review on failures of PV modules based on literature and site visits [39]. The treatise discusses at length inspection tools, observed failures observed and new proposed test methodologies. Different failure modes are classified by safety and their different time series behaviour, eg, linear versus nonlinear. In practice it is often difficult to distinguish between different time series behaviour and estimating the impact of a failure mode for several reasons. Different products may be susceptible to different failure modes at different times in their life cycle. PV modules often display several failure modes at the same time making it difficult to isolate a certain power loss to a specific failure mode. The use environment may accelerate different failure modes differently.

Instead, a simpler strategy may be to classify failures by their modes, mechanisms, severity and detectability, as it is commonly done in other industries. Hence, failure modes are typically defined as the effect by which failures are observed [40]. In contrast, a failure mechanism is the cause that leads to failure. In Table 3.1 we summarize typically seen failures and categorize them by failure mode, failure mechanism, the severity, the onset and the detectability. Detectability frequently relies on visual inspection, yet visually observed failure modes are often subject to the inspector's knowledge and experience. A standardized protocol for visual inspection has been developed and is provided as a resource in Section 3.7 at the end of the chapter. The onset of the failure relates to different behaviour in time; it can be a gradual process or a sudden process. Yet, even this classification may be limited due to the synergistic nature of many internal and external variables. Examples of the most commonly observed failure modes and mechanisms observed in the literature, Fig. 3.9A, and visits of several hundred systems, Fig. 3.9B, will be discussed in the following section.

3.3.2 Failure modes

Discolouration

The discolouration of the ethylene vinyl acetate (EVA) encapsulant was observed in the power loss of the Carrizo Plains installation in the 1990s [42] (Fig. 3.10). Subsequently, the discolouration was detailed and appeared to have been solved [43]. However, Fig. 3.9 implies that it is still a relevant topic today, although this may be aided by the fact that discolouration is most notable by visual inspection. Discolouration typically does not cause failure, but causes decreased power output and may be a key contributor to the slow, ~0.5%/year degradation typically observed for PV modules. The impact is the loss of light transmission that can be readily detected in a decreased short-circuit current (Isc) [44] (Fig. 3.11).

Table 3.1 Commonly observed PV module failure modes, mechanisms, severity, time onset and detectability

Failure mode	Failure mechanism	Severity	Onset	Detectability
Discolouration (EVA)	Chemical alteration of cross-linking	Low	Gradual	Evident, visual
Delamination (front side)	Adhesion, loss of elastomeric properties	Low to high	Gradual	Evident, visual
Backsheet failure	Adhesion, fracture	Low to high	Gradual, sudden	Evident, visual
Internal circuitry discolouration	Corrosion	Low to high	Gradual	Evident, visual
Internal circuitry interruption (solder bonds, ribbons)	Corrosion, fatigue	High	Sudden	Visual, requires detailed inspection
Glass breakage	Fracture	High	Sudden	Evident, visual
Cell breakage	Fracture	Low to high	Sudden	Evident, visual
Hot spots	Fracture, electrical mismatch	Low to high	Gradual	Hidden, need IR
Burn marks	Fracture, electrical mismatch	High	Gradual	Evident, visual
PID	Electrical stress, sodium migration	High	Gradual	Hidden, need IR, EL, performance data
External circuitry disruption (j-box, bypass diode, cables)	Wear, electrical stress	Low to high	Gradual	Evident, visual
Structural failures	Deformation (frame from snow load)	High	Sudden	Evident, visual
AR coating delamination	Adhesion	Low	Gradual	Evident, visual
Soiling	Wear, etching of glass	Low	Gradual	Evident, visual
LID	Chemical, oxygen-boron complex	Low	Gradual	Hidden, need performance data

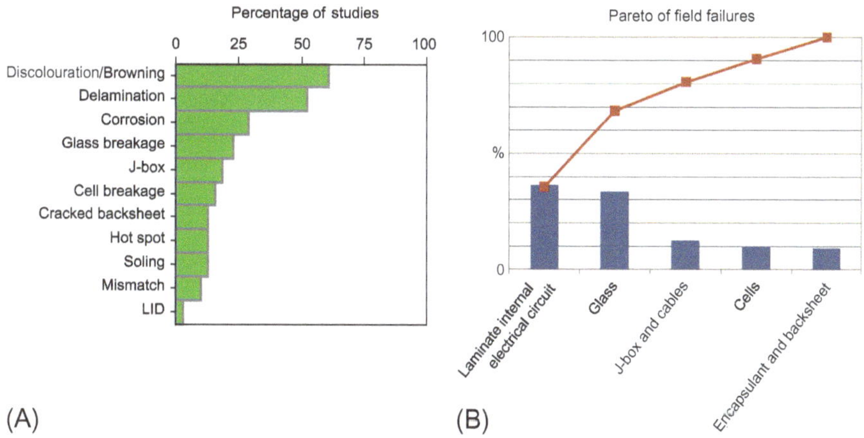

Fig. 3.9 Failure modes observed as a percentage of literature studies (A) [20] and Pareto chart of failure modes observed in a single study of multiple manufacturer (B) [41].

Fig. 3.10 Crystalline silicon module with significant encapsulant discolouration.

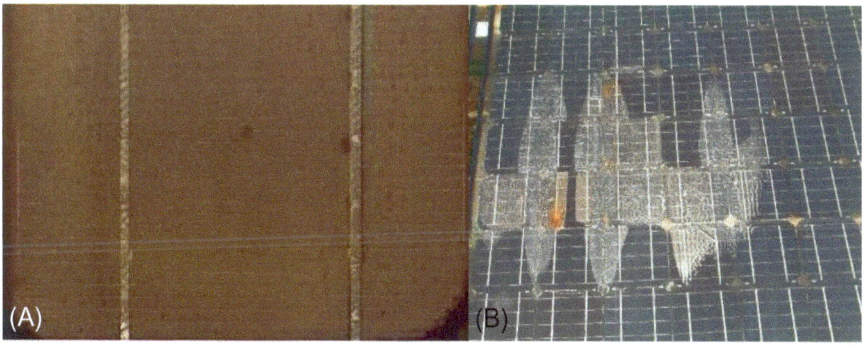

Fig. 3.11 Severe delamination over an entire cell (A) [45], and large-scale delamination across multiple cells (B) [46].

Delamination

Similarly to discolouration, delamination at or near the front surface can be seen in a decreased Isc, at least in the initial stages, as an additional interface develops and loss of transmission occurs. If delamination continues and becomes more severe, moisture may ingress and lead to corrosion of the internal circuitry that can be observable in decreased fill factor (FF). Furthermore, delamination can also occur behind the x-Si cells and may manifest itself as bubbling of the backsheet or complete delamination (Fig. 3.12).

Metallization corrosion

Discolouration of the internal circuitry or corrosion is often preceded by delamination. Initially, the power loss is caused by the added interface. However, over time with increasing delamination, moisture can penetrate and lead to corrosion of the metallic circuitry. Visually corrosion is easily detectable, as shown in Fig. 3.11. Electrically, the power loss is typically associated with FF loss through increased series resistance (Fig. 3.13).

Fig. 3.12 Delamination of the backsheet causing bubbles (A) and complete delamination for a frameless module (B) [26,41].

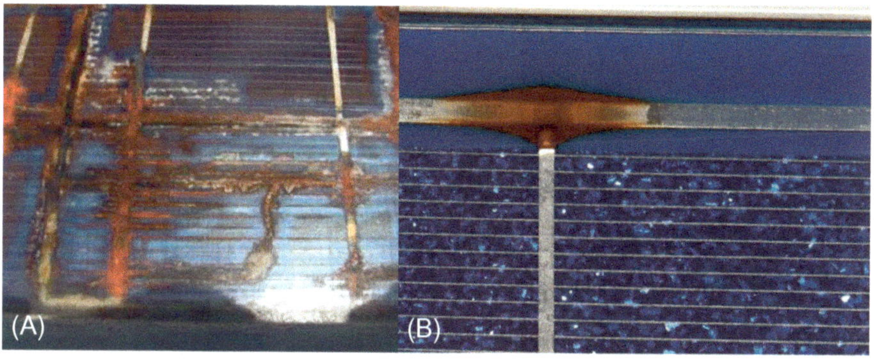

Fig. 3.13 Corrosion of the metallization (the busbar and fingers) (A) [59] and interconnect (B) [60].

Potential-induced degradation

Module degradation as a function of the position in the string was first reported on high-efficiency n-type Si modules [47]. The initial confusing effect was that modules that were exposed to more positive bias displayed higher degradation. Although not fully understood at the time, the remedy was grounding and avoidance of positive bias. Only a few years later a similar effect was discovered for p-type x-Si modules placed in negative bias [48]. While more is being learned in this rapidly developing field, the cause of the effect appears to be the migration of sodium ions into the semiconductor junction [49]. This potential-induced degradation (PID) and the partial reversal of the degradation have been of significant interest in the last few years because of the rapidity and severity of the effect in some cases. Module manufacturers routinely label their products as 'PID resistant' or 'PID free', though it is not always clear what tests were performed that led to said label. Two standard test methods have been defined in IEC/TS 62804.

Cracked cells and snail trails

Another common cause of loss of Isc can be broken cells. Even though cracked cells have become more common as cells have been thinned in recent years, there are few papers written specifically about cracked cells. Publications more commonly describe 'snail trails' (observed when the encapsulant has a different colour near the crack than elsewhere, Fig. 3.14) and conclude that these decorate cracked cells that lead to substantial power loss [51]. However, instead of identifying the cracked cells as the problem that needs to be fixed, often the snail trails are identified as 'the problem'. We have not seen a report that directly links the snail trails to performance loss, but note that the snail trails provide a simple way to identify cracked cells. Some PV customers wish to eliminate all cell cracks, but if a cell cracks in such a way that every

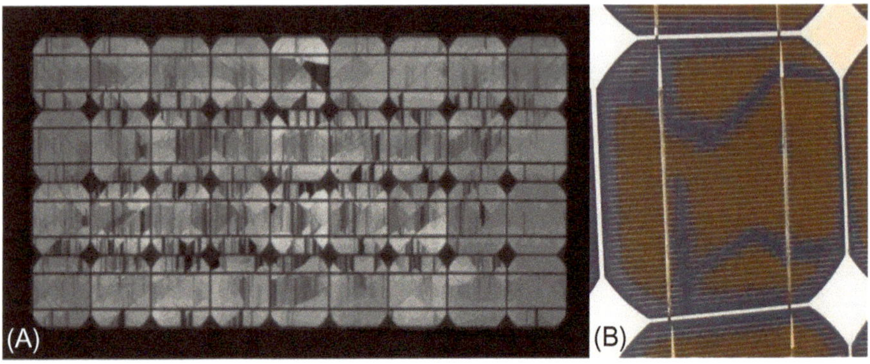

Fig. 3.14 Electroluminescence image of a module with many cracked cells (A) [50], and optical image of a cracked cell (B); the discolouration reveals the underlying crack.

Fig. 3.15 Image showing 'snail trails' that decorate cracked cells [55].

fragment remains connected, there may not be power loss. In order to reduce the silicon usage, and, therefore, module cost, we need to better understand how to avoid cracks, how to test for them, and how to identify whether some cracks are acceptable [52]. Electroluminescence is a useful tool for identifying cracked cells [53] (Fig. 3.15).

Hot-spots

As individual cells are serial connected in a module substring, the current flowing through the cells is the same. If one cell is shaded, that cell is reverse-biased, dissipating power and therefore leading to heating effects. Bypass diodes are connected in parallel and in the opposite direction to the PV cells limiting the power dissipating effect (Fig. 3.16).

Partial shading is not the only situation that can lead to reverse bias of one or more cells. Mismatch between individual cells, eg, one cell producing less current than the

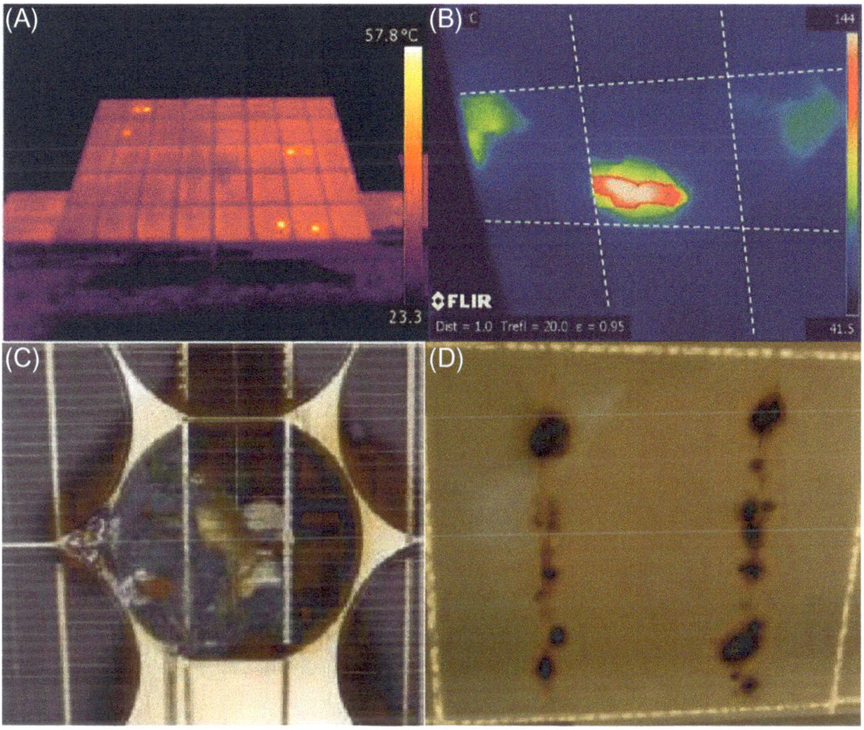

Fig. 3.16 Hot-spots observable in infrared imaging (A, B) [61,62] and damage caused by hot-spots on front side (C) and backside of module (D) [18,41].

other cells in a substring, can have a similar effect. The mismatch may increase during many years of field exposure as the individual cells degrade differently, exacerbating the effect. If the bypass diode fails to protect the substring, the heating can rapidly increase. Another ageing-related cause is the weakening connections of the internal module circuitry, such as solder bonds, through thermal cycling during field exposure. Finally, cracked cells may lead to hot-spots, as part of the cell is isolated and behaves similarly to a partially shaded cell. As the temperature difference between the hot-spot and the ambient area increases, safety becomes the primary concern. Significant damage to the front or the back of the module can occur, leading to elevated fire risk.

Glass

Glass breakage (Fig. 3.17) in thin-film modules is more common than in crystalline silicon because many thin-film products have annealed glass while most c-Si modules use tempered or heat-strengthened glass. Some arrays suffer from more glass breakage than others. Such glass breakage has been attributed to poor support structure design, inappropriate maintenance (eg, using a weed whacker to cut vegetation), poor

Fig. 3.17 Cracked front cover glass for silicon (A) and thin-film (B) modules [50,54].

Fig. 3.18 PV module frame damage because of snow load [55].

installation and/or handling practices (eg, a small nick on the edge of a module can nucleate a crack) and finally manufacturing defects.

Frame

Bent frames are common (Fig. 3.18) in climates with snow and ice. The snow may partially melt and then refreeze along the bottom edge of the module, causing stress on the frame. Joerg Althaus has led an effort [56] to define IEC 62938 'Non-uniform snow load testing for PV modules', planned for publication in 2016. This test will be valuable for those choosing modules for cold (icy) locations.

3.3.3 Standards development

Bankable PV (PV installations in which the investor has confidence in their reliable and long-term operation) benefits from international standards that address the module design, consistent manufacturing and system verification. The International PV Quality Assurance Task Force (PVQAT) was created to develop a set of universally accepted accelerated stress tests to identify adequate durability for all climates and mounting configurations [57]. PVQAT and IEC efforts to define improved PV standards are moving forward quickly to support the maturation of the industry. The challenge is to identify rapid test methods that adequately assess durability in a range of use environments while recognizing that quantitative lifetime assessment must be implemented for a specific bill of materials with a defined quality management system that specifies the variability allowed for the product implementation. PVQAT investigations will provide a basis for implementation of both climate-specific qualification tests and more quantitative service life predictions. Table 3.2 provides a summary of the module-level issues that were identified and the near- and long-term plans by PVQAT and Working Group 2 (WG2) (Modules) of IEC Technical Committee 82 (TC82) on PV.

Table 3.2 International standards efforts to address issues prioritized in Qualification Plus (Q+) [58]

Priority	Current status	Near-term plan	Long-term plan
UV durability of polymeric components (encapsulants, backsheets, connectors and junction boxes) and as a requirement for insulation materials (see Q+ component tests 1–4)	Recent and planned changes will add more UV and other testing of polymeric materials	(1) Submit 62788-7-2; (2) Submit amendment to IEC 61730 referencing IEC 62788-7-2 for materials weathering conditions; (3) Submit additional IEC 62788 parts; publish in 2016	Define climate-specific UV exposures
Bypass diode and j-box thermal test (see Q+ component test 5)	Multiple documents are in progress to address thermal endurance, cycling, and runaway	Five actions are described in text reference [55]	Define climate-specific diode tests

Continued

Table 3.2 Continued

Priority	Current status	Near-term plan	Long-term plan
Electrical failures within the module (see Q+ module test 1)	Ribbon interconnects can be tested with cyclic mechanical loading, but thermal cycling (TC) is still needed		Define climate-specific thermal-cycle tests and implementation within QMS
Power loss from cracked cells (see Q + module test 2)	IEC 62782 TS: Cyclic (dynamic) mechanical load testing and IEC 62759-1 PV modules—Transportation testing were published	Submit amendment to IEC 61215 to add 1000 cycles from IEC 62782 before TC and humidity freeze	Define climate-specific tests for cracked cells, if needed
Susceptibility to PID (see Q+ module test 3)	IEC/TS 62804-1 Test methods for detection of PID—Part 1 Crystalline Silicon has been published	Define standard labels for PID susceptibility	Understand relationship between degradation in the test and in the field; Identify quick tests for screening cells and encapsulant materials
Susceptibility to hot spot degradation (see Q+ module test 4)	Revised hot-spot test is included in 2016 version of IEC 61215		Define improved hot-spot test for thin-film modules
Improve confidence in Quality Management System (QMS) (see Q+ description of sampling and QMS)	IEC/TS 62941 Guideline for increased confidence in PV module design qualification and type approval was published	Implement through IECRE	Facilitate adoption and assess value of extending to quantitative assessment

Continued

Table 3.2 **Continued**

Priority	Current status	Near-term plan	Long-term plan
Structural failure from snow and ice (based on experience in Europe and New England)	IEC 62938 Non-uniform snow load testing for PV modules Committee Draft has been reviewed	Complete IEC 62938; publish in 2017	Encourage use of the snow load test to differentiate modules
Faster degradation in hot climates	Delamination, encapsulant discolouration, and thermal fatigue are documented to increase with high temperature	Define difference in use environment	Use IEC 62892 to implement comprehensive tests at higher temperatures
Assessment of system functionality	Drafts completed for both energy and capacity tests	Publication of IEC 61724-2 and IEC 61724-3 in 2016	Implement as part of IECRE

3.4 Conclusion and future trends

Today, the PV industry is fairly knowledgeable about the most common causes of failures in the field, but a complete understanding of the acceptable process and design windows is challenging, especially as companies may make changes to the bill of materials frequently. PV reliability standards are maturing by:

- addressing failure mechanisms that have recently become problematic;
- differentiating the durability of PV module designs as a function of the use environment and
- improving manufacturing consistency toward a consistent implementation of product design, moving toward inclusion of service life prediction in the most mature quality management systems.

Today's most critical issues will be addressed by standards completed in 2016. Quantitative service life predictions will become practical when the industry is mature enough for product designs to stabilize enough to allow time for testing and model development and validation.

3.5 Resources

3.5.1 Visual inspection sheet

Documentation of module condition

Date ________________________ Name of data recorder________________________________

Location ____

Latitude____________________ Longitude _________________ Altitude ____________________

1. System Data

System design: ☐ single module ☐ multiple modules (a.) ☐ unknown

(a.) Multiple module system:

Module location/number in a series string (from negative) ______________

# of modules in series (string) _______ # of strings in parallel (array) _________

# of bypass diodes____________ # of modules per bypass diode ____________

System Bias: ☐ open circuit ☐ resistive load ☐ max. power tracked ☐ short circuit ☐ unknown

System Grounding: ☐ grounded (a.) ☐ not grounded ☐ unknown

(a.) ☐ negative ☐ positive ☐ center of string ☐ unknown

BEGIN INSPECTION AT BACK SIDE OF MODULE

2. Module Data

Technology: ☐ mono Si ☐ multi Si ☐ a–Si ☐ CdTe ☐ CIGS/CIS ☐ other: _____________

Certification: ☐ unknown ☐ UL 1703 ☐ IEC 61215 ☐ IEC 61646 ☐ IEC 61730 ☐ other: _____________

Estimated deployment date _________________________

Photo taken of nameplate: ☐ **yes** ☐ **no**

Manufacturer ____

Model # ____

Serial # ____

Installation Site/Facility Serial # ______________________________________

Width ______________cm Length ______________cm

Nameplate: ☐ nameplate missing

P_{max} _____________ V_{oc} _____________ I_{sc} ____________

Sys Volt__________ V_{max} ___________ I_{max} __________

Bypass diode, I_f _______________________________________

Series fuse ____

3. Rear-side Glass: ☐ **not applicable** ☐ **applicable**

Damage: ☐ no damage ☐ small, localized ☐ extensive

Damage Type (mark all that apply):
☐ crazing or other non–crack damage
☐ shattered (tempered) ☐ shattered (non-tempered) ☐ cracked (a.) ☐ chipped (b.)
(a.) Cracks (#):☐ 1 ☐ 2 ☐ 3 ☐ 4–10 ☐ >10
Crack(s) start from: ☐ module corner ☐ module edge ☐ cell ☐ junction box
☐ foreign body impact location
(b.) Chips (#): ☐ 1 ☐ 2 ☐ 3 ☐ 4–10 ☐ >10
Chipping location: ☐ module corner ☐ module edge

4. Backsheet: ☐ not applicable ☐ applicable

Appearance: ☐ like new ☐ minor discoloration ☐ major discoloration
Texture: ☐ like new ☐ wavy (not delaminated) ☐ wavy (delaminated) ☐ dented
Material quality -chalking: ☐ none ☐ slight ☐ substantial
Damage: ☐ no damage ☐ small, localized ☐ extensive
Damage Type (mark all that apply):
☐ burn marks (a.) ☐ bubbles (b.) ☐ delamination (c.) ☐ cracks/scratches (d.)

(a.) Burn marks (#):☐ 1 ☐ 2 ☐ 3 ☐ 4–10 ☐ >10
Fraction of area burned:
☐ <5% ☐ 5–25% ☐ 50% ☐ 75–100% (consistent overall)
(b.) Bubbles(#): ☐ 1 ☐ 2 ☐ 3 ☐ 4–10 ☐ >10
Average bubble dimension: ☐ <5 mm ☐ 5–30 mm ☐ >30 mm
Fraction of area with bubbles > 5 mm:
☐ <5% ☐ 5–25% ☐ 50% ☐ 75–100% (consistent overall)
(c.) Fraction of area delaminated:
☐ <5% ☐ 5–25% ☐ 50% ☐ 75–100% (consistent overall)
Fraction of delamination that exposes circuit or cell(s)
☐ <5% ☐ 5–25% ☐ 50% ☐ 75–100% (consistent overall)
(d.) Cracks/scratches (#): ☐ 1 ☐ 2 ☐ 3 ☐ 4–10 ☐ >10
Cracks/scratches location: ☐ random/no pattern ☐ over cells ☐ between cells
Fraction of area affected by cracks/scratches (approx.):
☐ <5% ☐ 5–25% ☐ 50% ☐ 75–100% (consistent overall)
Fraction of cracks/scratches that exoose circuit (approx.):
☐ 0% ☐ 25% ☐ 50% ☐ 75% ☐ 100%

5. Wires/Connectors:

Wires: ☐ not applicable ☐ like new ☐ pliable, but degraded ☐ embrittled
(mark all that apply): ☐ cracked/disintegrated insulation ☐ burnt
☐ corroded ☐ animal bites/marks
Connectors: ☐ not applicable ☐ like new ☐ pliable, but degraded ☐ embrittled
Type: ☐ unsure ☐ MC3 or MC4 ☐ Tyco Solarlok ☐ other ________

(mark all that apply): ☐ cracked/disintegrated insulation ☐ burnt ☐ corroded

6. Junction Box:

Junction box itself: ☐ not applicable/observable ☐ applicable

Physical state: ☐ intact ☐ unsound structure

(mark all that apply): ☐ weathered ☐ cracked ☐ burnt ☐ warped

Lid: ☐ intact/potted ☐ loose ☐ fell off ☐ cracked

Junction box adhesive: ☐ not applicable/observable ☐ applicable

Attachment: ☐ well attached ☐ loose/brittle ☐ fell off

Pliability: ☐ like new ☐ pliable, but degraded ☐ embrittled

Junction box wire attachments: ☐ not applicable/observable ☐ applicable

Attachment: ☐ well attached ☐ loose ☐ fell off

Seal: ☐ good seal ☐ seal will leak

other: ☐ arced/started a fire

7. Frame Grounding:

Original state: ☐ Wired ground ☐ Resistive ground ☐ No ground ☐ unknown

Appearance: ☐ Not applicable ☐ Like new ☐ Some corrosion ☐ Major corrosion

Function: ☐ Well grounded ☐ No connection

Photos taken of ☐ back, label, and junction box

CONTINUE INSPECTION ON FRONT SIDE OF MODULE

8. Frame: ☐ not applicable ☐ applicable

Appearance: ☐ like new ☐ damaged (a.) ☐ missing

(a.) (mark all that apply): ☐ minor corrosion ☐ major corrosion ☐ frame joint separation ☐ frame cracking ☐ bent frame ☐ discoloration

Frame Adhesive: ☐ like new/not visible ☐ degraded (a.)

(a.) (mark all that apply): ☐ adhesive oozed out ☐ adhesive missing in areas

9. Frameless Edge Seal: ☐ not applicable ☐ applicable

Appearance: ☐ like new ☐ discoloration (a.) ☐ visibly degraded

(a.) Fraction affected by discoloration:

☐ <5% ☐ 5–25% ☐ 50% ☐ 75–100% (consistent overall)

Material problems:

☐ squeezed/pinched out ☐ shows signs of moisture penetration

Delamination: ☐ none ☐ area(s) delaminated (a.)

(a.) Fraction Delaminated:

☐ <5% ☐ 5–25% ☐ 50% ☐ 75–100% (consistent overall)

10. Glass/Polymer (front):

Material: ☐ glass ☐ polymer ☐ glass/polymer composite ☐ unknown
Features: ☐ smooth ☐ slightly textured ☐ pyramid/wave texture
☐ antireflection coating
Appearance: ☐ clean ☐ lightly soiled ☐ heavily soiled
Location of soiling:
☐ locally soiled near frame:
☐ left ☐ right ☐ top ☐ bottom ☐ all sides
☐ locally soiled on glass /bird droppings
Damage: ☐ no damage ☐ small, localized ☐ extensive
Damage Type (mark all that apply):
☐ crazing or other non-crack damage
☐ shattered (tempered) ☐ shattered (non-tempered) ☐ Cracked (a.)
☐ Chipped (b.) ☐ milky discoloration (c.)
(a.) Cracks (#): ☐ 1 ☐ 2 ☐ 3 ☐ 4–10 ☐ >10
Crack(s) start from: ☐ module corner ☐ module edge ☐ cell ☐ junction box
☐ foreign body impact location
(b.) Chips (#): ☐ 1 ☐ 2 ☐ 3 ☐ 4–10 ☐ >10
Chipping location: ☐ module corner ☐ module edge
(c.) Fraction of area:
☐ <5% ☐ 5–25% ☐ 50% ☐ 75–100% (consistent overall)

11. Metallization:

Gridlines/Fingers: ☐ not applicable/barely observable ☐ applicable
Appearance: ☐ like new ☐ light discoloration (a.) ☐ dark discoloration (a.)
(a.) Fraction of discoloration:
☐ <5% ☐ 5–25% ☐ 50% ☐ 75–100% (consistent overall)
Busbars: ☐ not applicable/not observable ☐ applicable
Appearance: ☐ like new ☐ light discoloration (a.) ☐ dark discoloration (a.)
(a.) Fraction of discoloration:
☐ <5% ☐ 5–25% ☐ 50% ☐ 75–100% (consistent overall)
(mark all that apply): ☐ obvious corrosion ☐ diffuse burn mark(s) ☐ misaligned
Cell Interconnect Ribbon: ☐ not applicable/not observable ☐ applicable
Appearance: ☐ like new ☐ light discoloration (a.) ☐ dark discoloration (a.)
(a.) Fraction of discoloration:
☐ <5% ☐ 5–25% ☐ 50% ☐ 75–100% (consistent overall)
(mark all that apply): ☐ obvious corrosion ☐ burn marks ☐ breaks
String Interconnect: ☐ not applicable/not observable ☐ applicable
Appearance: ☐ like new ☐ light discoloration (a.) ☐ dark discoloration (a.)
(a.) Fraction of discoloration:
☐ <5% ☐ 5–25% ☐ 50% ☐ 75–100% (consistent overall)
(mark all that apply): ☐ obvious corrosion ☐ burn marks ☐ breaks
☐ arc tracks (thin, small burns)

12. Silicon (mono or multi) module: ☐ **not applicable** ☐ **applicable**

Number of:

Cells in module ___________

Cells in series/string _______

Strings in parallel __________

Cell size: Width ____________ cm Length ____________ cm

Distance between frame and cell: ☐ >10 mm ☐ <10 mm

Distance between cells in a string: ☐ >1 mm ☐ <1 mm

Discoloration: ☐ none/ike new ☐ light discoloration ☐ dark discoloration

Number of cells with any discoloration:_________________

of those, average % discolored area:

☐ <5% ☐ 5–25% ☐ 50% ☐ 75–100% (consistent overall)

Discoloration location(s) (mark all that apply):

☐ module center ☐ module edges ☐ cell centers ☐ cell edges

☐ over gridlines ☐ over busbars ☐ over tabbing ☐ between cells

☐ individual cell(s) darker than others ☐ partial cell discoloration

Junction box area: ☐ same as elsewhere ☐ more affected ☐ less affected

Damage: ☐ none

(mark all that apply): ☐ burn mark (a.) ☐ cracking (b.) ☐ moisture

☐ worm marks/snail tracks (c.) ☐ foreign particle embedded

(a.) Burns (#): ☐ 1 ☐ 2 ☐ 3 ☐ 4–10 ☐ >10

(b.) Number of cells cracked: ________________

(c.) Number of cells with worm marks/snail tracks: _____________

Delamination: ☐ none ☐ from edges ☐ uniform ☐ corner(s) ☐ near junction box

☐ between cells (a.) ☐ over cells (b.) ☐ near cell or string interconnect

(a.) Fraction delaminated between cells:

☐ <5% ☐ 5–25% ☐ 50% ☐ 75–100 (consistent overall)

(b.) Fraction delaminated over cells:

☐ <5% ☐ 5–25% ☐ 50% ☐ 75–100% (consistent overall)

Likely interface (choose 2):

☐ glass ☐ semiconductor ☐ encapsulant ☐ back sheet ☐ busbar

13.Thin film module: ☐ **not applicable** ☐ **applicable**

Number of cells:

Number of cells in module ___________

Number of cells in series/string _______

Number of strings in parallel __________

Cell size: Width ____________ cm Length ____________ cm

Distance between frame and cell: ☐ >10 mm ☐ <10 mm

Appearance: ☐ like new ☐ minor/light discoloration ☐ major/dark discoloration

Discoloration type (mark all that apply):

☐ spotted degradation ☐ haze (encapsulant browning) ☐ other

Discoloration location (mark all that apply):
☐ overall/no location pattern ☐ module center ☐ module edge(s)
☐ cell center ☐ cell edges ☐ near crack(s)

Damage: ☐ no damage ☐ small, localized ☐ extensive
Damage Type (mark all that apply): ☐ burn mark(s) ☐ cracking
☐ possible moisture ☐ foreign particle embedded

Delamination: ☐ no delamination ☐ small, localized ☐ extensive
Location: ☐ from edges ☐ uniform ☐ corner(s) ☐ near junction box ☐ near busbar
☐ along scribe lines
Delamination Type: ☐ absorber delamination ☐ AR coating delamination ☐ other

Photos taken of ☐ front and defects

14. Electronic Records ☐ not applicable ☐ applicable
Photographs and I–V curves recorded electronically–list file names in blanks
Photo files ____
I–V curve ____
Connector function: ☐ functions ☐ no longer mates ☐ exposed
Irradiance ____________________Sensor ________________
Temperature ____________________Sensor ________________
EL picture ______________________
IR picture ______________________
Bypass Diode Test: ☐ not applicable ☐ applicable
Number of diodes:
In total__________, shorted_____________, open___________

OTHER

Acknowledgement

This work was completed under contract no. DE-AC36-99GO10337 with the US Department of Energy.

References

[1] International Electrotechnical Commission, Crystalline Silicon Terrestrial Photovoltaic (PV) Modules—Design Qualification and Type Approval, first ed., IEC Central Office, Geneva, 2016. IEC 61215.
[2] D.C. Jordan, S.R. Kurtz, Photovoltaic degradation rates—an analytical review, Prog. Photovolt. Res. Appl. 21 (1) (2013) 12–29, http://dx.doi.org/10.1002/pip.1182.
[3] J. Newmiller, G.L. DNV, White paper on photovoltaic module degradation, RANA-WP-03-A, 2015.
[4] S. Ransome, Modelling inaccuracies of PV energy yield simulations, in: Proceedings of the 33rd PV Specialists Conference, San Diego, CA, USA, 2008, pp. 1–5. http://dx.doi.org/10.1109/PVSC.2008.4922864.

[5] D. Dirnberger, B. Müller, C. Reise, PV module energy rating: opportunities and limitations, Prog. Photovolt. Res. Appl. 1 (2015), http://dx.doi.org/10.1002/pip.2618.
[6] D. Dirnberger, Uncertainty in PV module measurement—part II: verification of rated power and stability problems, J. Photovolt. 4 (3) (2014) 991–1007.
[7] L. Ngan, N. Strevel, K. Passow, A.F. Panchula, D. Jordan, Performance characterization of cadmium telluride modules validated by utility-scale and test systems, in: Proceedings of the 40th IEEE Photovoltaic Specialists Conference, Denver, CO, USA, 2014, pp. 1957–1962.
[8] D.C. Jordan, B. Sekulic, B. Marion, S.R. Kurtz, Performance and aging of a 20-year-old silicon PV system, J. Photovolt. 5 (3) (2015) 744–751, http://dx.doi.org/10.1109/JPHOTOV.2015.2396360.
[9] M.G. TamizhMani, J. Kuitche, A. Mikonowicz, Nameplate, datasheet and sampling requirements for photovoltaic modules, in: Solar ABC Stakeholder Workshop, Las Vegas, 2014, 2014.
[10] A. Detrick, A. Kimber, L. Mitchell, Performance evaluation standards for photovoltaic modules and systems, performance evaluation standards for photovoltaic modules & systems, in: Proceedings of the 31th IEEE Photovoltaic Specialists Conference, Lake Buena Vista, FL, 2005, pp. 1581–1586.
[11] M. García, L. Marroyo, E. Lorenzo, J. Marcos, M. Pérez, STC power for 15 MW of PV comparing nameplate, initial power and power after 4 years, Prog. Photovolt. Res. Appl. 22 (2) (2012) 242–247.
[12] J. Leloux, L. Narvarte, D. Trebosc, Performance analysis of 10,000 residential PV systems in France and Belgium, in: Proceedings of the 26th European Photovoltaic Solar Energy Conference, Hamburg, Germany, 2011, pp. 3939–3946.
[13] T. Oozeki, T. Yamada, K. Otani, T. Takashima, K. Kato, Performance trends in grid-connected photovoltaic systems for public and industrial use in Japan, Prog. Photovolt. Res. Appl. 18 (8) (2010) 596–602, http://dx.doi.org/10.1002/pip.992.
[14] A.M. Reis, N.T. Coleman, M.W. Marshall, P.A. Lehman, C.E. Chamberlin, Comparison of PV module performance before and after 11-years of field exposure, in: Proceedings of the 29th PV Specialists Conference, New Orleans, LA, USA, 2002, pp. 1432–1435, http://dx.doi.org/10.1109/PVSC.2002.1190878.
[15] O. Coiante, S. Castello, PV module performance after ten years of continuous outdoor operations, in: ISES 1491—Solar World Congress, 1991.
[16] J.E. Granata, W.E. Boyson, J.A. Kratochvil, M.A. Quintana, Long-term performance and reliability assessment of 8 PV arrays at Sandia National Laboratories. in: Proceedings of the 34th PV Specialists Conference, Philadelphia, PA, USA, 2009, pp. 1486–1491, http://dx.doi.org/10.1109/PVSC.2009.5411336.
[17] P. Sanchez-Friera, M. Piliougine, J. Pelaez, J. Carretero, M. Sidrach de Cardona, Analysis of degradation mechanisms of crystalline silicon PV modules after 12 years of operation in Southern Europe, Prog. Photovoltaics: Res. Appl. 19 (6) (2011) 658–666, http://dx.doi.org/10.1002/pip.1083.
[18] T. Friesen, D. Chianese, A. Realini, G. Friesen, E. Burà, A. Virtuani, D. Strepparava, R. Meoli, TISO 10 kW: 30 Years Experience with a PV Plant, in: Proceedings of the 27th European Photovoltaic Solar Energy Conference, Frankfurt, Germany, 2012, pp. 3125–3131.
[19] B.T. Hamzavy, A.Z. Bradley, Safety and performance analysis of a commercial photovoltaic installation, Proc. SPIE 8825 (2013), http://dx.doi.org/10.1117/12.2023406.

[20] D.C. Jordan, J.H. Wohlgemuth, S.R. Kurtz, Technology and climate trends in PV module degradation, in: Proceedings of the 27th European Photovoltaic Solar Energy Conference, Frankfurt, Germany, 2012, pp. 3118–3124.

[21] J. Singh, J. Belmont, G. TamizhMani, Degradation analysis of 1900 PV modules in a hot-dry climate: results after 12 to 18 years of field exposure, in: Proceedings of the 39th IEEE Photovoltaic Specialists Conference, Tampa, FL, USA, 2013, pp. 3270–3275.

[22] R. Dubey, S. Chattopadhyay, V. Kuthanazhi, J.J. John, C.S. Solanki, A. Kottantharayil, B.M. Arora, K.L. Narasimhan, J.M. Vasi, A. Kumar, O.S. Sastry, Performance degradation in field-aged crystalline silicon PV modules in different indian climatic conditions, in: Proceedings of the 40th IEEE Photovoltaic Specialists Conference, Denver, CO, USA, 2014, pp. 3182–3187.

[23] J. Ying Ye, T. Reindl, A.G. Aberle, T.M. Walsh, Performance Degradation of Various PV Module Technologies in Tropical Singapore, J. Photovolt. 4 (5) (2014) 1288–1294.

[24] N. Strevel, L. Trippel, M. Gloeckler, Performance characterization and Superior energy yield of first solar PV power plants in high-temperature conditions, in: Photovoltaics International, 2012.

[25] Desert: BWh, BWk; Hot & Humid: Af, Am, As, Aw, Cwa, Cfa; Moderate: Cfb, Csa, Csb, BSh, BSk; Snow: Dfa, Dfb, Dfc, ET.

[26] K. Yedidi, S. Tatapudi, J. Mallineni, B. Knisely, J. Kutiche, G. TamizhMani, Failure and degradation modes and rates of PV modules in a hot-dry climate: results after 16 years of field exposure, in: Proceedings of the 40th IEEE Photovoltaic Specialists Conference, Denver, CO, USA, 2014, pp. 3245–3247.

[27] M.A. Mikofski, D.F.J. Kavulak, D. Okawa, Y.-C. Shen, A. Terao, M. Anderson, S. Caldwell, D. Kim, N. Boitnott, J. Castro, L.A.L. Smith, R. Lacerda, D. Benjamin, E.F. Hasselbrink Jr., PVlife: an integrated model for predicting PV performance degradation over 25+ years, in: Proceedings of the 38th IEEE Photovoltaic Specialists Conference, Austin, TX USA, 2012, pp. 1744–1749.

[28] D.L. Staebler, C.R. Wronski, Optically induced conductivity changes in discharge-produced hydrogenated amorphous silicon, J. Appl. Phys. 51 (6) (1980) 3262–3268.

[29] C.E. Chamberlin, M.A. Rocheleau, M.W. Marshall, A.M. Reis, N.T. Coleman, P.A. Lehman, Comparison of PV module performance before and after 11 and 20 years of field exposure, in: Proceedings of the 37th IEEE Photovoltaic Specialists Conference, Seattle, WA USA, 2011, pp. 101–105.

[30] L. Abenante, F. De Lia, S. Castello, Long-term performance degradation of c-Si photovoltaic modules and Strings, in: Proceedings of the 25th European photovoltaic solar energy conference, Valencia, Spain, 4023–4026.

[31] A. Kyprianou, A. Phinikarides, G. Makrides, G.E. Georghiou, Robust principal component analysis for computing the degradation rates of different photovoltaic systems, in: Proceedings of the 29th European Photovoltaic Solar Energy Conference, Amsterdam, Netherlands, 2014, pp. 2939–2942.

[32] G. Belluardo, P. Ingenhoven, W. Sparber, J. Wagner, P. Weihs, D. Moser, Novel method for the improvement in the evaluation of outdoor performance loss rate in different PV technologies and comparison with two other methods, Sol. Energ. 117 (2015) 139–152.

[33] D.C. Jordan, S.R. Kurtz , K.T. VanSant , J. Newmiller, Compendium of photovoltaic degradation rates, Progress in PV, 2016, http://dx.doi.org/10.1002/pip.2744/.

[34] A. Skoczek, T. Sample, E.D. Dunlop, The results of performance measurements of field-aged crystalline silicon photovoltaic modules, Prog. Photovolt. Res. Appl. 17 (2009) 227–240, http://dx.doi.org/10.1002/pip.874.

[35] L. Fanni, I. Pola, E. Burà, T. Friesen, D. Chianese, Investigation of annealing and degradation effects on a-Si PV modules in real operating conditions. in: Proceedings of 24th European Photovoltaic Solar Energy Conference, Hamburg, Germany, 2009, pp. 3596–3599, http://dx.doi.org/10.4229/24thEUPVSEC2009-4AV.3.92.
[36] D. DeGraaff, R. Lacerda, Z. Campeau, Z. Xie, How do qualified modules fail—what is the root cause? in: International PV Module Quality Assurance Forum, San Francisco, CA, 2011.
[37] J.H. Wohlgemuth, D.W. Cunningham, A.M. Nguyen, J. Miller, Long term reliability of PV modules, in: 20th European PV Solar Energy Conference, Barcelona, Spain, 2005, pp. 1942–1948.
[38] A. Golnas, PV system reliability: an operator's perspective. IEEE J. Photovolt. 3 (1) (2013) 416–421, http://dx.doi.org/10.1109/JPHOTOV.2012.2215015.
[39] M. Köntges, S. Kurtz, C. Packard, U. Jahn, K.A. Berger, K. Kato, T. Friesen, H. Liu, M. Van Iseghem, Performance and reliability of photovoltaic systems, subtask 3.2: review of failures of photovoltaic modules, International Energy Agency, Report IEA-PVPS T13-01:2014, 2014.
[40] SAE Standard SAE J1739, 2002.
[41] D. DeGraaff, R. Lacerda, Z. Campeau, Degradation mechanisms in Si module technologies observed in the field; their analysis and statistics, in: PV Module Reliability Workshop, Golden, CO, USA, 2011.
[42] A.L. Rosenthal, G.C. Lane, Field test results for the 6 MW Carrizo solar PV power plant. Solar Cells 30 (1–4) (1991) 563–571, http://dx.doi.org/10.1016/0379-6787(91)90088-7.
[43] W.W. Holey, S.C. Agro, Advanced EVA-based encapsulants, NREL/SR-520-25296, 1998.
[44] R.M. Smith, D.C. Jordan, S.R. Kurtz, P.V. Outdoor, Module degradation of current–voltage parameters, in: World Renewable Energy Forum, Denver, CO, USA, 2012.
[45] A. Skoczek, T. Sample, E.D. Dunlop, H.A. Ossenbrink, Electrical performance results from long-term outdoor weathered modules, in: 22nd European Photovoltaic Solar Energy Conference, Milan, Italy, 2007, pp. 2458–2466.
[46] M. Alonso-Abella, F. Chenlo, A. Alonso, D. González, Toledo PV plant 1 MWp—20 years of operation, in: Proceedings of the 29th European Photovoltaic Solar Energy Conference, Amsterdam, Netherlands, 2014, pp. 2728–2733.
[47] R. Swanson, M. Cudzinovic, D. DeCeuster, V. Desai, J. Jürgens, N. Kaminar, W. Mulligan, L. Rodrigues-Barbarosa, D. Rose, D. Smith, A. Terao, K. Wilson, The surface polarization effect in high efficiency silicon solar cells, in: Proceedings of the 15th International Photovoltaic Science and Engineering Conference, Shanghai, China, 2005.
[48] S. Pingel, O. Frank, M. Winkler, S. Daryan, T. Geipel, H. Hoehne, J. Berghold, Potentially induced degradation of solar cells and panels, in: Proceedings of the 35th IEEE PV Specialists Conference, Honolulu, HI, USA, 2010, pp. 2817–2822.
[49] P. Hacke, K. Terwilliger, R. Smith, S. Glick, J. Pankow, M. Kempe, S.K.I. Bennett, M. Kloos, System voltage potential-induced degradation mechanisms in PV modules and methods for test, in: Proceedings of the 37th IEEE PV Specialists Conference, Seattle, WA, USA, 2011, pp. 814–820.
[50] J.H. Wohlgemuth, S.R. Kurtz, PV module reliability: how can we improve it? in: 30th European Photovoltaic Solar Energy Conference, Hamburg, Germany, 2015.
[51] A. Dolara, S. Leva, G. Manzolini, E. Ogliari, Investigation on performance decay on photovoltaic modules: snail trails and cell microcracks, IEEE Journal of Phtovoltaics 4 (5) (2014) 1204–1211.

[52] M. Köntges, S. Kajari-Schroder, I. Kunze, Crack statistic for wafer-based silicon solar cell modules in the field measured by UV fluorescence, IEEE J. Photovolt. 3 (2013) 95–101.
[53] G. Mathiak, J. Beniaminova, L. Pohl, J. Sommer, W. Herrmann, F. Reil, Quantitative electroluminescence imaging applied to hail impact damages, in: 29th European Photovoltaic Solar Energy Conference, Amsterdam, Netherlands, 2014, pp. 3285–3287.
[54] M.A. Quintana, D.L. King, T.J. McMahon, C.R. Osterwald, Commonly observed degradation in field-aged photovoltaic modules, in: Proceedings of the 22nd IEEE Photovoltaic Specialists Conference, Las Vegas, NV, USA, 2002, pp. 1436–1439.
[55] S. Kurtz, T. Sample, J. Wohlgemuth, W. Zhou, N. Bosco, J. Althaus, N. Phillips, M. Deceglie, C. Flueckiger, P. Hacke, D. Miller, M. Kempe, M. Yamamichi, M. Kondo, in: Proceedings of the 42nd IEEE Photovoltaic Specialists Conference, New Orleans, LA, USA, 2015.
[56] F. Reil, G. Mathiak, K. Strohkendl, S. Raubach, C. Schloth, B.V. Wangenheim, Experimental testing of PV modules under inhomogeneous snow loads, in: 27th European Photovoltaic Solar Energy Conference, Frankfurt, Germany, 2012, pp. 3414–3417.
[57] http://www.pvqat.org/.
[58] S. Kurtz, J. Wohlgemuth, M. Kempe, N. Bosco, P. Hacke, D. Jordan, D.C. Miller, T.J. Silverman, N. Phillips, T. Earnest, R. Romero, Photovoltaic Module Qualification Plus Testing, NREL/TP-5200-60950, 2013.
[59] J. Wohlgemuth, T. Silverman, D.C. Miller, P. McNutt, M. Kempe, M. Deceglie, Evaluation of PV module field performance, in: Proceedings of the 42nd IEEE Photovoltaic Specialists Conference, New Orleans, LA, USA, 2015.
[60] C.E. Packard, J.H. Wohlgemuth, S.R. Kurtz, U. Jahn, K. Berger, T. Friesen, M. Koentges, Fielded PV module condition, in: 27th European Photovoltaic Solar Energy Conference, Frankfurt, Germany, 2012.
[61] M. García, L. Marroyo, E. Lorenzo, J. Marcos, M. Pérez, Observed degradation in photovoltaic plants affected by hot-spots. Prog. Photovolt. Res. Appl. 22 (2013) 1292–1301, http://dx.doi.org/10.1002/pip.2393.
[62] K. Kato, PVRessQ! in: PV Module Failures Observed in the Field, PV Module Reliability Workshop, Golden, CO, 2012.

Modelling and prediction of PV module energy yield

4

B. Goss, I.R. Cole, E. Koubli, D. Palmer, T.R. Betts, R. Gottschalg
Loughborough University, Loughborough, United Kingdom

4.1 Introduction

The performance prediction of PV modules is fundamental to the entire system model. This chapter will try to answer the question 'What is the best model?' It should become clear that there is no single perfect model and that better agreement of a given model with a particular measured dataset does not necessarily mean that the model is better. The uncertainties are often not considered and this would show that better agreement in many cases may well be a statistical fortuity. More importantly, these model uncertainties equate to financial risk when models are used for system planning, and thus the understanding of these really is crucial for system planning.

A model is a tool to answer a given question. The quality of the answer depends on the specificity of the question. In terms of photovoltaics (PV), one should first ask 'for what?' There are various levels of complexity in available models, each suited to a different application. Models typically consist of different building blocks, as described in Section 4.2, and the different elements are often interchangeable. Model elements can therefore be chosen specifically to meet the application requirements. The majority of modelling is currently performed to predict the energy yield of installed systems as part of a due diligence cycle, ie, as part of the financial analysis of planned systems. The implications of this are that model complexity is limited and the model uncertainty is of particular importance as it equates to financial risks in the given application. Hence the uncertainty should be included in any results given. It is shown in Section 4.6 that the actual PV model is usually not the main contributor to the modelling uncertainty, and disagreements with measurements are mostly within the measurement uncertainty. The main contributor to uncertainty is usually the input data and not the model itself. Thus, Section 4.4 takes a closer look at some of the challenges of data collection.

This chapter should give the basic understanding of how PV yield modelling works and what the challenges and limitations are. These limitations are present also in commercial software and should be taken into account when discussing yield forecasts.

4.2 Overview of the modelling of module performance

The basic modelling steps will be applicable to all device categories but the level of detail will change. A general overview is given in Fig. 4.1.

The Performance of Photovoltaic (PV) Systems. http://dx.doi.org/10.1016/B978-1-78242-336-2.00004-5

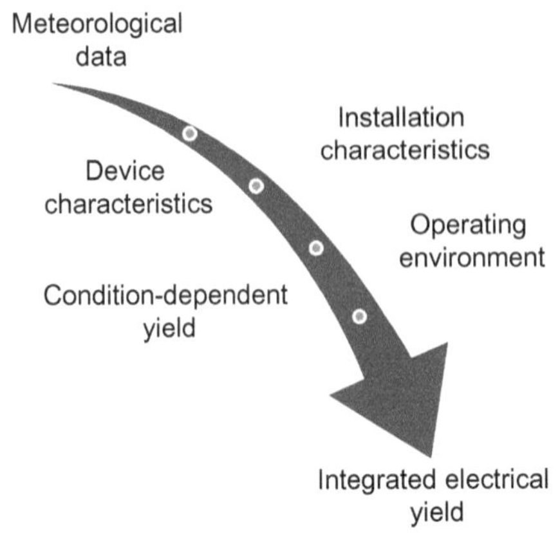

Fig. 4.1 Typical process of a PV yield modelling approach.

The type of modelling employed for a given scenario is dependent on the available data. Simulations are typically conducted for a new location for which no specific operating data exists. This often means only synthetic environmental data is available, typically interpolated meteorological data in the form of horizontal irradiance (G_h) and ambient temperature (T_A). However, this is not the environment to which the PV module will be exposed. The PV module will normally not be mounted horizontally and, as shown in Section 4.4.4, will not operate at ambient temperature. Therefore a device- and installation-specific translation of meteorological data to the actual operating environment is required. The determination of the operating environment has been noted as one of the main contributors to modelling uncertainty [1].

The other main contributor to modelling uncertainty is the actual device characterization [1]. Manufacturers' data sheets typically include values for the device operation at standard test conditions (STC) and the temperature coefficients for the different performance parameters of a single module. Commercial PV modelling software packages which include detailed device modelling will usually include a database of commercially available PV modules. The data included in the database is often limited to what is publicly available on the manufacturers' data sheets. Some software packages have an option for users to import module data from third-party module databases [2,3] or for users or manufacturers to upload module data [4].

A special case of energy modelling is the so-called energy rating (ER). PV modules are normally supplied with a power rating, which is the power measured at STC, as defined in IEC 61215 [5]. This rating provides a good comparison metric between PV devices of the same manufacturer and technology. However, STC is rarely met in realistic conditions. Thus, operating performance may vary, especially between devices of different technology due to variable environmental conditions. Energy yield prediction is a complex process that requires measured PV device characteristics under varying conditions as well as meteorological data from the site to be investigated. The effects to be assessed in terms of device modelling are predominantly the influences of varying irradiances and module temperatures. Typically devices lose efficiency with decreasing irradiance and increasing temperature, ie, the STC rating is often overoptimistic. The precise nature of this behaviour is estimated using temperature coefficients, ie, by measuring the I–V characteristics at different temperatures, and then extracting a linear relationship of short-circuit current, maximum power and open-circuit voltage. A better database is achieved by measuring an energy matrix as proposed in the standard [6]. Here, measurements are taken at a number of different irradiances and temperatures. A matrix of operating conditions is built up which allows the modelling of any nonlinearities. The next level of detail to be considered is the spectral response, angular response and operating response. These measurements are commonly not available, but this may be more common if the upcoming standard on ER [7] is internationally accepted.

The term ER is often applied to yield measurements. However, this is rather misleading as this is not a true metrological rating. A rating requires standardized measurements at standardized conditions and then a standardized method to present the energy yield. For example, the power rating of PV modules uses measurements at STC which are an irradiance of 1000 W/m^2, a cell temperature of 25 °C and an incident spectrum corresponding to air mass of 1.5 (which equates to a solar elevation angle of approximately 42°). Similarly, one would need standardized models (IEC61853-3) and standardized environments (IEC61815-4) to achieve a comparable rating. Industry-wide adoption of ER will require an IEC Standard (which is currently being developed) to be published and accepted. The standard is split into four parts, with parts one and two describing the required measurements, part three the calculation methodology and part four the reference datasets.

The advantage of using such a standard is that it provides a standard dataset and thus eliminates the rather large uncertainty in the microclimates described later. The downside is that, by its very nature, ER is not the simulation of a specific location and will be affected significantly by local effects.

4.3 Electrical PV module modelling

PV models are used to estimate an I–V characteristic as a function of irradiance and operating temperature. A large number of models are available, each of them having their own strengths and weaknesses. There are two main categories of PV models: those based on the equivalent circuit representation of the cell and those based on empirical data. Within these categories models are differentiated according to the complexity of the underlying physics and the fundamental parameters needed as input.

Empirical models include parameters that are fitted from data so they do not have a physical basis. Physical models have their basis in physical components with known behavioural characteristics. Within these categories models are differentiated mainly by the number and type of input parameters required, and whether the output is the full IV curve (parametric continuous), a set of I–V points (discrete), or just the maximum power value.

Typically, the basic unit of these models is the PV module and as such there is an underlying assumption that all cells in a module are equal and that there is no variation of the operating environment across the module. The importance of these assumptions will be discussed in Section 4.3.3.

4.3.1 Physical models

These models simulate the full I–V curve by means of parametric equations. The most common physical models take advantage of the similarity of solar cells and diodes; these are known as the diode models. One-, two- and three-diode models (which describe different charge carrier recombination mechanisms) are available. They are represented by equivalent circuits derived from physical principles [8]. Popular commercial software packages such as PVWATTS [9], TRNSYS [10], PV*SOL [11], PVSyst [12] and CECPV [13] employ one- or two-diode models to calculate the output of a PV system.

Equivalent circuit models define the entire I–V curve of a PV device (cell, module or array) as a continuous function for a given set of operating conditions. Assessment of the operation of solar cells and the design of power systems comprising solar cells should be based on their electrical characteristics, namely the voltage–current relationships of the cells under varying levels of irradiance and temperature [14]. The one-diode model is the simplest and longest established of the diode models and therefore the most commonplace. The equivalent circuit in this case may be represented as shown in Fig. 4.2.

For a given temperature and irradiance the current–voltage characteristic of the p–n junction is given by:

$$I = I_L - I_0\left(e^{\frac{q(V+IR_S)}{nkT}} - 1\right) - \frac{V+IR_S}{R_{SH}} \tag{4.1}$$

Where k is the Boltzmann constant. The other variables are described in the text below

Eq. (4.1) cannot be solved explicitly. It requires the extraction of five parameters: the light current I_L, the diode reverse saturation current I_0, the series resistance R_S, the shunt resistance R_{SH} and the diode ideality factor n. The physical boundaries of the ideality factor are one and two; these depend on material, fabrication processes and dominant recombination mechanisms. Values well outside this range are often

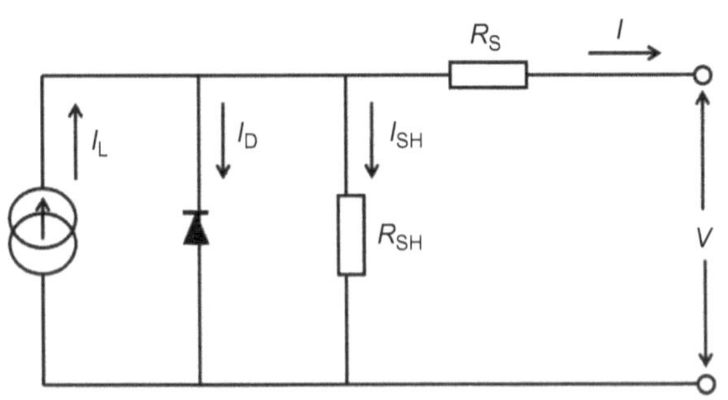

Fig. 4.2 Single diode equivalent circuit.

reported, which is due to this model being intended for single cells of (very) small dimensions and not full-scale modules as considered here.

These parameters are not provided in manufacturers' data sheets. Manufacturers typically provide the following information: open-circuit voltage (V_{OC}), short-circuit current (I_{SC}), voltage (V_{MP}), and current (I_{MP}) at maximum power point (MPP), temperature coefficient of open-circuit voltage and the temperature coefficient of short-circuit current. These values are measured at the STC of temperature (T_{STC} = 25 °C), solar irradiance (G_{STC} = 1000 W/m^2) and air mass (AM = 1.5). Some manufacturers also provide I–V curves at other environmental conditions. These are sufficient to extract the physical parameters.

The equation as given here does not describe the temperature behaviour. Auxiliary equations are needed to describe the parameters' dependence on cell temperature and irradiance. Eq. (4.1) is sometimes reduced into a less complex four-parameter model where it is assumed that $R_{SH} = \infty$. However, this model fails to predict the effect of high temperature on the current and typically fails to represent low light behaviour accurately. The reduced computational complexity does not justify the impaired prediction accuracy and thus it is not recommended to use this simplified version for modelling [15].

There are a plethora of proposed solutions for the five-parameter model starting as early as 1987 [16]. Solutions can be roughly classified into two main categories based on the approach applied to extract the five parameters, those that employ analytical expressions [16–21] and those that use numerical optimizations [22–27]. Analytical solutions often exploit mathematical tools such as the Lambert W function to solve for $I = f(V)$ [20]. Analytical solutions may be complicated and require simplifications or assumptions in order to be solved [18]. Numerical methods employ iterative optimization algorithms (for example, differential evolution [22]) which yield satisfactory results but often their performance is affected by the initial values of the parameters. Some models require knowledge of the principal I–V curve points given in the manufacturer's data sheet (for example, [17,19]), whereas many require measured I–V curves, often at different irradiances and temperatures.

Each of the diode models applies several mathematical and physical assumptions that slightly affect the result. Inaccuracies are primarily observed at low light or high temperature conditions [28] or for specific module technologies due to their different responses under given environmental conditions [26,29]. Crystalline silicon solar cells are generally described accurately by equivalent circuit models but various assumptions or modifications may be required for thin-film solar cells in order to account for device-specific behaviours (on shunt and series resistances) and different recombination mechanisms, as shown in early work on amorphous silicon [30,31] and CIS or CdTe solar cells [32,33]. Recent work by Mermoud and Lejeune [34] suggests an improved one-diode model which is more appropriate for amorphous, microcrystalline and CdTe modules. The more complex recombination mechanisms found in thin-film PV devices can be described by using numerical simulation programmes [32] or by employing the more accurate two-diode equation model [29,35], or simplified proposed models, for example [36], which considers very smooth I–V curves from various thin-film technologies. For some devices, such as dye-sensitized devices, the temperature coefficient is very different to conventional devices and thus equivalent circuit models fail completely and a different approach would need to be developed.

So far in this chapter, it has been implicitly assumed that all cells comprising a module are identical and thus a given diode model can be used for higher level devices (modules and arrays) as well as solar cells. This is a compromise between modelling accuracy and complexity. In reality, nonuniform conditions often occur across the area of a PV module due to the intrinsic variations of cell parameters due to fabrication processes and partial shading, which means that different areas across the module experience different conditions of irradiance and temperature. Mismatches in cell characteristics cause modules (and consequently entire arrays) to operate at nonideal conditions generally governed by the weakest cell. In the worst-case scenario, this can lead to damage to the module and thus protective measures are employed. Bypass diodes are used to protect the module by redirecting the current when the voltage across a substring is negative, to prevent individual cells from damage due to over-current.

In order for all these effects to be represented, circuit-based modelling is required. In these models, the module is resolved to smaller units of substrings or cells or even subcells where each one is represented by its I–V characteristics. Moreover, diode characteristics are also included. This increases the number of modelling input variables significantly. The single diode model-based simulation was first introduced by ref. [37] and was later followed by ref. [38]. Similar work has been done using the PSpice circuit modelling programme; recently developed models based on a subcell analysis and accelerated simulations have been introduced by ref. [39]. Cell level analysis of large arrays can be found in ref. [40].

4.3.2 Empirical models

Empirical models usually only estimate key points on the I–V curve (V_{MP}, I_{MP}, I_{SC}, V_{OC}). They are typically derived from numerical fitting of device measurements. As a consequence these models may vary significantly in complexity [41–43]. The power output is usually obtained by fitting a function of total irradiance and/or module temperature. In their simplest form, they can be described by a simple matrix of irradiances multiplied with a device descriptor matrix measured either in the laboratory or outdoors for a set of known conditions.

Eq. (4.2) describes a simplistic model based on a single equation for total production:

$$E_{\text{total}} = PR \cdot \eta \cdot A \cdot G_{\text{total}} \tag{4.2}$$

where E_{total} is the predicted total energy, PR the module performance ratio, η the nominal module efficiency, A the active surface area of the device and G_{total} the total irradiation received by the module. These models are commonly used in solar calculators, for annual performance estimation, but are unsuitable to simulate performance at a detailed level considering irradiance and temperature variations as well as other environmental and device-specific effects at higher resolution time series.

A set of simple mathematical expressions which correlate module efficiency and thus power output to incident irradiance and cell temperature, $\eta(G,T)$, is given in Skoplaki and Palyvos [44], for different mounting configurations (see Section 4.4.3).

One of the most commonly employed empirical models is the Sandia array performance model (SAPM) [45] which calculates power production with an algebraically simple method, using two extra points on the I–V curve (I_X and I_{XX} at $V_{OC}/2$ and $(V_{OC}+V_{MP})/2$, respectively). Individual equations contain coefficients which are determined empirically. Spectral and angle of incidence effects are also taken into account by the use of polynomial functions with empirically determined coefficients. Sandia provides an up-to-date library of all the modelling performance parameters of available commercial modules; an online array performance programme is also provided.

Simplified versions of the Sandia model have been employed in several publications concerning ER [46,47] and modelling of PV performance [48–51] and PV-GIS [52] software.

A popular version of the equation for MPP is shown in ref. [48,53]. It only requires in-plane irradiance and module temperature to predict maximum power output, according to Eq. (4.3)

$$P'(G',T') = G' \cdot \Big(1 + k_1 \ln(G') + k_2 \ln(G')^2 + k_3 T' + k_4 T' \ln(G') + k_5 T' \ln(G')^2 + k_6 T'^2\Big) \tag{4.3}$$

where $P' = P/P_{STC}$, $G' = G/G_{STC}$ and $T' = T - T_{STC}$. A comparison of this model to a number of other models [54] has shown good performance for crystalline silicon modules.

Empirical models which only use in-plane irradiance and module temperature as inputs and predict maximum power are also called single-point efficiency models or performance surface models, as their output can be plotted as a 3D surface (power vs. irradiance vs. temperature).

Further models taking into account device-specific behaviours, for example, low light nonlinearity, include the National Renewable Energy Laboratory (NREL) Module Energy Rating (MER). The MER model is based on I–V curve translations [55–57] and utilizes the translation equations from the ASTM E1036-96 standard but employs a different normalizing procedure to account for the changes in fill factor (FF). A matrix of I–V values from indoor measurements of 36 combinations of six irradiance (150–1000 W/m^2) and six temperature (5–60°C) values is used to provide three correction factors (α, $\delta(T)$ and $\beta(E)$) to account for the variations in short-circuit current with temperature, open-circuit voltage with temperature and solar irradiance from STC conditions (denoted with *zero* subscript), respectively. An I–V curve for a given set of conditions is then calculated using a reference I–V curve (denoted with an

R subscript) close to the desired condition. The translation equations are given by the following set of equations:

$$I_{SC} = \frac{E}{E_0} I_{SC_0} [1 + a(T - T_0)] \tag{4.4}$$

$$V_{OC} = V_{OC_0} [1 + \beta(E_0) \cdot (T - T_0)] \left[1 + \delta(T) \ln\left(\frac{E}{E_0}\right) \right] \tag{4.5}$$

$$I = I_R \frac{I_{SC}}{I_{SCR}} \tag{4.6}$$

$$V = V_R \frac{V_{OC}}{V_{OCR}} \tag{4.7}$$

where E is the useful portion of incident power given by:

$$E = \frac{\int_{\lambda_{min}}^{\lambda_{max}} E_{INC}(\lambda) SR(\lambda) d\lambda}{\int_{\lambda_{min}}^{\lambda_{max}} E_{REF}(\lambda) SR(\lambda) d\lambda} \cdot 1000\,\mathrm{W/m^2} \tag{4.8}$$

$SR(\lambda)$ is the spectral response of the device, $E_{INC}(\lambda)$ the incident spectral irradiance and $E_{REF}(\lambda)$ the spectral irradiance at AM = 1.5.

Another notable example is the loss factor model (LFM) described in Refs. [58,59,60]. More specifically, this model is based on fitting measured I–V curves from different PV module technologies under various environmental conditions (outdoor measurements). I–V curves at reference conditions are also required in order to derive the loss coefficients. The observed loss values are associated with specific independent normalized loss factors. These factors represent the percentage of power losses due to series resistances, shunt resistances and the FF. The result is given as the ratio of measured efficiency to efficiency at STC, characterized as performance factor and is given as the product of the calculated loss factors and a spectral mismatch correction factor.

4.3.3 Advantages of the different approaches

A detailed description of available PV software, employing either equivalent circuit or empirical expressions, is presented elsewhere in Klise and Stein [61]. These include the modelling of the components of the entire PV system, for example, inverter and other balance of systems components, as well as solar irradiance and temperature models. Some software, such as Solar Advisor Model (SAM) [62], also include financial scenarios. The capabilities and availability of each software package are discussed, but the aim of this chapter is not to prescribe a preferred package. This decision depends on the application and on the available data for modelling to minimize the modelling uncertainty as outlined previously.

There is no preferred model in either category of PV models, namely equivalent circuit or (semi)empirical models, since the speed and accuracy required depend on the application. The limitations of each category include computational intensity, model complexity and capability to represent specific aspects of PV module modelling. Accuracy is not discussed here since intercomparisons have shown that the difference between the models is within the measurement accuracy, so the difference is largely statistical noise [1].

Device physical models are generally implicit and their solutions can be quite complicated, time-consuming and computationally expensive. They also require corrections specific to different PV technologies. However, their advantage lies in the fact that they can be used to obtain the full I–V curve, which is of great importance for specific studies such as cell interaction modelling [39,40] to determine the mismatch effects on the PV devices. The calculation of irradiance on shaded arrays is described in Section 4.5.3.

Empirical models are generally more simplistic and less computationally intensive as well as more easily adjustable to different PV technologies. This makes them good candidates for PV performance mapping across large geographical areas and fast calculations on energy yield can be obtained. The accuracy depends on the data used to obtain the empirical coefficients for a PV module of specific technology. Also single-point efficiency models (where only maximum power is obtained) cannot be used to study in depth more complex scenarios such as shading effects or faults on the PV modules. A comparison between a one-diode model (CEC), SAPM and the LFM, carried out in Stein et al [63], points out the main aforementioned weaknesses of the different modelling approaches.

4.4 Modelling operating environments

The aim of this section is to give an overview of the general principles of modelling operating environments and enable the reader to assess the energy yield prediction. The steps to translate these to real operating environments are described. The starting point for any assessment of PV yield is finding available meteorological data. Section 4.4.1 gives an overview of some of the easily accessible sources. The next step in modelling the energy yield of modules is in translating existing environmental data to the actual operating environment. This is described in Sections 4.4.2–4.4.4. The section concludes with a demonstration of how this translation actually influences the energy yield; Section 4.4.5 discusses the differences in different orientations on the overall energy yield.

4.4.1 Available meteorological datasets

The largest functional impact on PV system output is the available solar resource. The usual practice in the PV industry is to use solar irradiance values from the internal databases of specialist software. Data is readily available in the form of monthly averages of 10- or 20-year periods. Meteonorm [64], PV-GIS [65] and PVSyst [66] are well-known packages which contain this type of information. These packages work on a long-term input dataset, sometimes several decades in the past, generate a statistically correct average and produce a forward prediction based on this. These approaches have the advantage that they smooth anomalies but annual variation and long-term trends are hidden. In order to study these, solar irradiance measurements at daily, hourly or per minute intervals are required for time periods of several decades.

National meteorological institutions often generate long-term datasets. The original intention was not for modelling PV, so detailed irradiance measurements may be sparse. Meteorological institutions have recently increased the number of measurement stations as well as the deployed instrumentation to meet demand. One example of such a large and detailed dataset is MIDAS from the UK Met Office [67]. This comprises both historical and current hourly and daily global radiation. This data collection began with just one recording station in 1947. At the time of writing (Nov. 2015), there are 88 operational weather stations measuring solar irradiance covering the whole of the UK.

Similar solar irradiation records may be found in various other countries; commonly the national irradiation archives commence with a small number of manual readings at an early date and gradually increase to an extensive network of fully automated observation posts with pyranometers.

However, not all countries have long-term historical records. In Brazil, for example, the SONDA project (Brazilian Management System of Environmental Data for the energy sector) commenced with eight ground sites in 2004 only. Prior to that, solar data was collated from airports, the Ministry of Agriculture, universities, utilities, agricultural research companies and regional weather services. This data is not standardized and problems arise with missing data and differing formats and measurement units.

For international studies, high quality but sparse data is available from the Baseline Surface Radiation Network (BSRN) [68]. This has 58 stations, covering the range of climatic zones from the Arctic Ocean to the coast of Antarctica. Data is logged by instruments of the highest available accuracy with time steps of between 1 and 3 minutes.

Developments in technology in recent years mean that official organizations are no longer the only source of meteorological data. The falling price of private weather stations and advancements in the Internet has allowed amateur enthusiasts and volunteer groups to set up their own networks. Studies are presently underway to assess the quality of this publicly contributed (crowd sourced) data. There are several such hubs worldwide. One major group receives solar radiation data from about one thousand member stations and publishes data from 2009 onwards [69].

An alternative source of long-term, low time interval solar irradiation data is satellite-derived values (rather than the ground-based measurements reviewed thus far). Satellites have been generating suitable images for decades but these require mathematical processing in order to extract irradiation values. Satellite data directly downloadable as ready-prepared solar irradiation data is available for a few regions only.

4.4.2 Translation of horizontal to in-plane irradiance models

As described in the previous section, irradiance data from weather stations is usually measured in the horizontal plane since it may be used for a variety of applications at differing tilts and azimuths. Therefore for the majority of PV systems, the horizontal irradiance (G_H) must be translated to in-plane irradiance.

Horizontal to in-plane translation of diffuse irradiance is more complex than for beam irradiance so the horizontal irradiance is first split into these two components. The beam component of the irradiance is translated using simple trigonometry. In-plane diffuse models can be divided into anisotropic models which take account of reduced diffuse irradiance closer to the horizon and isotropic models which do not. The flow of data between these models is shown in Fig. 4.3. A number of different algorithms have been developed for each of these three submodels; the accuracy of these models may vary depending on the type of climate. Several authors have compared the available models by validating against outdoor measured data [70–72]. The accuracy ranking of models varies considerably between different geographic locations.

4.4.3 Mounting configurations

Mounting configurations play an important role in module temperature as well as irradiance collection. The impact on operating temperature is due to different heat transfer mechanisms driving the thermal behaviour. The following mounting configurations need to be treated differently when modelling the device temperature:

- Building attached, sloped roof (BAPV)
- Free-standing
- Flat roof
- Building integrated (BIPV) including in facades and roofs

Building integrated PV (BIPV) is often confused with building attached PV (BAPV). BAPV comprises PV modules that have been installed after the construction of a

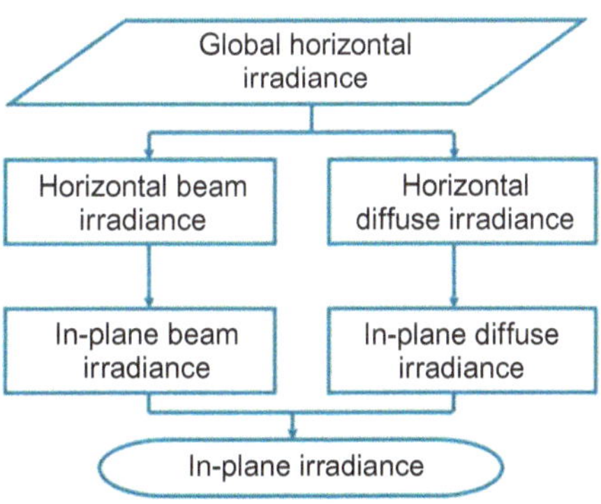

Fig. 4.3 Activity diagram showing calculation of irradiance components.

building, whereas BIPV comprises PV modules that have been used as part of the outer weatherproof layer of the building. Each configuration has a different effect on PV module efficiency due to different cooling mechanisms.

Cooling is the result of heat 'losses' from the module to the environment (sky and ground). These convective and radiative losses are each associated with a loss coefficient. The convective loss coefficient is highly dependent on the profile of the air near the surface of the module. While a free-standing module is able to cool more quickly, a building integrated module with a very thin air gap between its back and the material of the building not only cools more slowly but also heats up more quickly.

The differences in operating temperature between the different mounting configurations can exceed 20°C, which would equate to 5% or more variation in power for some devices. Thus, it is important to distinguish between these cases in energy yield forecasting.

4.4.4 Modelling operating temperature

The voltage of a module falls when its operating temperature rises, whilst its short-circuit current slightly increases. The result is a decrease in its output power. The operating temperature of a module determines its efficiency. As a rule of thumb, the power lost is slightly less than 0.5%/°C for crystalline silicon devices. A large number of thermal models which correlate environmental parameters to module temperature exist in the literature. Generally, thermal models can be split according to the simplicity or complexity of their expression. There are explicit equations which simply correlate module temperature to environmental parameters such as ambient temperature, solar radiation flux and/or wind speed. The first two parameters are found in all models (explicitly or implicitly), whereas wind speed is not always included in expressions of *steady state* models, for example [73], and nominal operating cell temperature (NOCT) [5]. NOCT is the temperature of a 45° South-facing module with an incident irradiance of 800 W/m^2, an ambient temperature of 20°C and a wind speed of 1 m/s. It is one of the parameters often provided in the PV module data sheet. A thermal model based on NOCT is given by:

$$T_{\mathrm{m}} = T_{\mathrm{a}} + \frac{G}{800\,\mathrm{W/m^2}}(\mathrm{NOCT} - 20°\mathrm{C}) \tag{4.9}$$

Ross's model is given by:

$$T_{\mathrm{m}} = T_{\mathrm{a}} + k \cdot G \tag{4.10}$$

where T_{m} (°C), T_{amb} (°C) and G (W/m^2) are module temperature, ambient temperature and in-plane irradiance, respectively. The k (K·m^2/W), or Ross coefficient, expresses the influence of the mounting configuration of the module [74] on its temperature; typical empirically derived Ross coefficients are given in Table 4.1.

Table 4.1 **Empirical Ross factor coefficients [75]**

Mounting type	k (°C·m^2/W)
Free-standing	0.021
Flat roof	0.026
Sloped roof: well cooled	0.020
Sloped roof: highly integrated	0.056

Other explicit expressions that also take into account wind speed include the Sandia thermal model [76]. Wind speed and a mounting configuration coefficient model are presented in Skoplaki et al. [77]. An extensive documentation of explicit as well as implicit correlations that covers models of about 25 years of literature is found in Skoplaki and Palyvos [78].

Implicit expressions not only account for ambient temperature, in-plane irradiance and wind speed but also the energy balance between the module and its environment. Therefore, they may include a number of intrinsic factors such as heat transfer coefficients, cell absorption coefficient, glass transmittance, etc. [79–82]. An example is the [81] model based on a simple energy balance. Module temperature is given by:

$$T_m = \frac{U_{pv}T_a + G[(C_\alpha \tau) - \eta_r - \gamma\eta_r Tr]}{U_{pv} - \gamma\eta_r G} \tag{4.11}$$

where U_{pv} is the heat exchange coefficient of the module and is dependent on wind speed, C_α is the cell absorption coefficient, τ is the glass transmittance, η_r is the module efficiency at 25°C reference temperature and 1000 W/m^2 irradiance and γ is the absolute value of the variation coefficient of the power with module temperature expressed as %/°C.

These models are generally more sophisticated and they describe module temperature better in high resolution time series since they account for non-steady-state phenomena and thermal lag. Simpler models are accurate when hourly (or lower resolution) measurements are used [83].

4.4.5 The influence of module orientation on yield

The tilt and azimuth of a PV system have two influences on performance. The first is an increase or decrease in annual energy yield depending on the suitability of the roof pitch and azimuth. The second is that the daily and/or seasonal peak energy generation is time shifted.

The sun path is a function of terrestrial location. The sun reaches its greatest elevation above the horizon on the summer solstice and falls to its lowest on the winter solstice. An example sun path is given in Fig. 4.4.

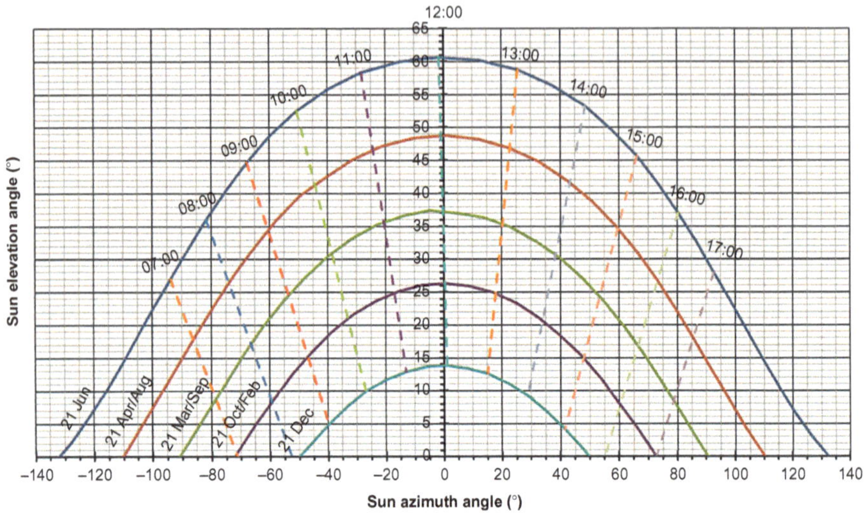

Fig. 4.4 Sun path diagram for Loughborough, UK. Latitude 52.76 degree, longitude −1.24 degree, altitude 80 m. Times in Greenwich Mean Time (GMT). Positions calculated using CREST implementation of Reda and Andreas [84].

The optimum module inclination angle for maximized net annual solar energy yield varies according to latitude. The reduction in yield resulting from suboptimal module orientation is a location-specific phenomenon. An example of net annual yield variation for varying azimuth and inclination angles in Loughborough, UK, is given in Fig. 4.5. This was calculated using the 'solarscene.xyz' optimized solar energy ray

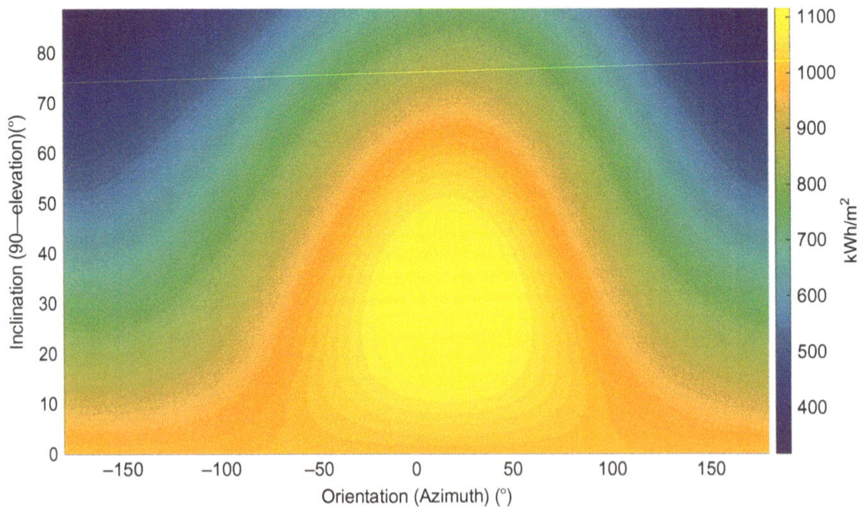

Fig. 4.5 Influence of module orientation on net annual irradiation harvest for Loughborough, UK, in 2014. Inclination angles from 0° (horizontal) to 90° (vertical), orientation angles N/E/S/W are ±180/−90/0/+90°, respectively.

tracing model developed at CREST, as compared with alternatives in ref. [85] and building on work of refs. [86–88].

The isolated effects of varying inclination angle for a south-facing plane for a typical meteorological year (TMY) using data from NCAS British Atmospheric Data Centre [89] are shown in Fig. 4.6. The TMY used here is the 10-year average from 2005 to 2014. More information regarding time-dependent meteorological variation can be found in ref. [90].

Steeper tilt angles result in higher energy harvests in the winter. For instance, a tilt angle of 60° has a greater yield between Mar. and Oct. than the net optimum of 30–40° in the UK. Shallower tilt angles result in higher energy harvests in the summer. For example, 15° may exceed 30° in energy production for several weeks in midsummer. This is due to increased collection from the lower and higher solar elevation sun paths in the winter and summer, respectively.

In a country like the UK, there is a greater energy demand in the winter due to shortened daylight hours and ambient temperature reductions below comfort. However, in a country like Australia there is a greater energy demand in the summer due to ambient temperature increases above comfort. Localized analysis of energy demand requirements can lead to subjective optimization of installation angles, reducing the need for expensive storage or backup generation solutions. Varying month by month average daily generation profiles for a range of module inclination angles in Loughborough, UK, are given in Fig. 4.7.

Array orientation may also be governed by other constraints. Systems on pitched roofs usually follow the shape and orientation of the existing roof. Systems on flat roofs are usually installed in rows of tilted modules so there is more flexibility for the designer to optimize tilt and azimuth. Small changes in tilt angle do not have a major impact on PV yield. For example, increasing the inclination from zero to optimum tilts for the UK only increases the annual irradiation harvest by approximately 10% (from 1.0 to 1.1 MWh) for a fixed south-facing array (Fig. 4.7).

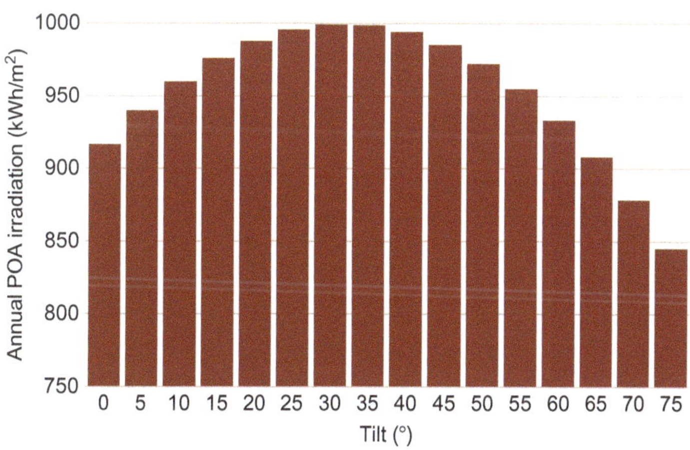

Fig. 4.6 Influence of tilt on irradiation: Annual average plane of array (POA) irradiation for a TMY dataset for Loughborough, UK. South-facing, tilt angle varied in bands of 5°.

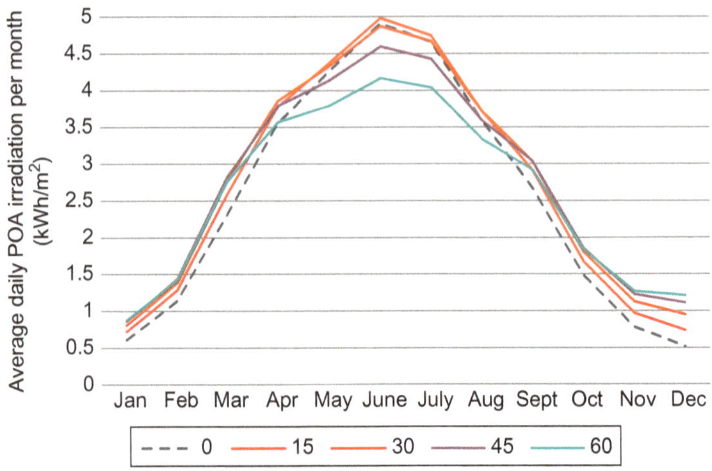

Fig. 4.7 Average daily POA irradiation for a TMY dataset for Loughborough, UK. Azimuth constant, south-facing, five tilt angles (0°, 15°, 30°, 45° and 60°) illustrated.

The effect of azimuth on net PV yield is shown in Fig. 4.8. Multiple azimuths and tilts were calculated using the solaR package [91] of the R language which utilizes the [92] equation to separate irradiation into its beam and diffuse components and transposes the irradiation constituents to the required tilt angle via the [93] equation with [94] correction. Less suitable orientations for PV (for example, east and west) receive an average of 15% less plane-of-array irradiation than the most advantageous aspect. This site described in Fig. 4.8 shows a positive bias for west-facing installations. This may be

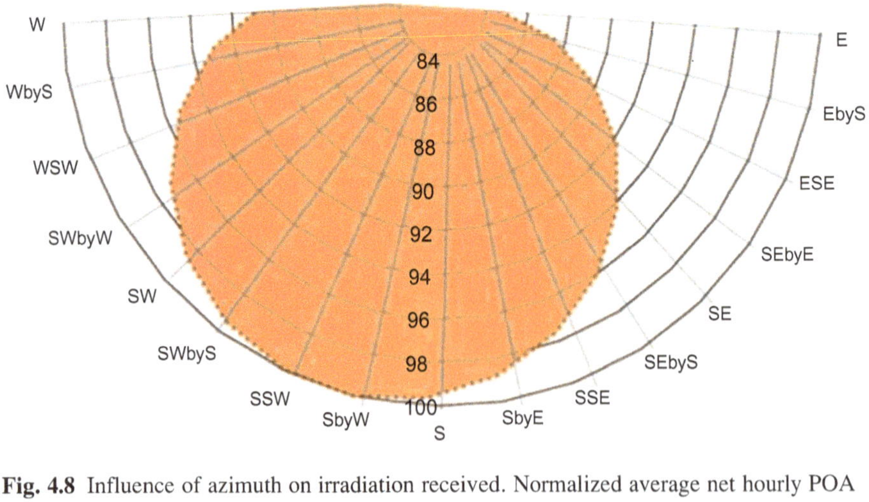

Fig. 4.8 Influence of azimuth on irradiation received. Normalized average net hourly POA irradiation per compass point for a TMY dataset for Loughborough, UK at a constant inclination of 38°.

unusual, but has been observed for a number of locations in the UK and has been associated with hazy mornings.

The optimum aspect for maximized irradiation harvest in the northern hemisphere is theoretically due south, resulting from sun path symmetry. However, this may vary slightly due to local climatic conditions such as morning mist. Fig. 4.8 illustrates that the most favourable azimuth for Loughborough is actually slightly to the west of south.

Azimuth influences the timing of peak energy production (Fig. 4.9). The deployment of a range of PV azimuths between east and west widens the time duration of solar energy generation, increasing production before 09:30 and after 14:30 at the expense of reduced annual yield. Demand from both residential and commercial users is still high before 09:30 and after 14:30, so widening the time duration of PV generation reduces the requirement for dispatchable power generation sources on the electricity network to be available for generation at times of higher demand but lower PV output.

In summary, although optimum orientations can be calculated for a given setting, installation for optimum orientation is often not possible in BAPV installations due to building features. Compromises in deployment are necessary. Furthermore, reductions in net annual yield can have positive side effects such as higher generation during times of higher demand. In order to fully understand yield variation complexity, it is necessary to have knowledge of the roof characteristics of current buildings. Moreover, optimization of generation profiles for energy demand matching requires knowledge of the local and/or regional demand profiles.

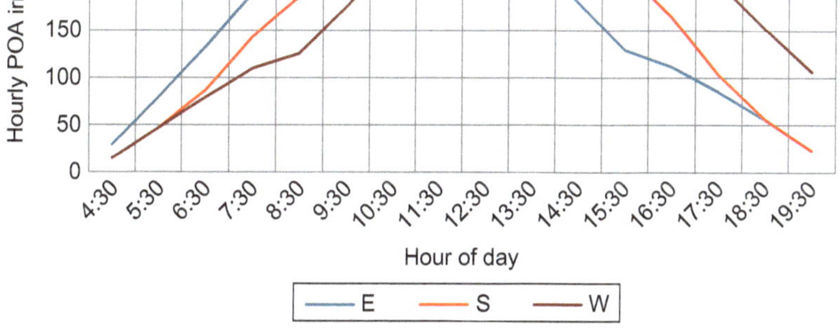

Fig. 4.9 Comparison of average hourly POA irradiation for a constant inclination of 38° for east, south and west-facing planes in Loughborough, UK (TMY).

4.5 Advanced considerations

4.5.1 The local environment for PV

The majority of PV arrays are installed on the ground, on flat roofs in rows or on pitched roofs. Systems mounted on the ground or on flat roofs can be designed with a specified angle of tilt and row spacing, whereas arrays on pitched roofs are usually restricted to the azimuth and tilt of the existing roof, except in the case of buildings which are new or undergoing major renovation.

The roof area suitable for PV may be influenced by roof style (gable ends, hips, lean-to). Roof features (including chimneys, vent pipes, aerials, roof lights, cross-gables, and dormer windows) may reduce the available roof area, especially once a suitable boundary has been allocated to limit the impact of shading. Shading by trees, adjacent buildings and terrain features must also be considered in a performance model (4.10).

Roof pitch is typically linked to building age, roofing material and geographic region. Materials used in older buildings, such as slate and flat clay tiles, required a steep roof pitch in excess of 40° to prevent ingress of water. Modern interlocking roof tiles are usually installed with bitumen or polymer based sarking layers, and therefore can be installed on much shallower pitches of 20–30°. The availability of heat-welded sheet materials has enabled the growth of 'flat' roofs with a pitch of less than 5°; these are found most commonly in urban areas. Systems installed on flat roofs or on the ground are usually installed in rows of one or two modules.

Building orientation (azimuth) is dictated primarily by road layout and local topography, such as slopes, rivers, coastline and valleys.

Fig. 4.10 Photograph showing a PV array with partial shading from an adjacent tree. Residential house, Beeston, UK.

Methods of collecting array spatial data

Various approaches can be used to survey potential PV array locations with varying accuracy. The most common methods are shown in Table 4.2.

Publicly available data is limited. Architects' building plans may occasionally be available. In the absence of existing information, data may be collected manually on-site with surveying instruments. LiDAR (Light Detection and Ranging) is becoming more widely used to automate the capture of roof characteristics over large areas. LiDAR data is collected by pulsing a laser from an aircraft flying at constant altitude and timing the return pulses to obtain height information in the form of a grid. Research is ongoing to extract roof characteristics from raw height values. Most methods begin by separating the roof into panes, often by detecting roof ridges. Canny edge detection is one such technique. This algorithm selects local maxima in the LiDAR as edges and rejects those grid cells which are out-of-line to define a thin ridge. The Douglas Peucker line simplification algorithm (1973) is similar.

4.5.2 Shading

The reduction in PV system performance due to shading can be subdivided into four distinct loss factors:

- A reduction in irradiance incident on the PV modules
- Difference in spectrum between beam and diffuse irradiance
- Mismatch losses due to varying irradiance and temperature of each cell
- A reduction in inverter MPP tracking, because the I–V curve has false maxima

The irradiance incident on a shaded cell is the sum of the visible portions of the diffuse fraction of the irradiance and the ground reflected irradiance, whereas the irradiance incident on unshaded cells also includes the beam irradiance. Irradiance reduction causes a reduction in current and therefore power generated by the affected cells. However, the current in any string of solar cells is limited to the current of the cell with the lowest irradiance; therefore the I–V curve mismatch between cells has an impact on cells which are not shaded. The modelling of module, string and array performance must therefore take into account the variations in irradiance across the array. Modelling of mismatched cells is discussed in Section 4.3.1.

4.5.3 Modelling of irradiance losses due to shading

The irradiance reduction due to shading can be modelled by considering potential rays of light from all possible directions and therefore poses a problem that is well solved by the application of ray tracing techniques.

Shading models consider this problem from two perspectives:

(a) viewing a surface from the point of view of the sky
(b) viewing the sky from the point of view of a surface

Table 4.2 **Surveying methods used to assess photovoltaic systems**

	Object		Structures and trees		PV array	Near horizon		Far horizon	
	Coordinate type	**Publicly available data**	**X, Y**	**Height**	**Tilt and azimuth**	**Elevation +Azimuth**	**Distance from array**	**Elevation +Azimuth**	**Distance from array**
a	Estimation from aerial photography	✓	✓				✓	✓	✓
b	Electronic maps or GIS data	✓	✓	✓			✓		✓
c	LIDAR device from an aircraft [95]	✓	✓	✓	✓			✓	✓
d	Electronic CAD (computer aided design) drawings		✓	✓	✓	✓	✓		
e	PanoramaMaster [96] or Meteonorm horicatcher [97]			✓		✓		✓	
g	Compass, inclinometer and tape measure		✓		✓	✓	✓	✓	
h	Total station [98]		✓		✓	✓	✓		
i	GPS surveying antennae		✓		✓	✓	✓		

Models in category (a) commonly use rendering to generate a three-dimensional view of a building or district, with shading used to indicate zones of varying irradiation [99–102]. A key challenge of this approach is the computational time required to model irradiance for each surface segment for all sun positions. A logical optimization is to bin sun positions into a finite number of zones responsible for similar annual irradiation, for example, from 4000 hourly sun positions above the horizon into 250 bins [102]. This approach is typically used in simulations at neighbourhood level using building coordinates from GIS or LiDAR coordinates.

Category (b) models use the concept of the sunpath diagram to analyse whether a given surface (or a portion of it) is shaded for a given time step [103–108].

PV design tools such as PV*Sol and PVSyst model shading in PV arrays with strings or modules as the smallest unit [11,109]. Modelling the energy yield cell-by-cell is considered too slow with currently available methods [110]. Simulation time has been identified as a key factor for users, after complexity, cost and CAD integration [111].

The challenge of modelling shading is that to model the beam and diffuse irradiance from every valid sun position onto every solar cell and for every time step (usually hourly) of a reference year is a time-consuming simulation [112]. The modelling time increases exponentially with the number of modules in the array. Publicly available modelling packages reduce the modelling time by taking computational shortcuts, for example, by assuming that the irradiance is the same across areas of the array larger than one cell to reduce the number of areas for which shading needs to be calculated.

4.6 Uncertainties in yield modelling

The modelling uncertainty has a number of components, not all of them related to the model itself. This section will look at the key contributors. The aim of an energy yield prediction is to understand the financial return of an investment that typically is made for 15–25 years. Several assumptions affect the accuracy of the yield prediction, including the input data and the model itself.

Model quality is assessed against some validation dataset. Typically the quality of the validation dataset is taken as 'perfect' and its measurement uncertainty not considered. However, if this set has a bias (as all measurement systems will to some extent; this is termed type 1 uncertainty) the result regarding the model quality will also be biased. Some of these issues are discussed in this section.

4.6.1 Uncertainty in modelling assumptions

The assumptions are not seen as physics-based assumptions. This section considers the general modelling assumptions affecting the energy yield modelling of modules.

All publicly available PV performance models assume that modules of one type are identical. This is not true, as demonstrated by the power ranges in manufacturers' data.

In simple terms, the bandwidth of the power rating determines the accuracy of the modelling. The impact of this will vary in magnitude depending if the model is used to model individual modules or systems. In terms of a single module simulation, the full bandwidth of power distribution affects the model accuracy. Consider that a typical variation of ±3% would add 3% to the model uncertainty. For larger systems this is somewhat reduced due to statistical smoothing and interconnection issues and half the bandwidth would be a good assumption [40]. Following on from the previous example, this would mean an uncertainty in the order of 1.5%. Further to the module-to-module variation in production batches, the batch-to-batch variation should also be considered. Module manufacturers continuously improve their production processes and thus a module produced as little as 6 months later than the one used for the model validation might behave differently.

Any energy yield model does not normally consider variations in the operating environment. There are many reasons why modules mounted next to each other will experience different operating environments. The irradiance can be affected by, for example, ground reflection or shading obstacles. The operating temperature can easily vary 2–3°C across a single rack since the convective cooling at the edges is more effective or there may be a heat source underneath some part of the rack (on roof systems). This will add an uncertainty in the same order of magnitude as the device binning to the overall modelling uncertainty.

The dominant uncertainty in the modelling of operating yield is the assumption of nonageing modules, inferring that the performance in year 20 equals that of year 1. This adds tens of percent of uncertainty and thus is discussed in Section 4.6.2.

4.6.2 Modelling uncertainty

PV Energy yield modelling depends on the input data as well as the model itself. Round robin comparisons of models tend to show that the agreement between the models is better than the uncertainty of the measurements used for the validation; thus in most cases model variations can be considered statistical noise. Larger deviations can, for the most part, be attributed to the input data.

The main uncertainty may be the underlying environmental data. The greatest contribution is the generation of data representing a realistic operating environment from historic data. There are several issues which affect the choice of data. The common challenge of forecasting is that it is not possible to predict the weather for the next 20 years. A number of publications have reported that solar irradiation is increasing over the decades [113,114]. There is also a major uncertainty in translation of horizontal irradiance to in-plane irradiance and interpolation from nearby weather stations. Overall, this should be quantified with an uncertainty of about 10%.

The measurement and parametrization of the actual PV module is often overlooked as an uncertainty. There are module-to-module variations, usually within manufacturers' defined tolerances, in addition to the accuracy of the measurements of a sample (or samples) of modules. The associated uncertainty depends on the model being employed, so not all the issues will be described here. The ability to describe modules can be evaluated by checking the agreement between good measurement laboratories. Recent round robin intercomparisons of the power determination at STC found a lab-to-lab disagreement of less than 1.5%. However, the temperature coefficients of power disagreement exceeded 20% and low light behaviour disagreements exceeded 10%. These measurements were conducted on the same module, so these uncertainties must be added to module-to-module variations. This dramatically increases the uncertainty (and thus risk) of energy forecasting.

4.6.3 Ageing

The common assumption for long-term behaviour of modules is that the device will meet its warranty conditions, which would equate to an annual degradation rate of 0.7%/annum. More optimistic approaches assume a lower degradation rate of 0.5%/annum or less, but often lack the scientific justification for this. The literature suggests that values up to 3% are highly likely [115,116]. The distributions supplied in these articles indicate that in about a third of reported cases, degradation rates were greater than the 0.7%/annum stated in warranties. There are many reasons for this which are described elsewhere [117]. A cumulative reduction in energy yield by 2%/annum reduces the energy yield over 20 years by 30%. This is an order of magnitude that could nullify the financial viability of the most prudently planned system.

Considerable effort is currently being expended to model this ageing behaviour, but no clear answers are yet apparent. There is a lack of understanding of the stress-climates, device response and appropriate accelerated ageing testing. A number of qualification tests are currently being undertaken but these are not suitable for a lifetime prediction and even less so for a lifetime energy prediction.

The correct approach is to isolate the dominant failure modes and model their impact over a lifetime. The present consensus is that this could be modelled using an Arrhenius relationship, where temperature accelerates the ageing due to a specific stress. This should be evaluated in accelerated tests, but there is no appropriate predictive test available at present. One major obstacle is the achievable acceleration. The issue is that PV is expected to have a lifetime of 20 years and, typically, acceleration factors beyond 10 years are unrealistic. The testing time frame discussed currently by research laboratories is a time frame not acceptable to the industry.

There is no clear quantification of ageing contribution possible at present, but it is thought to be in the two digit percentages.

4.7 Summary and conclusions

The general approach to modelling the energy yield of PV modules has been discussed. The modelling of devices receives the most attention in yield forecasting but this may not impact on the major contribution in the overall modelling uncertainties. The key factors to be considered are:

- There are empirical or physics-based PV output models which, for the most part, perform equally well within their statistical variations.
- Appropriate climatic information is important to accurately model the energy yield of PV devices; a number of different data sources were described.
- It is important to understand not only the model but also to assess the risks/uncertainty in the data.
- The main uncertainty is in the assumption of constancy of module-to-module behaviour, environment or long-term performance of devices.

Conventional modelling ignores key aspects of the performance, such as mismatch. Depending on the scenario, this is a major contribution to modelling quality. The level of detail required depends on the uses of the modelling.

References

[1] G. Friesen, S. Dittmann, S.R. Williams, R. Gottschalg, H.G. Beyer, D.E. Guerin, A. Montgareuil, N. Van Der Borg, A.R. Burgers, R. Kenny, T. Huld, B. Müller, C. Reise, J. Kurnik, M. Topic, Intercomparison of different energy prediction methods within the European Project "Performance"—results of the 2nd Round Robin, in: Proceedings of the 24th European Photovoltaic Solar Energy Conference, Hamburg, 2009.

[2] A. Mermoud, How to import data from the PHOTON database [Online], pvsyst.com: PVsyst. Available: http://forum.pvsyst.com/viewtopic.php?f=19&t=21, 2013 (accessed 14.11.15).

[3] Solar Verlag Annually, Photon's online module & inverter database, The Photovoltaic Magazine, Photon International. Solar Verlag, Aachen.

[4] C. Laughton, PV*SOL Frequently Asked Questions [Online], The Solar Design Company, Machynlleth. Available: http://www.solardesign.co.uk/faqpvsol.php#pvsolfaq_25, 2015 (accessed 14.11.15).

[5] IEC 61215, Crystalline terrestrial photovoltaic (PV) modules, Design Qualification and Type Approval, International Electrotechnical Commission, Geneva, 2005.

[6] IEC 61853-1, Photovoltaic (PV) Module Performance Testing and Energy Rating—Part 1: Irradiance and Temperature Performance Measurements and Power Rating, IEC (International Electrotechnical Commission), Geneva, 2011.

[7] IEC 61853-2, Performance Testing and Energy Rating of Terrestrial PV Modules—Part 2: Spectral Response, Incidence Angle and Module Operating Temperature Measurements (FDIS), IEC, Geneva, 2011.

[8] J.L. Gray, The Physics of the Solar Cell, John Wiley & Sons, Chichester, UK, 2011.

[9] A.P. Dobos, PVWatts Version 1 Technical Reference, National Renewable Energy Laboratory (NREL), Golden, CO, 2013.

[10] TESS. TRNSYS—The Transient Energy System Simulation Tool [Online], Thermal Energy System Specialists, LLC, Madison, WI. Available: http://www.trnsys.com/green.htm, 2009 (accessed 30.11.09).

[11] G. Valentin, PV*SOL®Expert:3D Visualization. User Manual, Dr Valentin EnergieSoftware Gmbh, Berlin, Deutschland, 2013.

[12] A. Mermoud, Modeling systems losses in PVsyst, in: SANDIA Photovoltaic Modelling Meeting, 2010.

[13] California Energy Commission, CECPV Calculator Versions 4.1 and 5.0, (Online), California Energy Commission, Sacramento, 2015. Available: http://www.gosolarcalifornia.org/tools/nshpcalculator/download_calculator.php (accessed 15.11.15).
[14] J.A. Duffie, W.A. Beckman, Solar Engineering of Thermal Processes, Wiley, Hoboken, NJ, 2006 (Chichester: John Wiley, distributor).
[15] A.N. Celik, N. Acikgoz, Modelling and experimental verification of the operating current of mono-crystalline photovoltaic modules using four- and five-parameter models, Appl. Energy 84 (2007) 1–15.
[16] D.S.H. Chan, J.C.H. Phang, Analytical methods for the extraction of solar-cell single- and double-diode model parameters from I–V characteristics, IEEE. Trans. Electron. Dev. 34 (1987) 286–293.
[17] W. De Soto, S.A. Klein, W.A. Beckman, Improvement and validation of a model for photovoltaic array performance, Sol. Energy 80 (2006) 78–88.
[18] K. Ding, J. Zhang, X. Bian, J. Xu, A simplified model for photovoltaic modules based on improved translation equations, Sol. Energy 101 (2014) 40–52.
[19] G. Farivar, B. Asaei, Photovoltaic module single diode model parameters extraction based on manufacturer datasheet parameters, in: PECon2010–2010 IEEE International Conference on Power and Energy, 2010, pp. 929–934.
[20] C.W. Hansen, Parameter estimation for single diode models of photovoltaic modules, SAND2015-2065, Sandia National Laboratories, Albuquerque, NM, USA, 2015, 68.
[21] A. Ortiz-Conde, F.J.G. Sánchez, J. Muci, New method to extract the model parameters of solar cells from the explicit analytic solutions of their illuminated characteristics, Sol. Energy Mater. Sol. Cells 90 (2006) 352–361.
[22] K.M. El-Naggar, M.R. Alrashidi, M.F. Alhajri, A.K. Al-Othman, Simulated annealing algorithm for photovoltaic parameters identification, Sol. Energy 86 (2012) 266–274.
[23] R. Gottschalg, M. Rommel, D.G. Infield, M.J. Kearney, Influence of the measurement environment on the accuracy of the extraction of the physical parameters of solar cells, Meas. Sci. Technol. 10 (1999) 796–804.
[24] R. Gottschalg, M. Rommel, D.G. Infield, H. Ryssel, Comparison of different methods for the parameter determination of the solar cells double exponential equation, in: Proceedings of the 14th European Photovoltaic Solar Energy Conference, Barcelona, 1997, pp. 321–324.
[25] K. Ishaque, Z. Salam, S. Mekhilef, A. Shamsudin, Parameter extraction of solar photovoltaic modules using penalty-based differential evolution, Appl. Energy 99 (2012) 297–308.
[26] E.Q.B. Macabebe, C.J. Sheppard, E.E. Van Dyk, Parameter extraction from I–V characteristics of PV devices, Sol. Energy 85 (2011) 12–18.
[27] H. Tian, F. Mancilla-David, K. Ellis, E. Muljadi, P. Jenkins, A cell-to-module-to-array detailed model for photovoltaic panels, Sol. Energy 86 (2012) 2695–2706.
[28] G. Ciulla, V. Lo Brano, V. Di Dio, G. Cipriani, A comparison of different one-diode models for the representation of I–V characteristic of a PV cell, Renew. Sust. Energ. Rev. 32 (2014) 684–696.
[29] F. Attivissimo, F. Adamo, A. Carullo, A.M.L. Lanzolla, F. Spertino, A. Vallan, On the performance of the double-diode model in estimating the maximum power point for different photovoltaic technologies, Measurement 46 (2013) 3549–3559.
[30] J. Merten, J.M. Asensi, C. Voz, A.V. Shah, R. Platz, J. Andreu, Improved equivalent circuit and analytical model for amorphous silicon solar cells and modules, IEEE Trans. Electron. Devices 45 (1998) 423–429.

[31] U. Stutenbaeumer, B. Mesfin, Equivalent model of monocrystalline, polycrystalline and amorphous silicon solar cells, Renew. Energy 18 (1999) 501–512.
[32] M. Burgelman, J. Verschraegen, S. Degrave, P. Nollet, Modeling thin-film PV devices, Prog. Photovolt. Res. Appl. 12 (2004) 143–153.
[33] R. Gottschalg, B. Elsworth, D.G. Infield, M.J. Kearney, Investigation of the back contact of CdTe solar cells, in: Proc. of the ISES Solar World Congress, Jerusalem, Israel, 1999, pp. 124–128.
[34] A. Mermoud, T. Lejeune, Performance assessment of a simulation model for PV modules of any available technology, in: 25th European Photovoltaic Solar Energy Conference, Valencia, Spain, 2010.
[35] B. Werner, W. Kolodenny, M. Prorok, A. Dziedzic, T. Zdanowicz, Electrical modeling of CIGS thin-film solar cells working in natural conditions, Sol. Energy Mater. Sol. Cells 95 (2011) 2583–2587.
[36] R. Miceli, A. Orioli, A. Di Gangi, A procedure to calculate the I–V characteristics of thin-film photovoltaic modules using an explicit rational form, Appl. Energy 155 (2015) 613–628.
[37] J.W. Bishop, Computer simulation of the effects of electrical mismatches in photovoltaic cell interconnection circuits, Solar Cells 25 (1988) 73–89.
[38] V. Quaschning, R. Hanitsch, Numerical simulation of current–voltage characteristics of photovoltaic systems with shaded solar cells, Sol. Energy 56 (1996) 513–520.
[39] X. Wu, M. Bliss, A. Sinha, T.R. Betts, R. Gupta, R. Gottschalg, Accelerated spatially resolved electrical simulation of photovoltaic devices using photovoltaic-oriented nodal analysis, IEEE. Trans. Electron. Dev. 62 (2015) 1390–1398.
[40] J.P. Vargas, B. Goss, R. Gottschalg, Large scale PV systems under non-uniform and fault conditions, Sol. Energy 116 (2015) 303–313.
[41] K. Ding, Z. Ye, T. Reindl, Comparison of parameterisation models for the estimation of the maximum power output of PV modules, Energy Procedia 25 (2012) 101–107.
[42] B. Marion, Comparison of predictive models for photovoltaic module performance, in: 33rd IEEE Photovoltaics Specialists Conference (PVSC), San Diego, California, 2008.
[43] A. Massi Pavan, A. Mellit, V. Lughi, A.M. Pavan, A. Mellit, V. Lughi, Explicit empirical model for general photovoltaic devices: experimental validation at maximum power point, Sol. Energy 101 (2014) 105–116.
[44] E. Skoplaki, J.A. Palyvos, On the temperature dependence of photovoltaic module electrical performance: a review of efficiency/power correlations, Sol. Energy 83 (2009) 614–624.
[45] D. King, J. Kratochvil, W. Boyson, Photovoltaic Array Performance Model, Sandia National Laboratories, Albuquerque, New Mexico, 2004.
[46] T. Huld, E. Dunlop, H.G. Beyer, R. Gottschalg, Data sets for energy rating of photovoltaic modules, Sol. Energy 93 (2013) 267–279.
[47] R.P. Kenny, G. Friesen, D. Chianese, A. Bernasconi, E.D. Dunlop, Energy rating of PV modules: comparison of methods and approach, in: 3rd World Conference on Photovoltaic Energy Conversion, Osaka, Japan, 2003, pp. 5–8.
[48] T. Huld, M. Šúri, E.D. Dunlop, Geographical variation of the conversion efficiency of crystalline silicon photovoltaic modules in Europe, Prog. Photovolt. Res. Appl. 16 (2008) 595–607.
[49] T. Huld, R. Gottschalg, H.G. Beyer, M. Topic, Mapping the performance of PV modules, effects of module type and data averaging, Solar Energy 84 (2) (2010) 324–338.
[50] E. Koubli, D. Palmer, P. Rowley, T.R. Betts, R. Gottschalg, Assessment of PV system performance with incomplete monitoring data, in: 31st European Photovoltaic Solar Energy Conference and Exhibition, Hamburg, 2015.

[51] E. Koubli, D. Palmer, P. Rowley, R. Gottschalg, Replenishing deficient datasets in PV system monitoring, in: 11th Photovoltaic Science Application and Technology (PVSAT-11), Leeds, UK, 2015, pp. 37–40.
[52] M. Šúri, T.A. Huld, E.D. Dunlop, H.A. Ossenbrink, Potential of solar electricity generation in the European Union member states and candidate countries, Sol. Energy 81 (2007) 1295–1305.
[53] T. Huld, G. Friesen, A. Skoczek, R.P. Kenny, T. Sample, M. Field, E. D. Dunlop, A power-rating model for crystalline silicon PV modules, Sol. Energy Mater. Sol. Cells 95 (2011) 3359–3369.
[54] S. Dittmann, G. Friesen, S. Williams, T.R. Betts, R. Gottschalg, H.G. Beyer, A.G.D. Montgareuil, N.V.D. Borg, A.R. Burgers, T. Huld, B. Müller, C. Reise, J. Kurnik, M. Topic, T.F.F. Zdanowicz, Results of the 3rd Modelling Round Robin Within The European Project "Performance"—comparison of module energy rating methods, in: 25th European Photovoltaic Solar Energy Conference and Exhibition/5th World Conference on Photovoltaic Energy Conversion, Valencia, Spain, 2010, pp. 4333–4338.
[55] B. Kroposki, W. Marion, D.L. King, W.E. Boyson, J.A. Kratochvil, Comparison of module performance characterization methods, in: Photovoltaic Specialists Conference, 2000. Conference Record of the Twenty Eighth IEEE, IEEE, Anchorage, 2000, pp. 1407–1411.
[56] B. Marion, A method for modeling the current–voltage curve of a PV module for outdoor conditions, Prog. Photovolt. Res. Appl. 10 (2002) 205–214.
[57] B. Marion, B. Kroposki, K. Emery, J. Del Cueto, D. Myers, C. Osterwald, Validation of a photovoltaic module energy ratings procedure at NREL, Paper No. NREL/TP-529-26909, 1999.
[58] S. Sellner, J. Sutterlüti, L. Schreier, S. Ransome, Advanced PV module performance characterization and validation using the novel loss factors model, in: 2012 38th IEEE Photovoltaic Specialists Conference (PVSC), 2011, pp. 2938–2943.
[59] J. Sutterlueti, S. Ransome, R. Kravets, L. Schreier, Characterising PV modules under outdoor conditions: what's most important for energy yield, Channels 48 (2008) 24.
[60] J. Sutterlueti, S. Ransome, R. Kravets, L. Schreier, Characterising PV modules under outdoor conditions: what's most important for Energy Yield, in: Proc. 26th EUPVSEC, 2011, pp. 3608–3614.
[61] G.T. Klise, J.S. Stein, Models used to assess the performance of photovoltaic systems, SAND2009-8258, Sandia National Laboratories, Alberqueque, NM, USA, 2009.
[62] National Renewable Energy Laboratory, System Advisor Model (SAM) User Guide (Online), National Renewable Energy Laboratory, Golden, CO, 2008 Available: https://sam.nrel.gov/ (accessed 15.11.15).
[63] J.S. Stein, J. Sutterlueti, S. Ransome, Outdoor PV performance evaluation of three different models: single-diode, SAPM and loss factor model, in: 28th European Photovoltaic Solar Energy Conference, Paris, France, 2013, pp. 2865–2871.
[64] Meteotest. Meteonorm–Global Solar Radiation Database [Online], Meteotest. Available: http://www.meteonorm.com/pages/en/meteonorm.php, 2015 (accessed 25.11.15).
[65] M. Šúri, T.A. Huld, E.D. Dunlop, PV-GIS: a web-based solar radiation database for the calculation of PV potential in Europe, Int. J. Sustain Energ. 24 (2005) 55–67.
[66] A. Mermoud, PVSYST: a user-friendly software for PV systems simulation, in: 12th European Photovoltaic Solar Energy Conference, James and James Science Publishers, Amsterdam, Netherlands, 1994.
[67] THE MET OFFICE, Global Radiation Observations, Part of the Met Office Integrated Data Archive System (MIDAS), NCAS British Atmospheric Data Centre, Exeter, 2006.
[68] World Radiation Monitoring Center, Baseline Surface Radiation Network World Radiation Monitoring Center, 2015.

[69] Citizen Weather Observer Program (CWOP), CWOP Information [Online]. Available: http://wxqa.com/, 2015 (accessed 17.11.15).
[70] C.A. Gueymard, Direct and indirect uncertainties in the prediction of tilted irradiance for solar engineering applications, Sol. Energy 83 (2009) 432–444.
[71] J. Roy, T. Betts, R. Gottschalg, Validation of the energy yield prediction of photovoltaic modules, in: 7th Photovoltaic Science Application and Technology (PVSAT-7) Conference and Exhibition, Heriot-Watt University, Solar Energy Society & PVNet-UK, Edinburgh, Scotland, 2011.
[72] E.B. Serra, M.G. Gomez, M. Hartung, Evaluation of the energy yield prediction error depending on the methodology to estimate global tilted irradiance from measured global horizontal irradiance, in: 39th IEEE Photovoltaic Specialists Conference (PVSC), IEEE, 2013, pp. 0746–0750.
[73] J.R. Ross, Interface design considerations for terrestrial solar cell modules, in: 12th IEEE Photovoltaic Specialists Conference (PVSC), Baton Rouge, LA, USA, 1976, pp. 801–806.
[74] T. Nordmann, L. Clavadetscher, Understanding temperature effects on PV system performance, in: Proceedings of the Third World Conference on Photovoltaic Energy Conversion, Osaka, Japan, May 11–18, 2003 (poster), pp. 2243–2246.
[75] S. Krauter, A. Preiss, Comparison of module temperature measurement methods, 34th IEEE Photovoltaic Specialists Conference (PVSC), 2009, pp. 000333–000338.
[76] J.A. Kratochvil, W.E. Boyson, D.L. King, Photovoltaic array performance model, SAND2004-3535, Sandia National Laboratories (SNL), New Mexico, USA, 2004.
[77] E. Skoplaki, A.G. Boudouvis, J.A. Palyvos, A simple correlation for the operating temperature of photovoltaic modules of arbitrary mounting, Sol. Energy Mater. Sol. Cells 92 (2008) 1393–1402.
[78] E. Skoplaki, J.A. Palyvos, Operating temperature of photovoltaic m\odules: a survey of pertinent correlations, Renew. Energy 34 (2009) 23–29.
[79] M.K. Fuentes, A simplified thermal model for flat-plate photovoltaic arrays. SAND-85-0330, Sandia National Laboratories, New Mexico, USA, 1987.
[80] D. Infield, L. Mei, U. Eicker, Thermal performance estimation for ventilated PV facades, Sol. Energy 76 (2004) 93–98.
[81] M. Mattei, G. Notton, C. Cristofari, M. Muselli, P. Poggi, Calculation of the polycrystalline PV module temperature using a simple method of energy balance, Renew. Energy 31 (2006) 553–567.
[82] A.J. Veldhuis, M. Nobre, I.M. Peters, T. Reindl, R. Ricardo, An empirical model for rack-mounted PV module temperatures for Southeast Asian locations evaluated for minute time scales, IEEE J. Photovolt. 5 (2015) 774–782.
[83] P.M. Segado, J. Carretero, M. Sidrach-De-Cardona, Models to predict the operating temperature of different photovoltaic modules in outdoor conditions, Prog. Photovolt. Res. Appl. 23 (2014) 1267–1282.
[84] I. Reda, A. Andreas, Solar position algorithm for solar radiation applications (revised), National Renewable Energy Laboratory (NREL), Golden, CO, 2008.
[85] D. Palmer, B. Goss, I. Cole, T. Betts, R. Gottschalg, Detection of roof shading for PV based on lidar data using a multi-modal approach, in: Proceedings of the 31st Photovoltaic Solar Energy Conference, Hamburg, 2015. WIP.
[86] I.R. Cole, T.R. Betts, R. Gottschalg, Solar profiles and spectral modeling for CPV simulations, IEEE J. Photovolt. 2 (2012) 62–67.
[87] I.R. Cole, R. Gottschalg, Improved model for circumsolar irradiance calculation as an extended light source and spectral implications for high-concentration photovoltaic devices, IEEE. J. Photovolt. (2015).

[88] I.R. Cole, R. Gottschalg, Optical modelling for CPV systems: insolation transfer variations with solar source descriptions, IET Renew. Power Gen. 9 (2015) 412–419.
[89] NCAS British Atmospheric Data Centre, Met Office Integrated Data Archive System (MIDAS) Land and Marine Surface Stations Data (1853-current) [Online], UK Meteorological Office, Exeter. Available: http://badc.nerc.ac.uk/view/badc.nerc.ac.uk__ATOM__dataent_ukmo-midas, 2012 (accessed 21.09.14).
[90] T.R. Betts, R. Gottschalg, Expected variability of monthly and annual energy yield over PV system lifetime, in: Proceedings of the 9th Photovoltaic Science, Application and Technology (PVSAT) Conference, Swansea, UK-ISES, 2013.
[91] O. Perpiñán Lamigueiro, solaR: solar radiation and photovoltaic systems with R, J. Stat. Softw. 50 (2012) 1–32.
[92] B. Ridley, J. Boland, P. Lauret, Modelling of diffuse solar fraction with multiple predictors, Renew. Energy 35 (2010) 478–483.
[93] J. Hay, D. McKay, Estimating solar irradiance on inclined surfaces: a review and assessment of methodologies, Int J Sustain Energ 3 (1985) 203–240.
[94] D.T. Reindl, W.A. Beckman, J.A. Duffie, Evaluation of hourly tilted surface radiation models, Sol. Energy 45 (1990) 9–17.
[95] Bluesky International Limited, Airborne LiDAR [Online], Bluesky International Limited, Coalville. Available: http://www.bluesky-world.com/products/lidar/, 2013 (accessed 29.10.13).
[96] Solar Design Company, PanoramaMaster Bracket and Software [Online], The Solar Design Company, Machynlleth. Available: http://www.solardesign.co.uk/panoramamaster.php, 2015 (accessed 14.05.15).
[97] Meteotest. Meteonorm Horicatcher [Online], Meteotest, Bern. Available: http://meteonorm.com/products/meteonorm-horicatcher/, 2013 (accessed 20.10.13).
[98] South Survey Ltd, Total Stations [Online], South Survey Ltd, Clitheroe, UK. Available: http://www.totalstations.co.uk/, 2013.
[99] R. Compagnon, Solar and daylight availability in the urban fabric, Energ. Buildings 36 (2004) 321–328.
[100] R. Levinson, H. Akbari, M. Pomerantz, S. Gupta, Solar access of residential rooftops in four California cities, Sol. Energy 83 (2009) 2120–2135.
[101] N. Lukač, D. Žlausa, S. Seme, B. Žalika, G. Štumberger, Rating of roofs' surfaces regarding their solar potential and suitability for PV systems, based on LiDAR data, Appl. Energy 102 (2013) 803–812.
[102] J. Mardaljevic, M. Rylatt, Irradiation mapping of complex urban environments: an image-based approach, Energ. Buildings 35 (2003) 27–35.
[103] M. Cellura, A. Di Gangi, A. Orioli, A photographic method to estimate the shading effect of obstructions, Sol. Energy 86 (2012) 886–902.
[104] H.D. Cheung, T.M. Chung, Analyzing sunlight duration and optimum shading using a sky map, Build. Environ. 42 (2007) 3138–3148.
[105] S.J. Crocker, C.R. Sullivan, Predicting shading of photovoltaic systems with cell-level resolution, IEEE Energy Conversion Congress and Exposition (ECCE), Denver, Colorado, USA, 2013, pp. 4488–4495. 15–19 Sept.
[106] M. Drif, P.J. Pérez, J. Aguilera, J.D. Aguilar, A new estimation method of irradiance on a partially shaded PV generator in grid-connected photovoltaic systems, Renew. Energy 33 (2008) 2048–2056.
[107] V. Quaschning, R. Hanitsch, Irradiance calculation on shaded surfaces, Sol. Energy 62 (1998) 369–375.

[108] D. Robinson, A. Stone, Solar radiation modelling in the urban context, Sol. Energy 77 (2004) 295–309.
[109] A. Mermoud, T. Lejeune, Partial shadings on PV arrays: by-pass diode benefits analysis, in: Proceedings of the 25th European Photovoltaic Solar Energy Conference, WIP—Renewable Energies, Valencia, Spain, 2010.
[110] S. Macalpine, M. Brandemuehl, R. Erickson, Beyond the module model and into the array: mismatch in series strings, in: 38th IEEE Photovoltaic Specialists Conference (PVSC), IEEE, Austin, Texas, 2012, pp. 003392–003396.
[111] M. Horvat, M.-C. Dubois, Tools and methods for solar design—an overview of IEA SHC Task 41, Subtask B, Energy Procedia 30 (2012) 1120–1130.
[112] B. Goss, I. Cole, T. Betts, R. Gottschalg, Irradiance modelling for individual cells of shaded solar photovoltaic arrays, Sol. Energy 110 (2014) 410–419.
[113] T. Betts, R. Gottschalg, Expected variability of monthly and annual energy yield over PV system lifetime, in: 5th Photovoltaic Science, Application and Technology Conference. Swansea, UK, 2009.
[114] B. Goss, T.R. Betts, R. Gottschalg, Uncertainty analysis of photovoltaic energy yield prediction, in: 6th Photovoltaic Science Application and Technology (PVSAT-6) Conference and Exhibition, PVSAT, Newcastle, England, 2012.
[115] D.C. Jordan, S.R. Kurtz, Photovoltaic degradation rates—an analytical review, Prog. Photovolt. Res. Appl. 21 (2013) 12–29.
[116] D. Wu, J. Zhu, T.R. Betts, R. Gottschalg, Long term degradation of photovoltaic devices under real outdoor operating conditions, in: Proceedings of the 21st Photovoltaic Science and Engineering Conference, Fukuoka, 2011.
[117] IEA, Review of Failures of Photovoltaic Modules, Performance and reliability of photovoltaic systems, International Energy Agency, Paris, France, 2014.

Part Two

Photovoltaic system performance

Photovoltaic (PV) balance of system components

5

Basics, performance

F. Baumgartner
Zurich University of Applied Sciences ZHAW, Winterthur, Switzerland

Over the last few years, balance of system (BOS) costs have become the crucial factor for overall system costs of photovoltaic (PV) electricity production. Thus, substitution of fossil and nuclear electricity generation by photovoltaics is no longer preferentially limited by the PV module costs, but by the requirement for substantial further reduction of BOS costs. Typically the PV module costs are lower than half the total investment costs for the system (Fig. 5.1), except for very large PV plants. BOS costs together with the operation and maintenance costs (O&M) lead to the total plant lifetime costs, with only roughly one-third belonging to PV modules at the end [2]. In Switzerland the very high O&M costs [3] in 2013 of PV plants was about 4€cents/kWh and thus in the range of the European Energy Exchange (EEX) [4] large-scale electricity generation costs.

The performance of the BOS components of a grid-connected PV system is described typically by their annual losses, as given in Table 5.1. Improvements in losses are possible by selecting more optimized components, such as more efficient inverters and more copper due to increased wiring cross-sections. Additional improvements may be obtained by mounting the modules in a higher position on the roof and reducing the average module temperature. Cable-based mounting (discussed further in

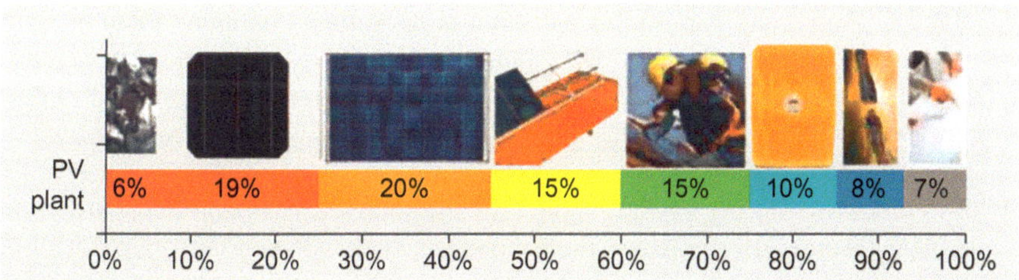

Fig. 5.1 Typical cost share of PV plants of total installation cost in Switzerland in 2013 for 200 kWp scale roof top system [1]. From left: wafer 6%; cell 19%; module total 45%; mounting installation total 30%; inverter 10%; electrical wiring 8%; others like planning 7%.

The Performance of Photovoltaic (PV) Systems. http://dx.doi.org/10.1016/B978-1-78242-336-2.00005-7

Table 5.1 Typical annual losses in standard good quality AC grid connected plants compared to best practices on the market (MW plant without transformer)

BOS losses in the PV plant relative to production in % (best practice today)	Small roof top <10 kW	PV plant 1 MW scale
Inverter [5]	3 (2)	2 (1.5)
DC wiring	1 (0.5)	1 (0.5)
AC wiring	–	1 (0.5)
Soiling	1	1

Section 5.2) several metres above the ground [6] may reduce the temperature by about 5°C and thus gain performance of about 2%, although it would be less on the roofs even if the spacing is increased to about half a metre. Thus, optimized module mounting may have the same effect as optimizing both inverters and cabling together. The customer decides which solution should be chosen, mostly on the sensitivity of performance gain and present price levels, which are also dependent, for example, on the raw material prices of, say, copper, aluminium or steel.

Finally, using the Zurich location as an example, DC performance ratios (for good quality standard plants with polycrystalline silicon modules) of 93% were measured on the DC module level [7]. A typical AC performance ratio of 88% is measured including the inverter losses, wiring losses, module mismatch losses and soiling, strongly dependent on the location (see Chapter 1 for a discussion of the calculation of the performance ratio).

5.1 Overview of PV system types and BOS components

The pricing of PV modules with different efficiency values is typically related to the cost share of the area-related BOS costs such as mounting structure, manpower and cost of land. If we compare two PV plants with the same total investment costs and different module efficiency of 16% and 20%, the latter modules will allow higher market prices due to the lower area-related costs (Table 5.2).

Apart from that, reduced mounting costs are needed to get economical prices for thin-film modules offering lower module efficiency values. However, there are limits to this. At 6.4% module efficiency, the PV modules would have to be offered for free to get the same total plant price for the same nominal power compared to the 16% efficient module, assuming the cost shares are as in Fig. 5.1.

On the other hand, this simple comparison shows that the market will offer only a small increase of €/Wp prices for very highly efficient PV modules. Thus BOS prices will dominate the sensitivity of specific PV module €/Wp prices versus module efficiency. Substantial BOS cost reduction is also needed to reach generation costs below 10€cents/kWh for small-scale PV plants on the roofs of family houses (Fig. 5.2).

Table 5.2 Cost share of PV modules and BOS components shown for different PV module efficiency and prices to get the same total PV plant prices per nominal system power equipped with 16% PV modules at 2013 market prices of 0.60€/Wp (see Fig. 5.1)

Module efficiency	PV module cost share	BOS area related costs mounting (material + labour)	Others BOS inverter, planning
20%	51% (0.68€/Wp)	24%	25%
16%	**45% (0.60€/Wp)**	**30%**	25%
12%	35% (0.47€/Wp)	40%	25%
8%	15% (0.20€/Wp)	60%	25%

In a recent market survey in Germany, the total installation cost in Q1/2015 was 1300€/kWp with a share of 52% for the BOS costs for 10–100 kW PV plants [10]. Other markets may differ significantly due to wages, experience and regulations, such as was published for France with BOS costs between 1 and 2€/Wp in 2011 [11].

For large-scale, ground-mounted PV plants, a detailed analysis of BOS costs and their reduction potential of about 9% annually between 2010 and 2016 is discussed by Ringbeck and Sutterlueti [12], including module design and mounting and electrical wiring structures.

The International Technology Roadmap for Photovoltaic (ITRPV) [8] gives the trends for the future development of installation costs of a PV plant larger than 100 kW relative to the system price of 1000€/kWp for Europe and the United States in 2014, without the additional costs of regional permits. Hence, the levelized cost of PV electricity (LCOE) production is below 8€cents/kWh for locations higher than

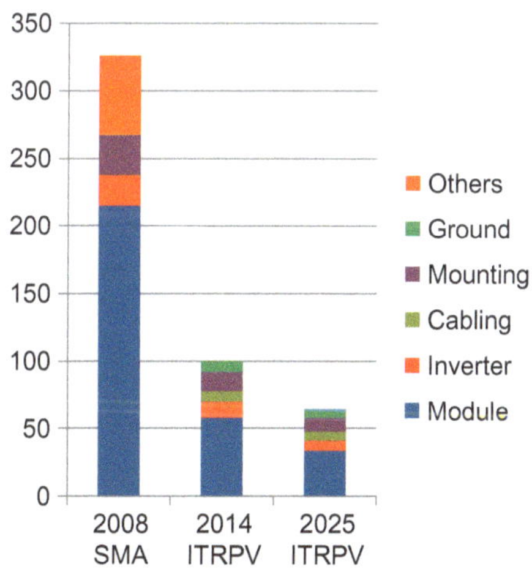

Fig. 5.2 Development of cost share in percent relative to the costs according to ITRPV in 2014 Ref. [8] of PV components of PV plants of about 1 MW range (see SMA Ref. [9]; Table 5.3).

Table 5.3 **Relative development of PV module and BOS costs for large systems greater than 100 kWp in Europe, the United States and Asia [8] and for SMA [9]**

	2008	2014	2025
% relative to 2014 costs	SMA	ITRPV_ed.2015	ITRPV_ed.2015
Module	215	58	33
Inverter	23	12	8
Cabling	–	8	7
Mounting	29	14	10
Ground	–	8	6
Others	59	–	–
Total	326	100	65

1000 kWh/kWp annual yield. They expect a cost reduction in PV inverters of about one-third in the coming decade.

Back in 2008, the specific system price was 3260€ per each kWp for a 1.4 MW-sized PV plant with crystalline silicon PV modules with a cost share for BOS of only 24%, including 9% for installation and 7% for a central inverter.

In 2012 the US National Renewable Energy Laboratory (NREL) published a survey about the nonhardware BOS costs (SOFT BOS) accounting for about half of the installation costs of a small residential PV plant. The module manufacturing costs were equal to the installation labour costs. In 2013 about 70% of the US residential PV plants were realized by third-party installers [13].

Even in Germany, with a sizable experience in residential PV installations, the working hours for the installation of a 4 kW plant are nearly equal to all the other tasks involved to realize the PV plant, such as marketing, planning together with an appointment at the house, ordering the products, software installation for monitoring, services to get funding or the final handover to the customer.

To maximize long-term reliability of the plant over two decades, sufficient working hours for the installation have to be provided to allow experienced craftsmen to carefully handle the PV module during installation, especially for PV on/in buildings. All additional paperwork and quality tests concerning the installation process after starting production withdraw funds from the installation itself and should be minimized.

Rough handling of the PV modules may increase the risk of power losses and thermal hot spots in the coming years, because microcracks, present in the wafer at the manufacturer gate before shipping, will further extend. Detecting these failures after years of plant operation, by infrared imaging or other testing methods to identify power losses in PV modules, will not allow them to be clearly assignable to the faulty mounting process itself, because other possible causes could have occurred during operation. It is recommended to improve training of the installers and apply random site visits during the construction phase.

5.2 Mechanical mounting and performance

In 1990 Germany started a subsidy programme to install 1000 PV roofs and continued with a 100,000 roof programme in 1999, which was finally replaced by the very successful EEG feed-in programme in 2004. This continuous and dependable German market growth led to the development of several efficient PV mounting solutions. One of today's market leaders started to launch their German fastening systems under the brandname Schletter Solar Mounting Systems in 2001 and have mounted 15 GW PV modules with their products in the past 15 years worldwide. Also well-known technology providers from the general construction industry entered the PV mounting solution market successfully, for example the Hilti Corporation in 2006.

A variety of products exist to mount, mainly framed, PV modules, whether on tilted or flat roof systems or even for green-field installations. Each PV module is typically fastened by four anchoring supports, located on the module frame about one-quarter of the distance to each of the smaller sides of the module, by standard top-down clamps to the metal framing channels. Today the dominant products still use metal fastening materials, primarily anodized aluminium or steel. The amount of material needed to clamp modules has slightly reduced over the past years. Other fastening materials such as plastics or wood could not be established with a relevant market share.

By establishing low elevation systems between 10 and 30 degrees module inclination on large-scale commercial flat roof systems (Fig. 5.3), the mounting material costs could be cut in half. These systems incorporate wind shields in the frame to minimize the uplift forces imposed and reduce the required ballast weight by more than a factor of ten. Because of that, the ballast to withstand the wind load was reduced from up to 200 kg of concrete to about 6 kg for each module, saving material and costs. The light metal mounting frame itself comes with about 4–8 kg of aluminium for each square meter of module area, or about 25 kg metal per kWp, including the steel screws. Thus about 2% of the overall PV system price is accounted for by the associated stock metal exchange prices of aluminium. The product price of these mounting racks without labour costs is about 10% of total PV system costs for several 100 kW plants, roughly equal to the inverter prices.

Fig. 5.3 Low-slope flat rooftop south-facing PV mounting: additionally to the optimum 15- or 10-degree module tilt angle, this shows a particularly low-weight roof load per unit including module <11.5 kg/m^2 (www.schweizer-metallbau.ch). Wind tunnel tested according to Wind technische Gesellschaft (WtG, http://www.wtg-dach.org/) standards.

Fig. 5.4 Inroof system using frameless PV modules to replace the standard roof (Solrif product of www.schweizer-metallbau.ch, European Patent EP 1,060,520).

A leak-free PV roof mounting is indispensable and these lightweight systems all have the benefit of not penetrating the existing roof membrane, because they need no permeation by additional building elements to transmit mechanical forces between PV modules and the original roofs below, and this is very important for the owner of the building.

In 2011 a market survey discussed products from 24 manufacturers of these low-slope commercial roof PV applications, about 8 years after the first appearance of this type of product [14]. The typical 10-year warranty may be extended individually to 25 years for an extra fee. Structural engineering requirements vary significantly from project to project, between ground-mount PV racks and their foundations and low-slope commercial systems which may need adapted ballast designs. Highly cost-efficient solutions of about 15-degree tilted angle use preassembled racks and come with optimized logistics for shipping.

After a quarter of a century, innovative solutions such as replacing the standard roof by PV in roof systems (Fig. 5.4), PV facade solutions or cable-based solutions are still very restricted niche markets.

5.2.1 Static requirement—building codes

All roof structures and similar supporting structures of buildings have to accommodate the mechanical load from wind and snow. Hence, the PV mounting structures have to be in accordance with the local building codes: for example, in Europe EN 1990:2010-04 general building and EN1991-1-3 for snow and EN1991-1-4 for wind load; in Germany the DIN 1055-4 for wind and -5 for snow load; in Switzerland SIA 261; and in the United States the international building code 7-05, developed by the American Society of Civil Engineers (ASCE) [15]. All these codes are adapted periodically. Most of them revised their standards in the last few years to accommodate the special requirements for the different load conditions of square edges of buildings highly relevant to flat-roof PV mounting systems.

In most of the building codes the calculation method to find the mechanical load relevant for structural engineering is similar. Different local wind and snow zones and addition factors and coefficients are applied, developed mainly by the local architects and engineering associations and standards committees.

Therefore, the mechanical load depends on several boundary conditions which the PV mounting structure must be designed to safely resist. The static load increases linearly with the snow depth due to the weight, but increases by a factor of four if the wind speed doubles.

- The wind load is calculated based on the wind zone. In Switzerland, for example, seven wind zones are defined, with the levels of 0.9, 1.1 and 1.3 kN/m^2 in the typical residential areas of Switzerland rising to 3.3 kN/m^2 in the most exposed Alpine locations. In Germany, four wind zones are described, with up to three subcategories showing static pressure values between 0.5 and 1.4 kN/m^2, the latter on small islands in the North Sea.
- Wind load increases roughly with the square of the height of the building. The resulting static wind pressure is given by a tabulated coefficient according to the height exposure category, which has to be multiplied by the preceding wind zone static pressure.
- A second wind coefficient has to be multiplied to account for the surroundings of the building. In Germany four zones are given: for a city, a village, flat country with trees and only flat country.
- The final wind exposure depends on the roof zone placement. For large roofs, the distance to the edge of the roof is relevant due to separation of strong vortices that then interact with the array's panel edges, especially for lightweight, low-slope, flat-roof PV systems. Again, relevant coefficients are given in the building codes to quantify these effects. Several manufacturers test their structures via experiments in wind tunnels and offer calculation methods and coefficients or web-based tools to address this complex issue (see Fig. 5.5).
- The total wind load is then split into the vertical load and the horizontal component which is relevant for the resistance-against-drag coefficient for movement of the whole structure.
- Snow load increases with the altitude of the location above sea level and depends on regional snow zone maps for each country. Recently, the Swiss SIA adapted their SIA 261:2014 snow codes for large flat roofs.
- The vertical load component has to be calculated from the overall wind and snow load together with the weight of the PV module itself. The values have to be well below the acceptable structural roof load of the building. On the other hand, a negative value, for example near the corner of a roof, may lead to lift-off damage of the PV modules. Thus the output of these static calculations may give different values of required ballast weight for different locations, for instance on flat roofs (see Fig. 5.5).

Polysun, a software tool to design and plan PV and solar thermal systems, at the time of this writing planned to offer as an additional feature a software-assisted calculation of the wind statics for several regions, including detailed suggestions for placing ballast elements, for example on a flat-roof PV installation [16].

5.2.2 New development and challenges of module mounting

Power losses and thermal hot spots may occur by unintentional growth of plants in between the rows of low-elevation PV mounting structures. To prevent these effects,

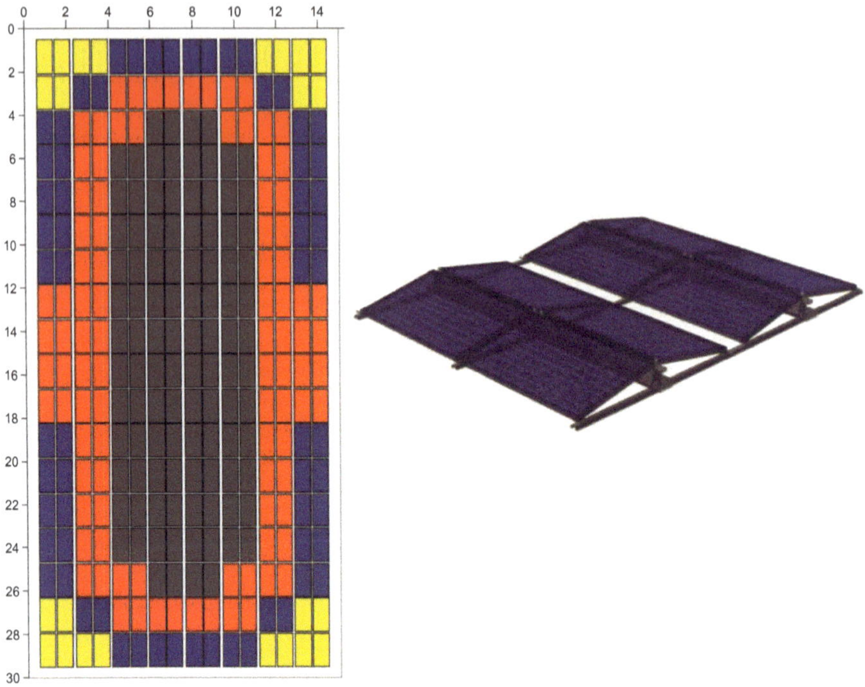

Fig. 5.5 The required ballast needed for a low-elevation lightweight mounting structure is higher close to the corner of a flat roof building. Shown for 22 m height building in Winterthur, Switzerland in the city centre and a Schletter EAST/WEST mounting solution with 15-degree inclination. For each PV module the required additional ballast at different roof zones is given by different colours: yellow 119 kg, blue 68 kg, red 33 kg and grey no ballast (wind zone Winterthur 0.9 kN/m^2, basic snow zone value 0.9 kN/m^2, weight of the PV module 0.15 kN/m^2). Sizes of the grid are given in metres, with each rectangle showing one PV module.

the operation and maintenance cost will significantly rise or the costly insertion of an additional roof membrane will be needed (Fig. 5.6).

In some city centres, such as Chicago or Zurich, there are regulations to try to increase the number of green roofs to improve the ecosystem and incorporate water retention in case of heavy rains. In the future, more projects will arise to find synergies between green and PV roofs. Combining both the benefits of PV electricity production with green roofs, the need for higher PV mounting structures emerges, which will increase the costs relative to the common systems (see Fig. 5.7).

Other new approaches such as the mounting of bifacial modules [18] need a requisite amount of light collected by the backside of the PV module, which again leads to higher mounting structures compared to conventional low-cost, low-elevation systems such as are shown in Fig. 5.8. An interesting approach is to mount bifacial modules in a vertical position facing to EAST/WEST. This includes the benefit of making the gardener's task easier to perform, at reduced O&M costs. Another advantage of the bifacial approach is that the vertical PV modules do not generate the maximum power

Fig. 5.6 Thermal hot spot due to shading of foliage plant growing on a flat-roof EAST/WEST mounting of crystalline standard PV modules in Winterthur—second PV module in the shaded area on the right side. Two photographs: shading led to hot-spot temperature of around 90°C [17].

at noon compared to conventional south mounted PV modules (see blue curve in Fig. 5.8). Two production peaks, appearing in the morning (see red curve in Fig. 5.8) and the afternoon, lead to a better matching to the typical load profile. In future, the late afternoon load peak in particular will further increase due to the rising demand for charging of electrical vehicles. Thus costs for the daily peak shift of energy by the use of external batteries are reduced by the use of vertically mounted bifacial modules.

Fig. 5.7 Left: PV green roof project Winterthur on the flat roof of Werkhof Winterthur (left photograph) using Zinco mounting structure (www.zinco-greenroof.com, Project 2014, ZHAW, Solarspar, Klimafond Stadtwerk Winterthur). Right: Standard, not green, roof 30 degrees using Hilti MSP mounting structure: EKZ ZHAW Reference Test Field on the flat roof with concrete ballast each 320 kg for each standard crystalline Si 230 W PV module.

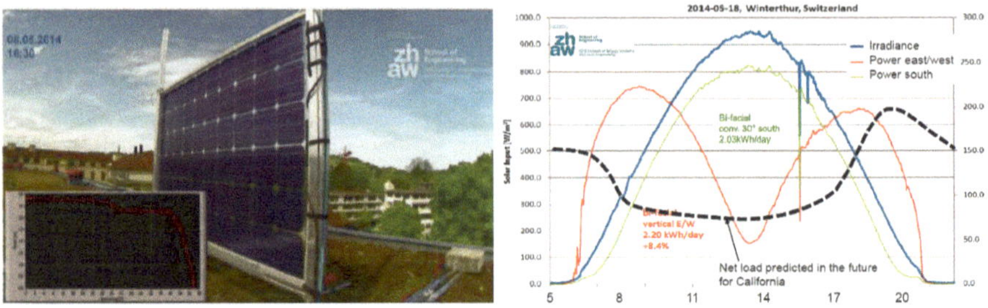

Fig. 5.8 Daily course of PV production of a vertical mounted bifacial module facing EAST/WEST (azimuth 90 degrees, Pn_back = 255.6 W and Pn_front = 232.6 W) and blue inclination 30 degrees south (azimuth 0 degrees, Pn_back = 255.3 W and Pn_front = 231.7 W) (on the ZHAW roof in Winterthur) [19].

5.2.3 Orientation and yield of the fixed PV generator

Large MW parks or flat-roof systems are built as rows of PV modules oriented East/West and each module oriented south, with a typical inclination angle between 25 and 35 degrees in central Europe [20]. The design of the BOS component, the mounting structure, is relevant in regard to several percent of the final yield. The classical design rule assumes that shading starts to come from the PV module row in front towards south at noon in the wintertime, which results in a so-called shading angle of 20 degrees in central Europe [21].

Detailed measurement analyses performed in Zürich showed that, due to the foggy conditions in the morning and evening, standard commercial PV planning tools like PVSyst [22] overestimate the shading losses by up to a factor of 2. A grid with a mapping interval of 2.5 degrees for elevation and 5 degrees for azimuth is built and the amount of measured annual yield in each raster element is calculated (Fig. 5.9). Additionally, the shading limit (red curve) is drawn for a shading angle of 20 degrees in that reference test field.

The result of Fig. 5.9 was used to calculate a potential shading impact for each grid element. Consequently, the shading losses for different shading angles could be calculated using different models of how partial shading affects power losses. In the EKZ model A, the total yield is lost if the shadow hits the reference module. That means that partial shading has the same effect as full shading. This ends up with an overestimation of the losses. The EKZ model B delivers different yield losses depending on the sector where the shadow occurs. There are three sectors according to the bypass diodes. If the shadow line is located in the lower third, the losses are 1/3 of the yield. The losses are 2/3 in the middle third and 3/3 in the upper third. This loss classification into three areas is also used in model C. For shadow in the lower third, the losses are still 1/3. In all remaining cases, the yield is reduced to 10% assuming that 10% of the global irradiation is diffuse irradiation. This situation is more realistic due to the limited voltage range of each inverter.

Finally, these detailed loss analyses are compared in Fig. 5.10 with the findings of the commercial PVSyst software tool using different. There are large deviations of the shading losses with respect to the same angles. The losses for the standard shading

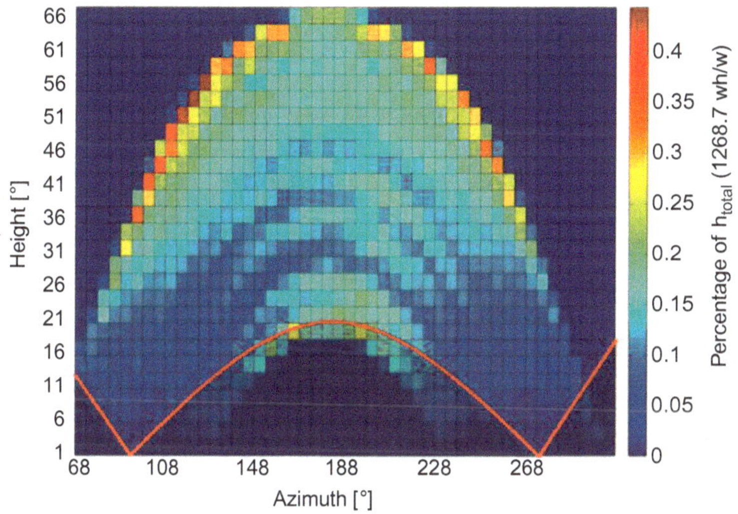

Fig. 5.9 Mapping of annual measured PV electricity production of a 30-degree tilted south-oriented multicrystalline silicon reference module depending on the direction of the direct sunbeam (location Dietikon, Zurich in 2011). The amount of annual yield in each raster element is given in nominal hours kWh/kWp and marked in colours according to the scale on the right side. Shading limit due to shading angle of 20 degrees is shown in red. PV electricity mapping of a 30-degree inclined south-oriented polycrystalline silicon module mounted in Zürich [23].

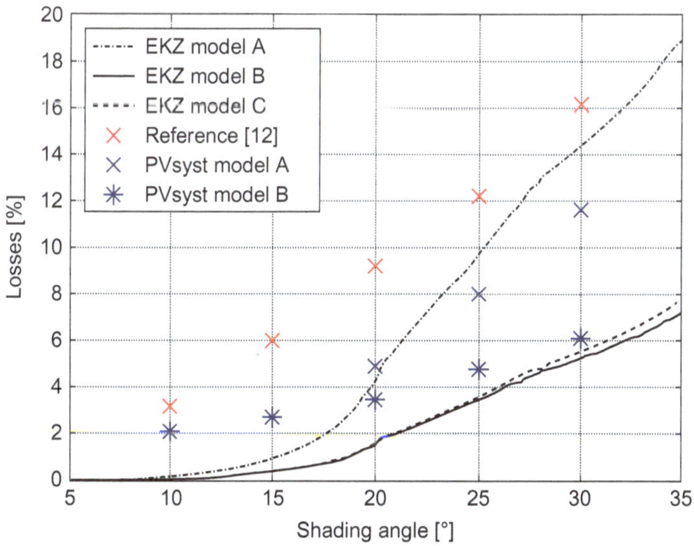

Fig. 5.10 Shading losses of flat-roof or green-field PV plants with different types of arrangement of the mounting structure with rows oriented East/West and the modules towards south, at different distances of the rows described with the shading angle. Standard crystalline silicon module mounted portrait with three bypass diodes [23].

angle used of 20 degrees vary between 1.5% and 4.9% and even 9% for another reference used (for further details see Ref. [23]).

Assuming shading losses of 2%, a shading angle of only 10 degrees is proposed by PVSYST while the detailed analyses based on real measurements as shown in Fig. 5.9 are based on a 20-degree shading angle. The latter offers an excessively higher total plant power output for the same standard flat-roof space.

5.2.4 Performance increase through mechanical tracking

Mechanical tracking of PV modules increases the electricity production of PV plants by about 20–30% relative to the annual yield of fixed installations, depending on location. Before the dramatic cost reduction of PV modules starting in 2009, from a level of about 3€/W to about 20% of that level today, several one-axis and two-axis tracking systems were developed and absorbed by the markets. At that time the additional costs of mechanical tracking systems relative to fixed mounting solutions increased the total plant costs by a percentage smaller than the increase of electricity yield due to tracking. This resulting economic benefit is reduced by the higher O&M costs associated with all the moving parts in the system (Fig. 5.11). Another reduction is needed due to the higher land requirement, because of the smaller capability of power extracting per square (compare Fig. 5.12).

A more theoretical approach showed that 41% more sunlight is available by mechanical tracking the PV module to follow the daily course of the sun, relative to fixed installations, if there is a very high amount of direct sunlight available, as in the most 'suntrap' locations on earth [24]. Calculated values between 31% and 36% of the two-axis gain in the Mediterranean region, compared to fixed installations, are achieved and predicted via public web calculation tools and also commercial PV planning tools [25]. Huld showed [26] that in Northwest and Central Europe, the relative

Fig. 5.11 Left: standard south-oriented tilted 30 degrees (1.6 MW Park CdTe, Rickelshausen, Germany); right: east/west-oriented as one part of the 20 kW reference plant on the roof of ZHAW Winterthur University building replacing the PV production peak at noon of south mounted modules by one peak in the morning and a second peak in the afternoon and equipped with 10 different module technologies, each connected by an individual 2 kW inverter to the AC grid.

gain of two-axis tracking is higher than in the south. In commercial PV plants, additional losses will occur relative to the single mechanical tracked module by shading losses of the nearby rows of also mechanically tracked PV modules on the limited available ground area of the plant. During the first three years of operation in Switzerland near Basel in a 650 kW PV plant, 23.5% gain due to one-axis tracking by solar wings with cable-based mounting solutions was measured [27]. Prototypes of such a cable-based tracker were installed in 2011 on top of a ski-lift in the Swiss Alps and on top of an outdoor storage facility in Flums in 2010, also in Switzerland (see Fig. 5.12).

At today's low PV system prices for large-scale fixed ground mounted plants, the 25% tracking increase in electricity yield will be covered mostly by the raw material stock exchange price of 200 kg steel needed per kWp to build the two-axis mechanical tracker [29]. New developments were performed optimizing one-axis systems to a very low demand of 100 kg/kW metal needed for the tracker. The mechanical PV tracker market accounted for only 3% of worldwide installed PV modules of about 1 GW in 2012 and about 4 GW were installed in 2014. In 2014 the world's largest single-axis tracking plant with a power of 206 MW started operation on an area of 801 ha on Mount Signal in southeastern California. Still, global annual installations of single axis trackers are forecast to reach over 9 GW in 2019, driven mainly by the growth of utility-scale installations, as predicted by an IHS tracker market outlook [30].

Mechanical tracking still offers the benefit of increasing PV production in the morning and in the afternoon at a period when other south-mounted PV plants will not be able to supply the electricity grid in future (see Fig. 5.13). In the coming years, these extra costs of tracking may be economically more feasible than an electrical battery to close the gap between production of PV electricity and consumption.

Fig. 5.12 Two-axis tracking solar wings in Flums and a one-axis tracker at the Solar Ski Lift Tenna, both in Switzerland, as examples of lightweight tracker solutions [28].

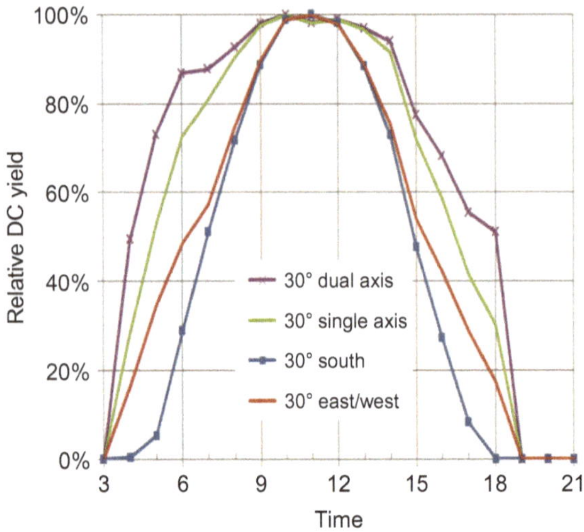

Fig. 5.13 Yield of PV plants fixed mounted and mechanical tracked modules—see legend□—in Freiburg, Germany on a clear-sky day in July [31].

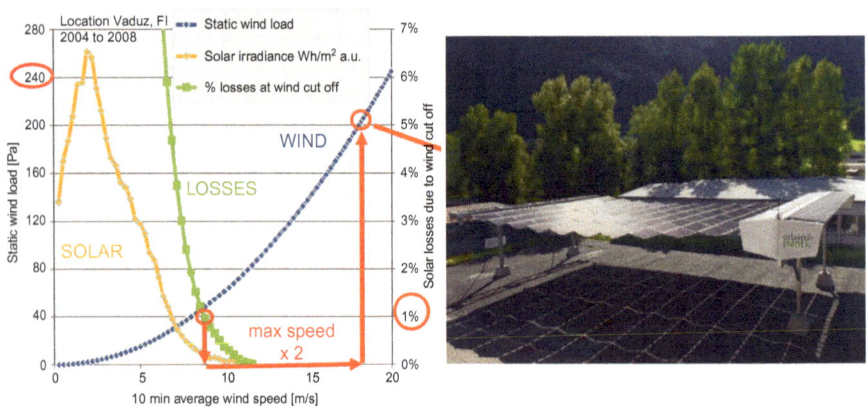

Fig. 5.14 Wind speed, wind pressure and histogram of solar irradiance below a certain wind speed for an annual period. Right-hand picture shows the 50 kW prototype of urban plant PV mounting system moving the module back into a safety box, for example at very high wind load, resulting in a few percent annual energy losses depending on the location (see Ref. [28]).

New innovative concepts such as mounting PV modules on top of car parking areas and moving them in a safety box before very high wind or snow load occurs are only demonstrated at the prototype stage (Fig. 5.14). Such a solution will offer an extremely low amount of mounting material, but on the other hand might cause higher O&M costs.

Today most of the systems are still mounted on the rooftop with several centimetres distance behind the PV panel to the roof to benefit from better cooling compared to roof integrated PV systems. Besides the higher electrical performance due to the negative temperature coefficient of power, the O&M costs are much lower if PV modules have to be replaced, due to the much easier handling.

5.2.5 Reliability aspect of mounting solutions and O&M costs

The service lifetime of ground-mounted solutions is improved by choosing the appropriate material to avoid corrosion and designing the systems to withstand wind and snow load, which may be crucial depending on the location. Tracking systems will improve specific yield per installed nominal PV power but show a drawback in higher service time due to the moving parts in the plant. To limit the latter, only a small number of electrical drives per megawatt are used, equipped with dry bearing and a smaller tracking accuracy of 2% together with a robust mechanical design.

One of the single-axis tracker market leaders provides the power for parallel rows of 700 kW modules from a single lifetime sealed 2 kW AC motor including gearbox. The supplier is claiming zero-scheduled maintenance to the lifetime sealed drive and a mechanical system that features passive wind management that does not require active wind stow control or an uninterrupted power source (tested to 250 km/h wind, 3-second gust). This supplier published 99.996% uptime for their present products, responsible for tracking about 3 GW PV modules worldwide [32].

Industrial suppliers from India, for example, are installing seasonally adjusted trackers, which allow the angle of trackers to be changed seasonally by humans rather than with motors, at lower installation costs.

5.3 Electrical MPP tracking principles and losses

Maximum power (MP) production of the PV generator is reached by applying the appropriate DC voltage V_{mp}. Ten percent higher voltage than V_{mp} shows a loss of 16% of power, while 10% lower voltage than V_{mp} shows only 5% in power losses (Fig. 5.15).

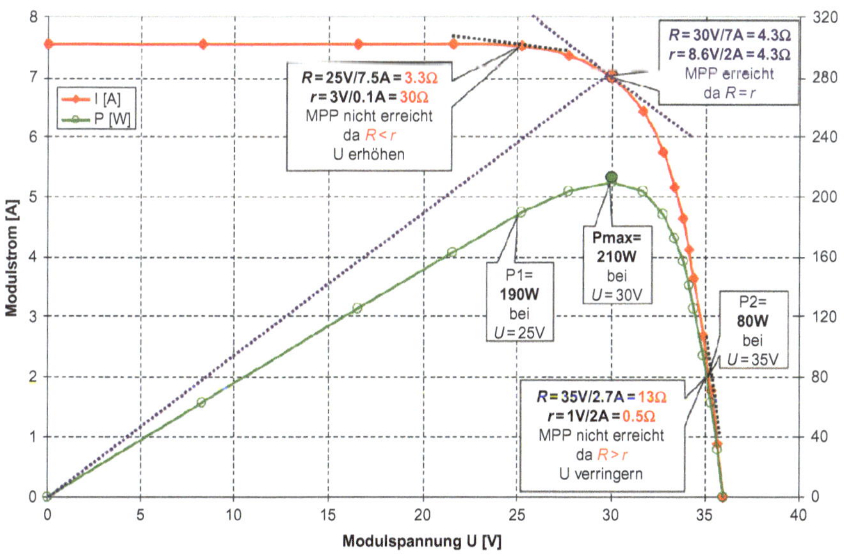

Fig. 5.15 *I–V* curve and power versus voltage characteristics together with the values of differential resistance *r* and absolute resistance *R* in MPP and two other operation points.

The topic of optimization of MPP tracking and related losses has been relevant in PV research and the PV industry for decades. The experimental findings are in accordance with the modelling output of the electrical current voltage characteristics of a solar cell, since the groundbreaking publication of Shockley in the middle of the last century [33] (see also Chapter 1).

In real operation, a deviation of the standard test conditions (STC) voltage given in the module datasheet occurs due to:

- Increasing solar *irradiance* P_{in} in the module plane will lead to a rise of V_{mp} as a function of the logarithm of the intensity ($V_{mp} = c \ln(P_{in}/P_0)$ with c and P_0 being constant factors) if the cell temperature is not changing; in detail it is not the power of irradiance that has a linear relationship to the short-circuit current but the number of photons with energies higher than the bandgap that are proportional—see Fig. 5.16.
- Increasing *temperature* T of the solar cell, the voltage V_{mp} will decrease by approximately 2 mV per K at constant current in standard crystalline silicon solar cells, as with silicon semiconductor diodes. The negative value of the temperature coefficient TC_V leads to the smaller voltage $V_{mp}(T) = V_{mp\_STC}(1 + TC_V(T - T_{STC}))$ compared to the value at STC temperature of 25°)—see Fig. 5.16.
- *Partial shading* conditions on a crystalline silicon module equipped with typically three bypass diodes may show one absolute and one additional local maximum of power versus voltage. This number of total maxima may be equal to the number of bypass diodes depending also on the shading conditions—see Fig. 5.17 (cause of shading losses may be

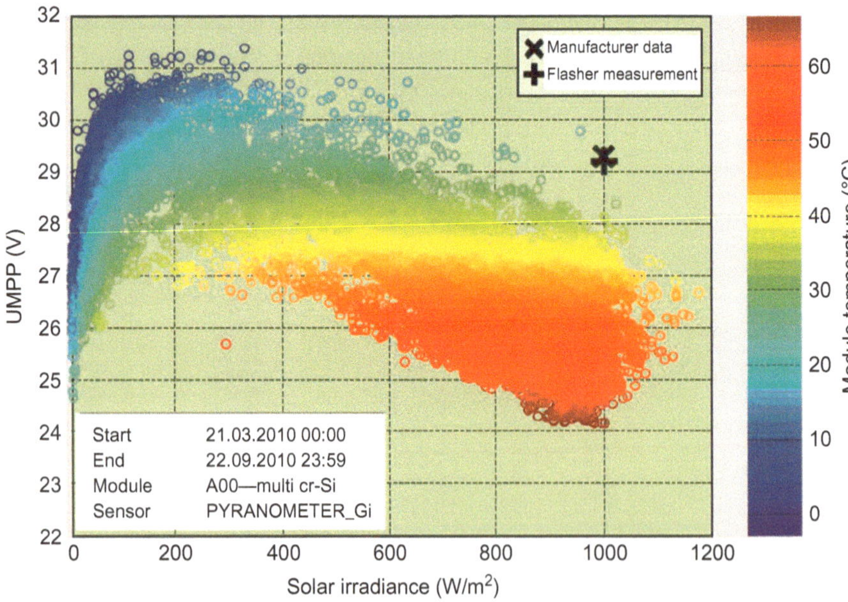

Fig. 5.16 Measured voltage at MP of a 30-degree south-mounted polycrystalline module versus irradiance in W/m^2 is shown for half a year during summer of 2012, based on 1-min intervals of a single reference module at four wire conditions as well as the indicated measured PV-module temperature in °C indicated by the colours. The V_{mp} condition of the PV module label as well as the flasher measurement result is also marked [34].

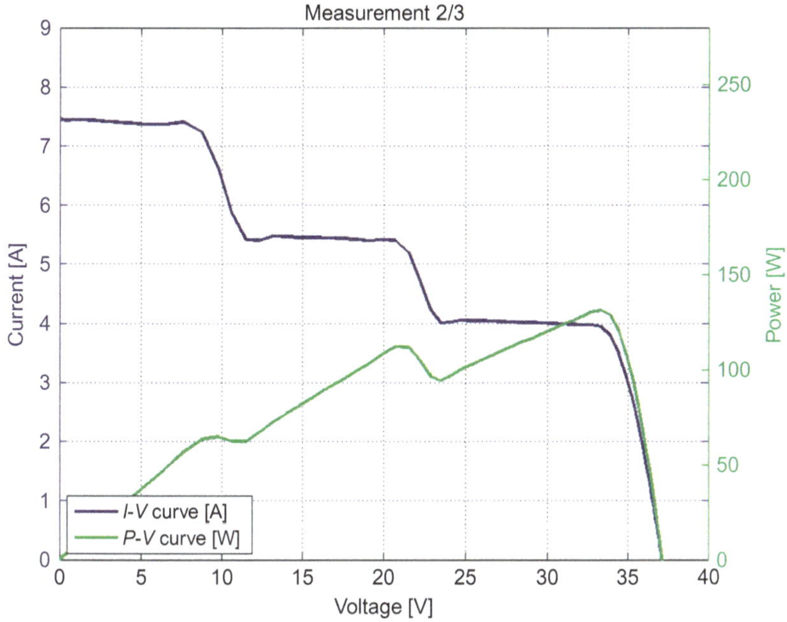

Fig. 5.17 Partially shaded PV module shows measured local power maxima together with the absolute power maxima versus module voltage depending on the number of active bybass diodes and the shading conditions [35].

nearby objects like houses and plants or building elements, mountains, the next row of PV modules or soiling, starting from the PV module edges).

- *Faulty conditions* in either the module wiring of the cell interconnectors and/or the junction box shunting or broken wires will show a fundamentally different power output than expected STC power, mostly zero power.
- *Degradation* of individual solar cells will lead to problems, for example the decrease of the fill factor of an amorphous silicon solar cell due to light-induced degradation of up to 10% [36].

The optimum voltage to reach MP will be a mixture of the real outdoor or module conditions described previously. The results of the occurrence of V_{mp} values of a polycrystalline silicon module, without major failures or shading, for the summer half-year operated in Zurich, with measurement intervals of one minute, are shown in Fig. 5.16. At 1000 W/m^2 irradiance, the V_{mp} in that practical application is 18% below STC value at a module temperature of 65°C. It is only 4% lower than V_{mp_STC} for the same irradiance at much lower ambient temperatures, leading to a final module temperature of only 35°C.

The measured real outdoor mapping for July 2015 of the MP operation conditions for voltage and power is given for the same reference installation for two different module technologies, polycrystalline silicon and heterojunction silicon, in Fig. 5.18. They use the same mounting conditions and have the same orientation. It is obvious that the voltage span at the same output power is much smaller for

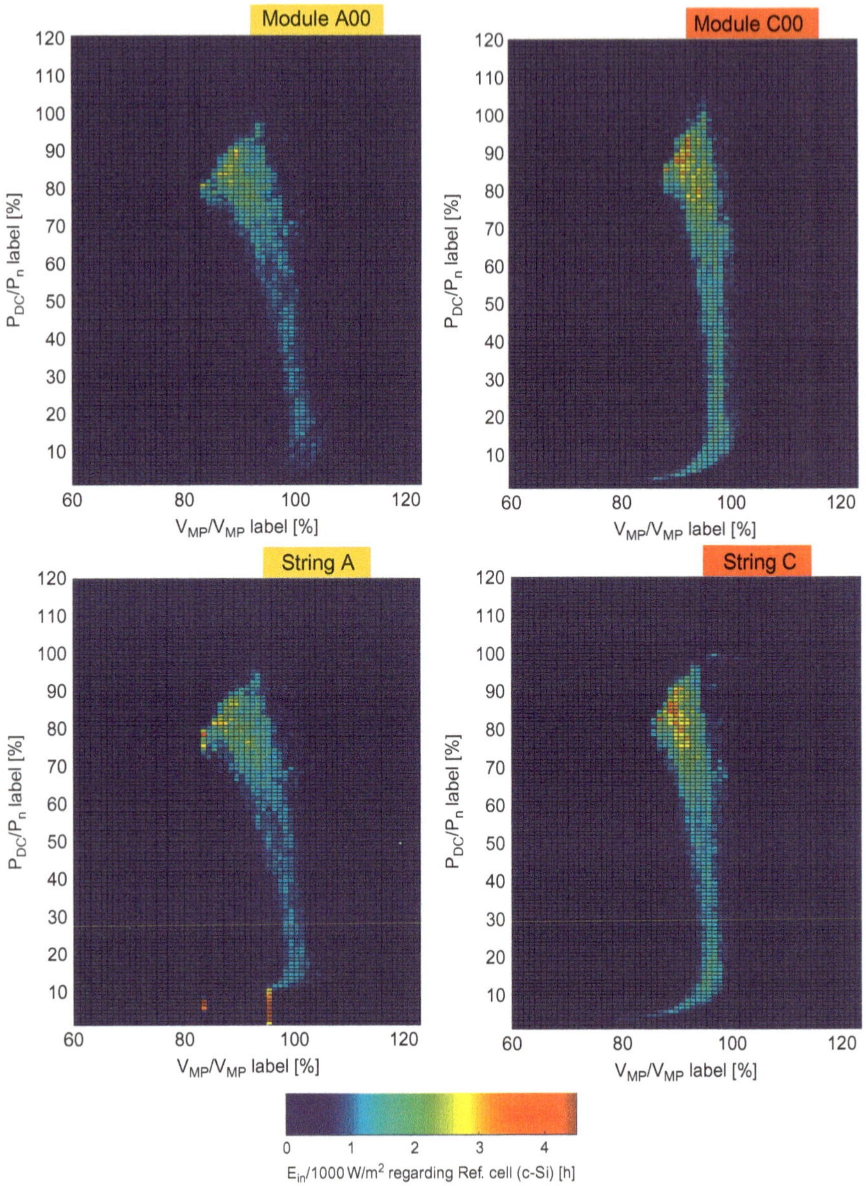

Fig. 5.18 Mapping of the MP power relative to MP voltage in July 2015 of a polycryst: 60 cell standard PV module (left: Module A00) and a heterojunction standard module (Module COO) at an inclination of 30 degrees south-mounted at the EKZ ZHAW Reference PV plant in Zürich. The preceding measurement values are based on a 1 min *I–V* curve scanning measurement of one single reference module at a total yield in that month of 181.3 Wh/Wp for Module A and 183.4 Wh/W for Module C. The two lower figures show the mapping of the measured performance of a 2 kW string of the same technology A and C, each powering a standard grid-connected inverter. See power limitation of the inverter in the lower right graph [37].

the heterojunction modules due to the smaller TC_V value. A further small increase of that voltage span is found by comparing the string voltage for serial connection of the modules in Fig. 5.18. This is due to the real imperfect MP tracker performance of the commercial DC/AC inverter behind the string date compared to the mapping picture of a single reference V_{mp} calculated from each I–V scan of the single reference PV module mounted in the same installation.

Due to the fact that not all serial connected modules offer the same current at the individual MP, mismatch losses will expand the voltage span of the string shown in the voltage power mappings.

Even if a deterministic model is used, appropriate for the relevant module technology, and the cell and the cell temperature are measured together with the irradiance, this highly sophisticated MP voltage predictor will fail if partial shading (Fig. 5.17) or faulty wiring conditions occur by accident in such a module string.

5.3.1 MPP tracking algorithm and performance

Over the last few decades several methods and algorithms have been developed to operate PV generators at MP. A fast-changing irradiance at days with conditions that alternate between cloudy and sunny every few seconds or minutes (see Fig. 5.19) is very challenging. Improved algorithms are needed to handle emerging local power voltage maxima due to partial shading or faulty conditions in the module wiring (Fig. 5.17). Switching power electronic converters (see Section 5.4) are used to adapt the DC voltage on the PV generator, enabling it to track the new optimized MPP, forming a maximum power point tracker (MPPT).

The highest MPPT efficiency values are reached with clear-sky conditions at smooth solar irradiance characteristics compared to cloudy days (Fig. 5.19). The measurement uncertainty is typically much higher than 1% compared to the indoor laboratory measurements which could reach uncertainty values below 0.5% [39]. To achieve low uncertainty measurements under real outdoor conditions, longer averaging periods are helpful but also limiting for fast irradiance changes. The annual outdoor measurement of the inverter efficiency of a 650 kW PV plant in Waldshut, Germany was in accordance with the 97% euro efficiency given by the manufacturer of the 30 kW inverter [40].

MPPT algorithm—perturb and operate

The most widely used MPPT method is to increase the voltage of the PV generator V_{mp} to reach the new operation point at higher maximum power. The resultant power before and after that small period of seconds or minutes is measured. Depending on the sign of the voltage change and the sign of the power change, an optimized next small change of the voltage, a voltage step is proposed by the algorithm. Four different logical input states are possible and the resultant change of the next voltage presetting is defined by a simple logical diagram. This algorithm is called the perturb and operate (P&O) or hill climbing MPP method (Table 5.4).

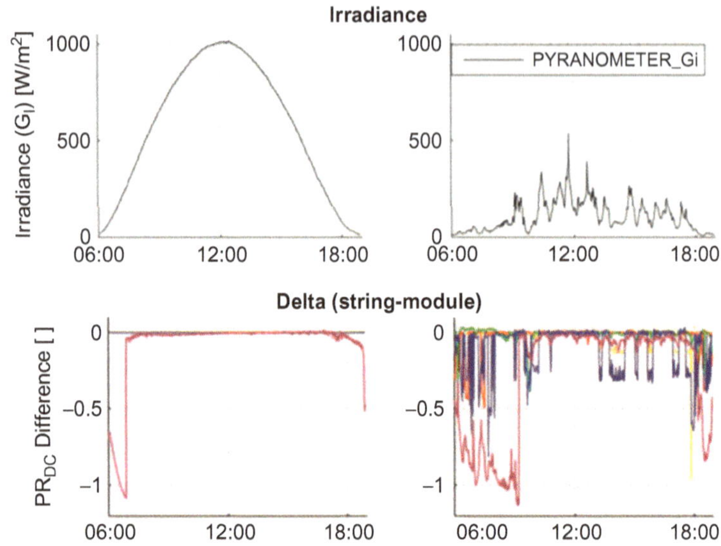

Fig. 5.19 The measured outdoor performance of the MPP tracking of a commercial inverter versus the measured MPP of a single reference PV module is shown for a clear-sky day (left) and a cloudy day (right) based on a 1 min interval [38].

Table 5.4 Logical rules/diagram according to the P&O MPP algorithm where V3 − V2 has an absolute constant value but changing sign and corresponds to the proposed new optimized voltage V3 which has to be applied to the PV generator as a result of the two measurements before the test period V1, P1 and during the test period V2, P2

Process	V2 − V1 > 0	P2 − P1 > 0	V3 − V2 > 0	
A	YES	YES	YES	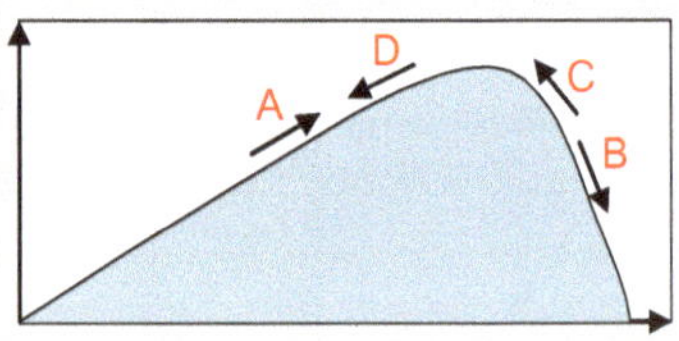
B	YES	NO	NO	
C	NO	YES	NO	
D	NO	NO	YES	

Disadvantages of the P&O method are:

- Oscillating around a constant MP voltage at very constant irradiance conditions leads to losses during suboptimal operating conditions besides MPP (a smaller voltage step could then be used).
- If process A is performed with starting voltage V1 below MP and V2 as shown, then the next voltage step point is in the wrong direction (this is solved by control or estimate of the irradiance); a similar problem occurs for process C around the MPP.

- Staying in submaxima at partial shading condition (this problem is solved by switching to larger steps of voltage changes).
- With very quick irradiance changes because of fast-moving clouds, the MPP tracker takes too long to find the MPP.

The modified P&O method was proposed to solve some of these problems by decoupling the PV power fluctuations caused by the hill-climbing process from those caused by irradiance changes [41]. This method adds an irradiance-changing estimate process in every perturb process to measure the amount of power change caused by the change of irradiance condition, and then compensates for it in the following perturb process. There are two operation modes named: Mode 1 for an estimate process and Mode 2 for the perturb process. Mode 1 measures the power variation due to the previous voltage change and irradiance change, and keeps the PV voltage constant for the next control period. Mode 2 measures the power variation and determines the new PV voltage based on the present and the previous power variations. The tracking accuracy is significantly improved by this estimate-perturb-perturb (EPP) method. The perturb process conducts the search over the highly nonlinear PV characteristic, and the estimate process compensates the perturb process for irradiance changing conditions. Further improvements are reached by estimating the expected slower voltage decrease after a fast and strong irradiance increase.

MPPT algorithm—incremental conductance

A very efficient, robust and fast tracking method uses the ratio of current and voltage and the differential conductance near the MP of the PV generator characteristics. The differential conductance is calculated by the small current change resulting from a small increase of the voltage and is reciprocal to the differential resistance. Additionally the absolute conductance, the ratio of the absolute value of current and voltage have to be calculated. In a very rough method only the sign of the difference of the absolute and incremental conductance may be used, positive below U_{mp} and negative above to control the PV generator voltage. It can be seen in Fig. 5.15 that the IV characteristic of a solar cell shows a smaller value of the differential conductance below V_{mp} compared to the absolute conductance. Only at the MPP does the difference between absolute and differential conductance diminish. The control strategy is to minimize the difference. This can be done using a constant period and constant voltage step as given in Table 5.5. This will result in an exponential convergence of the voltage to the optimum if the value of the step voltage is constant (see Fig. 5.20). Another advanced solution is to calculate the value of the voltage step as a function of the difference of both absolute and differential conductance values. Other faster and more efficient modified incremental conductance (IC) methods adapt the step voltage value depending on the difference of the absolute and incremental conductance.

Before using a micro controller, hardware analogue solutions of the IC method suffered from the ripple of the measured current and voltage and thus unrealistic jumps of the incremental conductance occurred in real outdoor operation. Similar problems exhibit at fast changes of irradiance. Still today a high quality high bit analogue digital converter has to be implemented to provide low error IC tracking results.

Table 5.5 **Principal of the IC MPPT algorithm based on the comparison of the absolute I/V and differential conductance $\Delta I/\Delta V$ of a solar cell characteristics near the MP voltage. V_{mp2} is the resulting proposition of the PV generator voltage for the next voltage step with a step size of V_{step}**

Sectors of the IV characteristic	Condition at Ump1: $I/V > \lvert\Delta I/\Delta V\rvert$	Next voltage: $V_{mp2} = V_{mp1} + \Delta V_s$
$V < V_{mp}$	Yes	$\Delta V_s = +V_{step}$
$V > V_{mp}$	No	$\Delta V_s = -V_{step}$

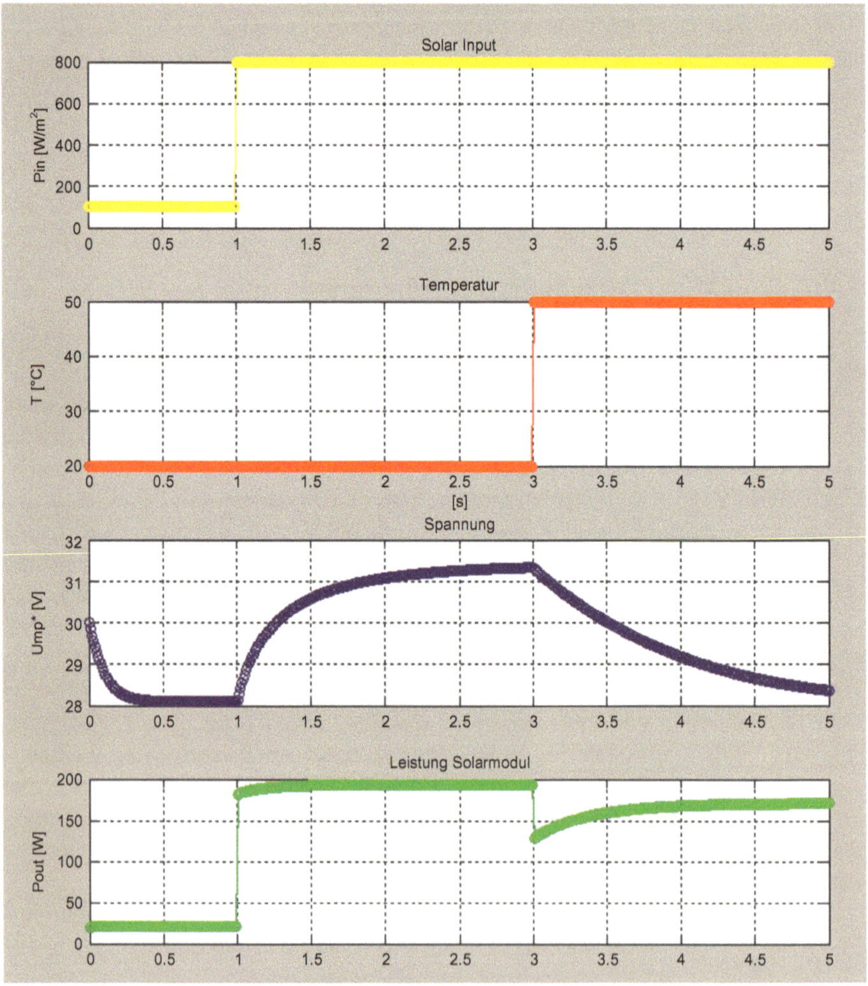

Fig. 5.20 Convergence of the IC MP Tracking method according to Table 5.5 for a step function of irradiance changes with the deviation to the optimum power of 200 W identified as the dynamic MPP tracking error (time scale in seconds).

MPPT algorithm—hybrids

More than six families of MPP tracking algorithms were reviewed by Onat [42]. The most powerful solutions in high-quality inverters are a combination of several methods implemented in the control software. To avoid, for example, being locked on a local maximum of the *I–V* curve, methods were developed to apply larger voltage steps in longer periods to test for the absolute maximum. In the IEC standards (see Section 5.5) procedures are defined to test the MPP tracking quality by defined ramps of irradiance for days with a fast change of cloudy and sunny conditions.

5.4 Basic power electronic concepts and losses

5.4.1 Basic principles and losses in power electronic circuits

The very first photograph advertising photovoltaic components shows a PV module by Bell Telephone Systems together with a lead acid battery and a smiling family back in the 1950s [43]. About 60 years later, innovative new products are coming onto the market offering to feed PV electricity into the grid and additionally offering the option to store electricity in so-called house storage batteries to use at night (see Section 5.5.2; Table 5.8).

A resistive load directly connected to the PV generator always finds changing operating conditions and thus the value of the load resistor needs to change to extract the optimum power at maximum power point. Such simple circuits are very cheap because they only use one switch to disconnect the load. However, directly connected applications deliver losses due to not operating the PV generator at MPP all the time, such as directly charging a battery. In the last decades, widely used PV chargers with typically 36 serial-connected crystalline silicon solar cells per module were installed to power standard 12 V nominal lead acid battery sets. These 36-cell PV modules were chosen due to the voltage drop in this stand-alone PV battery system from the MPP to the maximum charging voltage of close to 14 V on the battery terminals. Today's standard 72-cell modules would be appropriate to power 24 V lead acid systems in a similar approach for very simple and robust PV DC storage systems (Table 5.6).

DC/DC converters solve this problem by decoupling the voltage level at the PV generator from the output voltage at the load. For decades, switched power electronic

Table 5.6 PV inverter power classes for AC grid operation and worldwide market share [44]

Type of PV inverter	Power range	Efficiency	Market share
Central inverter	>100 kW	<98%	≈57.5%
String inverter	<100 kW	<98.5%	≈41%
Micro-inverter	<400 W		≈1.5%
DC/DC optimizer	Module range	98.8% without DC/AC	n.a.

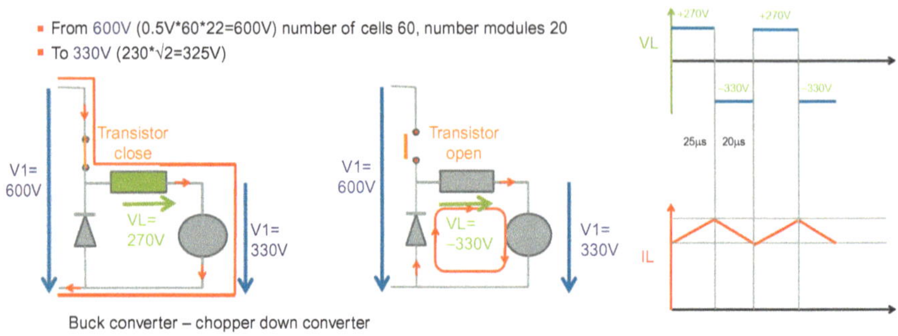

Fig. 5.21 Basic circuit of a DC/DC converter to transform 600 V DC voltage to a 330 DV level by PWM; in the right-hand picture, the increase and decrease of the inductor current is shown.

circuits were developed for up or down conversion for electrical drives, telecommunications and PV, decreasing the losses for the best products on the market from 8% in 1995 to about 1–2% today [45] (Fig. 5.21).

Typically the core of each switched circuit is to apply a different voltage level to an inductor, which carries the load current. A positive voltage for a short period of time is applied to the terminals of the inductor, leading to an increase in current followed by a small period with a negative voltage and a further decrease of the inductor current. By permanently adapting the length of the periods and the average value of the load current, the current of the PV generator is controlled. This method is called pulse-width modulation (PWM) and has been used since the beginning of semiconductor applications [46].

Different topologies are developed to place different semiconductor switches by means of appropriate DC voltage levels for the desired small period onto the inductor. This principle is also used in DC/AC inverters to directly connect the PV generator to the AC grid and thus control for a sinusoidal inductor current of 50 Hz (see Section 5.4.3).

5.4.2 Power electronic basics of grid-connected inverter

About 4600 PV inverter types have been offered on the international markets in the last few decades [47]. The lowest prices in 2014 showed the highly reliable central inverter at about 80€/kW and 110€/kW for the smaller string inverter used mainly for PV plants in buildings. Three phase string inverters account for one-third of the world market. The smaller micro-inverter, offered over the last two decades to directly connect a single PV module to the AC grid, has a very small market share, in the 1% range. Distributed DC/DC converters, coupling one or a few modules to track them at the optimum MPP, have gained a slight market share and are specially installed in buildings with a chance of partial shading conditions. Together with the DC/AC inverter, which groups several distributed DC/DC optimizers, the total efficiency is about 1–2% lower compared to a standard, highly efficient string inverter (Fig. 5.22).

Up to now, very little independent information about the reliability of these optimizers and their service life in real outdoor operation has been published. It is

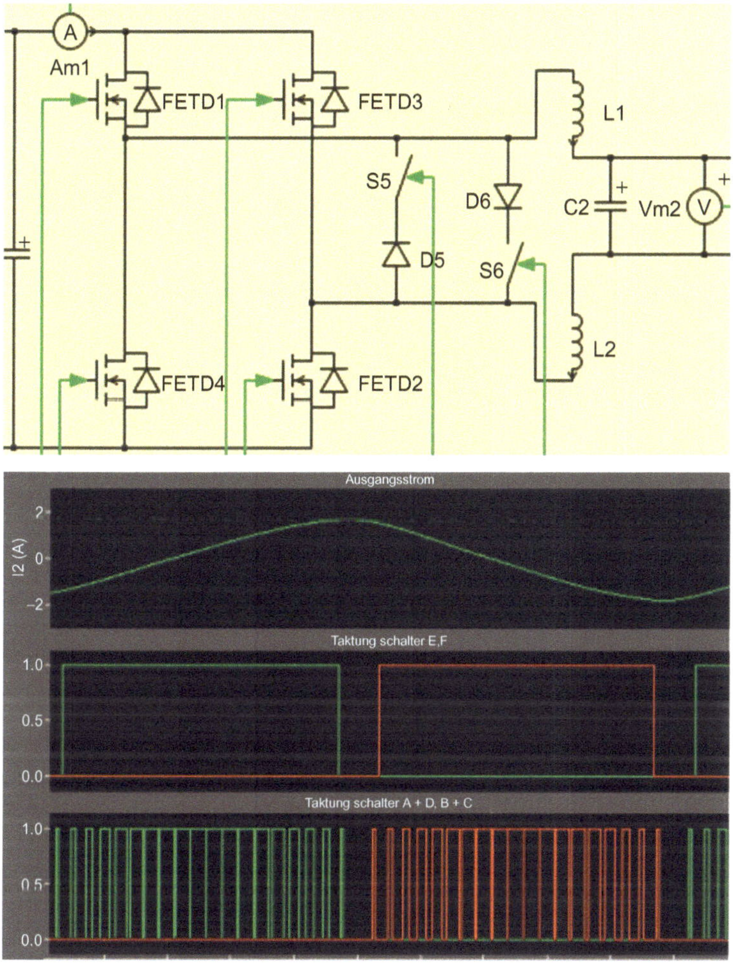

Fig. 5.22 Basic power electronic circuit of the ground-breaking HERIC topology invented in 2002. The graphs show the characteristic of the feed-in current into the AC grid over one grid period, while the second characteristic gives the switching intervals of 10 ms for semiconductor switches S5 and S6 and below the synchronous PWM signal of transistor pair FETD1 and FETD2, as well as FETD3 and FETD4. In total a minimum of only six power electronic switches are needed for this PV AC inverter topology. These simulations are performed by the use of PLECS software [48].

expected that the higher operation temperature of the electronic components used, such as the electrolytic capacitors mounted close to the PV module, will lead to a reduction of operation lifetime. There is a risk of significantly increasing O&M costs as a result.

A 1% increase in inverter efficiency improves the annual output of the PV plant by the same amount. Thus 1% higher total investment costs are allowable to reach the same levelized cost of electricity production. In that case, this higher efficiency inverter can achieve 10% higher prices per nominal power, due to the fact that the total cost share of the inverter is only 10% (see Fig. 5.1). These economic considerations have been the driver to improve PV inverter efficiency gradually from 90% in 1990 to 98% today.

In 2001 the average efficiency of a PV inverter smaller than 5 kW was 92% euro-efficiency (see Section 5.1) with a maximum efficiency of 96% for the transformer-less products, developed in the decade before [49]. These inverters offer the benefit of no galvanic coupling of the PV generator and the AC grid together with the disadvantage of about 3% higher losses in the transformer itself. Thus most of these transformer inverter systems were grounded at the minus pole of the PV generator, in accordance with the regulations in the United States at that time.

In May 2002 Heribert Schmidt applied for a patent [50] of his transformer-less HERIC topology, increasing the efficiency of the industrial products of the Sunway Company to 97% by using fewer components. Other related topologies such as the H5 from SMA followed, pushing the efficiency up to current values, slightly above 98% [51]. Twelve string inverters are in the 2014 top listing of a performance test survey with average European efficiency values between 97.5% and 98.5%, followed by another 117 products with tested efficiency values higher than 92% [52].

Transformer-less inverters with additional DC/DC converters also offer a large DC input voltage range, for example between 250 and 750 V. Some thin-film module products showed power degradation effects if the electrical potential of the generator dropped below earth potential. Thus some manufacturers of transformer-less PV inverters offered topologies equipped with an internal additional DC/DC converter to quarantee that the PV module was always above ground potential [53]. The downside of that approach is an approximately 2% lower inverter efficiency. In the last few years, the thin-film PV module manufacturers have improved their products and the latter type of inverter is not relevant on the market today.

However, string inverters offer new features to cut active power, for example, to 70% of nominal load if the grid operator requests that mode. Today this is, in most cases, the cheapest solution for a distribution grid operator facing voltage problems. As shown in Fig. 5.23, at a limitation of the AC power feed into the grid to 70% of nominal STC PV generator power, the annual losses are 4.4% for the location in Zürich, Switzerland.

The control of reactive power is another new feature of string inverters which has been implemented in the >100 kW inverter for several years, starting in the German markets. This allows the distribution grid operator to stabilize the local grid voltage. The arising design challenges for inverter manufacturers are to increase the value of nominal electrical capacitors and handle the higher current.

These recently implemented functional features of PV inverters are able to handle the feed-in of a high amount of PV electricity into a distribution grid of, for example, a small village. Thus, in such cases about 50% of the electricity consumption could be delivered by the PV inverters without additional investment in expensive smart grid technical solutions [54].

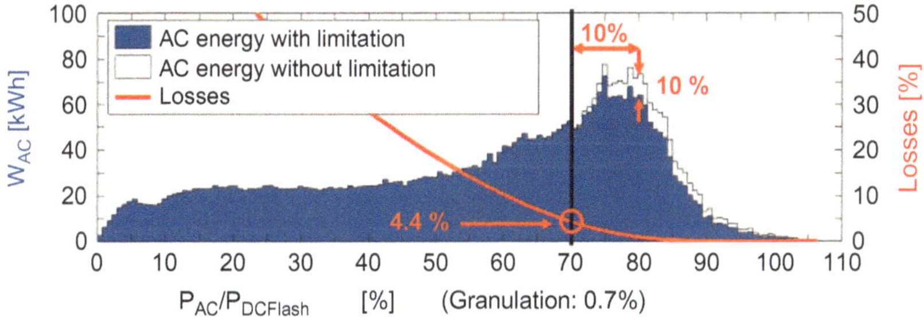

Fig. 5.23 Density plot of PV electricity production without limitation (*white* graph) and with power curtailment given as a ratio of AC power limit to nominal DC power PD_{CFlash} together with the total annual losses at certain values of power curtailment (*red* curve) for the location of Zurich, Switzerland [23].

A few inverter suppliers offer products on the market able to control a heat pump and the charging process of electrical vehicles using a weather algorithm at nearly no additional cost.

5.5 Results of BOS-related performance aspects

5.5.1 Performance of grid-connected PV systems

DC/AC inverter efficiency

The ratio of the DC input power reduced by the inverter losses and the DC input is called the conversion efficiency. Losses in power electronic devices may be separated into switching losses that are proportional to the inverter power P, resistive losses that are proportional to the square of the power and losses that are considered to be constant. Due to these different sources of losses and loss characteristics relative to the PV generator power and voltage, the inverter efficiency is not constant. The power of the sun changes during the day and thus the inverter efficiency depends on the individual share of the different loss mechanisms in the inverter. The measured losses of the PV inverter may be fitted to the following equation to get the relevant efficiency for a given power (see Fig. 5.24).

$$\eta_{\text{conv}} = \frac{P_{\text{ac}}}{P_{\text{dc}}} = \frac{P_{\text{dc}} - P_{\text{losses}}}{P_{\text{dc}}}$$

$$P_{\text{losses}}(P) = a_0 + a_1 P + a_2 P^2$$

The resulting fitting parameters a_0, a_1 and a_2 may be attributed to different loss mechanisms. The resistive losses are produced by semiconductor switches or inductors, the switching losses again by the power transistors and diodes and the constant losses come typically from the power supply and microcontroller or internal fan. Back in 1991, Hotopp proposed weighting factors to calculate the annual average inverter efficiency [56]

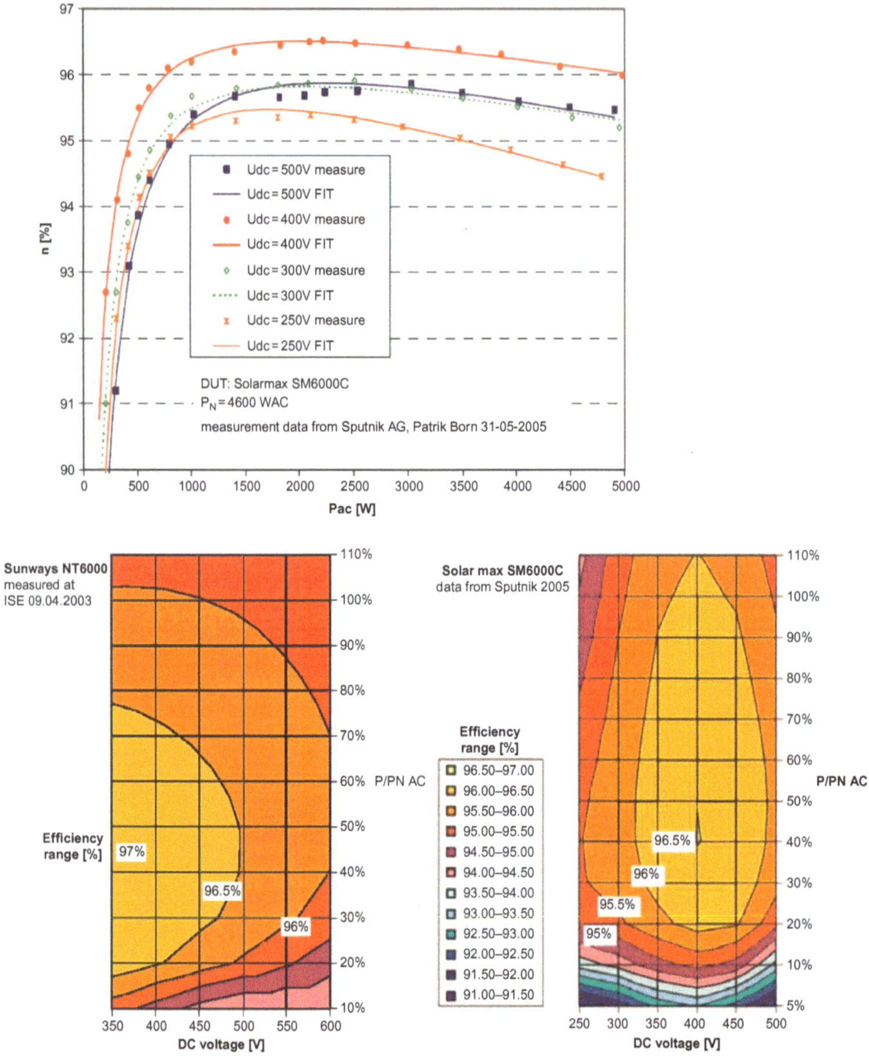

Fig. 5.24 The efficiency of a PV grid-connected inverter is a function of power but also applied DC voltage from the PV generator, as can be seen in the top graph of efficiency versus power with the DC voltage as a parameter or in the mapping graphs of efficiency in the power-DC voltage plane [55].

$$\eta_{\text{conv,stat,EURO}} = 0.03\eta_{5\%} + 0.06\eta_{10\%} + 0.13\eta_{20\%} + 0.10\eta_{30\%} + 0.48\eta_{50\%} + 0.20\eta_{100\%}$$

They are multiplied with the measured efficiency of the inverter under test in the laboratory at 5%, 10%, 20%, 30%, 50% and 100% of nominal power. The weighting factors were identified for the region of the Rhine valley in northern Germany, but have been established as a standard for the whole of Europe since then and are called the Euro efficiency value. Other weighting factors were proposed to have some additional values between 50% and 100% but the preceding coefficients are now established as the relevant norm [57] for testing inverters. The California Energy Commission (CEC) also defined weighting factors appropriate to the irradiance conditions at their locations:

$$\eta_{\text{conv,stat,CEC}} = 0.04\eta_{10\%} + 0.05\eta_{20\%} + 0.12\eta_{30\%} + 0.21\eta_{50\%} + 0.53\eta_{75\%} + 0.05\eta_{100\%}$$

These conversion efficiencies are measured in the laboratory at a static operation point of DC input and AC grid output and thus have the index $\eta_{\text{conv,stat}}$. Switching losses will increase with the DC voltage level. In the field the inverters are not operated just at a single DC voltage level, and thus $\eta_{\text{conv,stat}}$ also has to be measured at several DC voltage levels within the MPP range. Hence, the coefficients of the power losses a_0, a_1 and a_2 each must be enhanced by a polynomial term as a function of the DC voltage V:

$$\begin{aligned} P_{\text{losses}}(P, V) = {} & \left(c_{0,0} + c_{0,1}V + c_{0,2}V^2 + c_{0,3}V^3\right) + \\ & + \left(c_{1,0} + c_{1,1}V + c_{1,2}V^2 + c_{1,3}V^3\right)P + \\ & + \left(c_{2,0} + c_{2,1}V + c_{2,2}V^2 + c_{2,3}V^3\right)P^2 \end{aligned}$$

Ten years ago only one value of the Euro efficiency was relevant on the PV market for most of the installers and planners. Knowledge about the real efficiency, which also varied up to 3% with the DC voltage, was missing for products on the market at that time. It was evident that 2% higher efficiency affects the inverter prices significantly. At about 20%, the share of the inverter is about one-tenth of the total costs. In 2004, this voltage dependency was published [58] and a year later the voltage mapping as shown in Fig. 5.23 was published at a specialist photovoltaic conference.

In the same year, *Photon* magazine published, in their annual PV inverter market survey, Fig. 5.24 together with other detailed technical data of 234 inverters on the German market [59]. Several requests followed from planners and installers to the suppliers of the inverters regarding the DC voltage chosen to calculate their Euro efficiency value given in the data sheet. Several suppliers started to present the Euro efficiency, for example, at three DC voltage levels. *Photon* magazine started to discuss the performance including the power voltage mapping of a single inverter per issue and offered further detailed data to the customers. In the published ranking list of inverter Euro efficiencies, the resolution is given as 0.1%. It should be taken into account that the measurement uncertainty of the efficiency in the laboratory is typically between 0.6% down to 0.2%, at a confidence level of 95%, for high-quality equipment and an experienced operator [60].

Today, the relevant norms propose efficiency measurements at different DC voltage levels. The voltage variations of that static conversion efficiency tend towards values below 1% due to the higher overall efficiency level.

The European Standard EN 50530 [61] provides a procedure for the measurement of overall efficiency of grid-connected photovoltaic inverters. It includes the definition of a procedure to measure the static conversion efficiency and also dynamic efficiency, describing the accuracy of the MPP tracking (MPPT) of inverters. Several publications discuss the approach to total efficiency as presented in the following [39,62,63].

The overall efficiency η_t is defined as the ratio of the AC power output and the theoretical maximum available DC input power at MPP within a defined measuring period T_M; this is also valid for changing MPP values of voltage and current:

$$\eta_t = \frac{\int_0^{T_M} p_{\text{ac}}(t) \cdot dt}{\int_0^{T_M} p_{\text{mpp}}(t) \cdot dt}$$

The measurements of the final η_t are separated into the two categories of static conversion efficiency and dynamic MPPT efficiency:

$$\eta_t = \eta_{\text{conv}} \cdot \eta_{\text{MPPT}}$$

The static conversion efficiency η_{conv} is measured as the ratio of the AC energy output to the DC energy input over a measurement period of 10 min. Measurements must be taken at 5%, 10%, 20%, 25%, 30%, 50%, 75% and 100% of nominal power at three voltage levels [63] (see Section 4.3.1 in Ref. [63]). There are more sampling points of power compared to the Euro efficiency weighting factors. But in an annex to the standards the previous old weighting factors are given and the resulting Euro and CEC efficiency values finally have to be calculated.

$$\eta_{\text{conv}} = \frac{\int_0^{T_M} p_{\text{ac}}(t) \cdot dt}{\int_0^{T_M} p_{\text{dc}}(t) \cdot dt}$$

The MPPT efficiency η_{MPPT} is measured as the ratio of the energy drawn by the device under test to the energy provided theoretically by the PV simulator for operation at MPP.

$$\eta_{\text{MPPT}} = \frac{\int_0^{T_M} p_{\text{dc}}(t) \cdot dt}{\int_0^{T_M} p_{\text{mpp}}(t) \cdot dt}$$

where $p_{\text{mpp}}(t)$, $p_{\text{dc}}(t)$ and $p_{\text{ac}}(t)$ are the instantaneous values, respectively.

The given selected input parameters and definitions are necessary.

V_{dcmin}/V_{dcmax} is the minimum/maximum input voltage; $V_{dc,r}$ is the rated input voltage; V_{mppmin}/V_{mppmax} is the minimum/maximum MPP voltage; $I_{dc,max}$ is the maximum input current; $P_{dc,r}$; rated input power (or derived from AC power $P_{dc,r} = P_{ac,r}/\eta_{conv,min}$)

The requirements for the PV array simulator and the other test environment necessary for the measurements are defined in the standard. This includes tables of the required fill factor and voltage levels at MPP for the PV simulator distinguished between crystalline silicon and thin-film cell technologies.

To test the MPP tracker, ramp gradients ranging from 0.5 up to 100 W/m^2/s are used for the tests with two different irradiance levels, from 100 to 500 W/m^2 and from 300 to 1000 W/m^2.

PV inverter and PV generator sizing

One of the first tasks in the PV plant design is to find the correct number of serial connected PV modules to fit with the MPP tracking voltage range of the inverter. If the temperature coefficient of the PV module is not taken into account and the nominal STC module voltage was used only to reach the lower inverter MPP voltage range, very high power losses will occur in summer (see Fig. 5.25). This will limit the lowest possible voltage of the MPP inverter tracking to an individual module voltage of about 15% higher than the optimum V_{mp}. As was discussed in Section 5.3 (see Fig. 5.15), this much higher operation voltage relative to V_{mp} will result in roughly 50% losses compared to MPP.

Another output of the mapping is the appearance of a constant voltage fixed tracking, as can be seen in Fig. 5.25 in the maxima at low power and 97% of STC string voltage in the right-hand plot.

Quality of MPP tracking

Especially at irradiance ramps of larger than 10 W/m^2 per second, significant tracking losses are found [65]. Still today not all of the inverter manufacturers deliver products with robust MPP trackers, especially those suppliers who do not have many years of experience in the PV field (see Fig. 5.26).

The detailed test reports published by some technical photovoltaic magazines and performed in high-quality test laboratories provide an independent measured value of the Euro efficiency, which in most cases is in very good accordance with the manufacturer data sheets [66]. There is often helpful information as to whether there are problems regarding the practical quality of the MPP tracker, which otherwise a single user would be unable to detect. Again, in most of the reports, the MPPT efficiency is very high and overall efficiency is equal to the static conversion efficiency. But in some cases, these tests find faulty operating conditions. For example, an automatic disconnection from the grid was detected due to reducing the power factor from 1 to 0.7, and thus the resulting current value was above the shut-off limit [67].

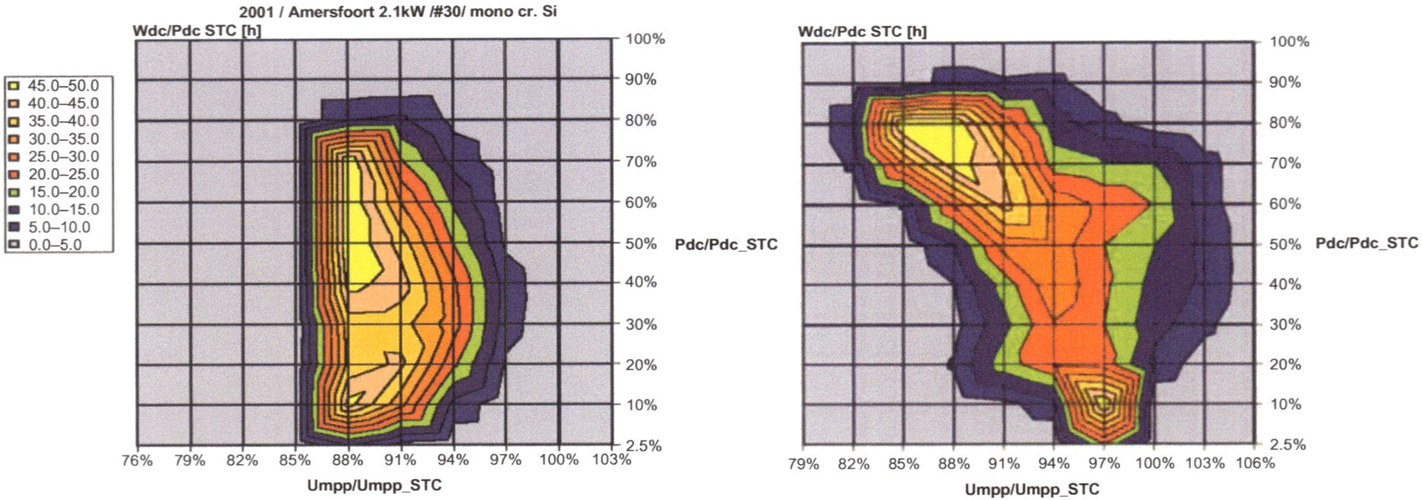

Fig. 5.25 Monitoring of PV plant operation in a summer month and analysis of the performance by voltage-power mapping. The colours give the measured amount of energy collected in each raster element defined by the interval of the DC voltage given in percent of nominal STC module voltage and the produced PV power given relative to the nominal STC PV module power. The left-hand plot shows a plant in the Netherlands with a clearly faulty plant design, having too few modules connected in series and thus operating the PV generator always at the lower limit of the inverter MPP voltage range. For comparison, the right-hand plot shows a well-designed plant in Switzerland, with the colours showing the amount of PV electricity produced in nominal hours Wh/Wp in that operating point respective raster element [64].

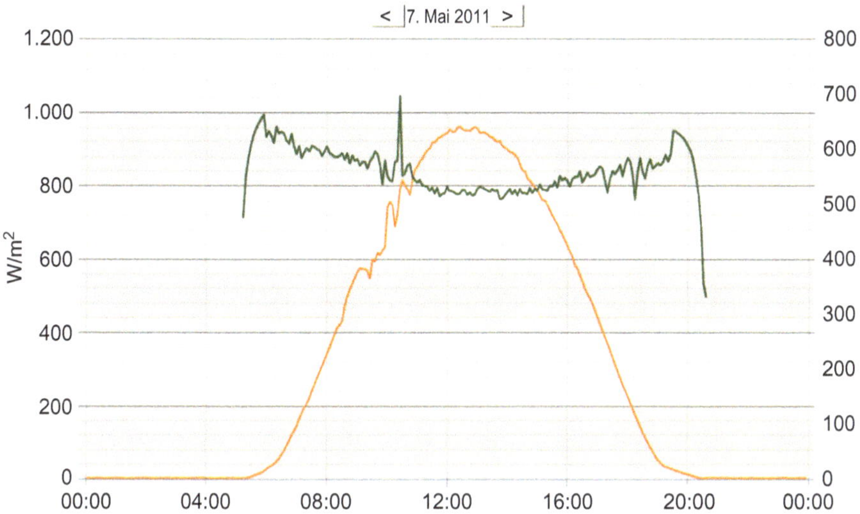

Fig. 5.26 Real outdoor monitoring results of measured irradiance (*orange* plot) and generator DC voltage (*green* plot) of a string inverter in a MW plant showing a voltage peak at about 10 am, where the absolute value of irradiance is not changing significantly but the voltage change leads to a dramatic drop of PV generator power (see Fig. 5.15); thus, a nonoptimized MPP tracker implementation is suspected here.

Performance and wiring losses

A very common design approach is to limit the ohmic losses of the DC wiring to 1% when the PV generator is operated at STC power. Due to annual distribution of the produced DC power (see Fig. 5.23), the final ohmic losses over one year depends on the location and is roughly between one- or two-thirds higher than the ohmic losses at STC. To control this in real outdoor operation, mapping of the V_{mp} may help (Fig. 5.25).

5.5.2 Performance of grid-connected and storage PV systems

The very first advertising photograph of a photovoltaic module was presented as the 'Bell Solar Battery' together with a lead-acid battery by the Bell Labs back in 1954 [68]. Today this combination has become more important for improving the rate of self-consumption and unburdening the electrical distribution grid.

In Germany about 30,000 PV AC systems equipped with battery storage were in operation at the end of 2015, mainly in private households [69]. Thus, PV electricity could be applied via the battery even after sunset. About a 59% share of global storage installations were accounted for by the United States, Japan and South Korea in 2015 and 2016, and the UK will add 1 GW of storage facilities by 2020 [69]. Some of the storage market actors, claiming cost parity for large storage in 2015 in the UK, were supported by a feed-in tariff (FIT); others are expecting cost parity for smaller residential building solutions till 2017 [70]. A detailed survey of the German markets in 2015 based on 75 products resulted in storage costs of 5 kWh in the range of 0.3–0.7€/kWh and 0.2–0.6€/kWh for 10 kWh systems without subsidy [71].

These residential PV storage systems have been under intensive development during the last decade [72–75]. Reliable data on the performance of these grid-connected PV storage devices, as well as the quality of control, will be expected in the coming years, due to the fact that in Germany today half of the installed devices have been in operation for only one or two years.

Fig. 5.24 reflects the corresponding energy transfer which has to be controlled by the central power electronic conditioner. It includes the PV DC/AC inverter (see Sections 5.4.1, 5.4.2 and 5.3) as well as the battery charge controller.

To reduce this power electronics complexity, recently devices have been launched to operate only in the electricity exchange between the AC grid and the battery. This subsystem working in parallel with the PV system circuits is illustrated in Fig. 5.27 (see energy flow marked *E7*). Leading PV power electronic companies started the market introduction of such AC battery storage systems in 2016, heading for the potentially large retrofit markets represented by existing PV single-house system owners with numbers of roughly one million in each of the countries of Germany, the UK and Australia [76]. The installation costs of such retrofits are reduced and they are open for several battery products on the markets.

The definitions in Table 5.7 are widely used to describe the several energy flows shown in Fig. 5.24. The first one is the self-consumption rate (SCR), which specifies the share of PV electricity that is either used immediately or via the battery. Without the use of the battery the key figure is called the immediate self-consumption rate

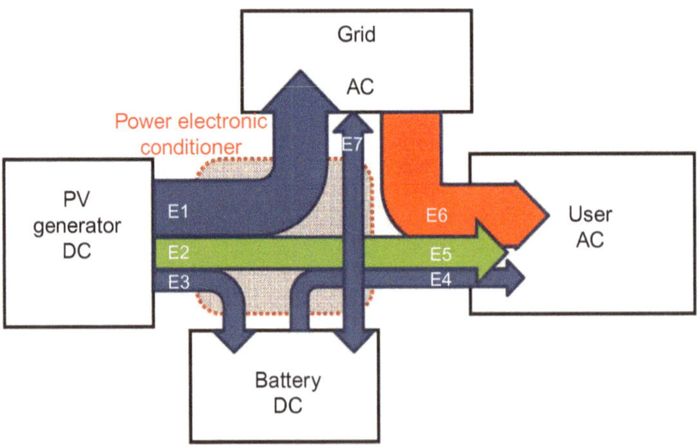

Fig. 5.27 Schematic of energy flow between PV generator, AC grid, battery storage and the user's plug controlled by the power electronic conditioner used to supply, for example, a single family house.

Table 5.7 Definition of key figures to describe the energy flow, typical on a yearly interval [77,78]

Key figure abbr.	Key figure	Formula (see Fig. 5.24)	Description
PVG	PV grid	E1	PV electricity feed into the public AC grid
PVS	PV stored	E4	PV energy consumed via the battery storage
SCR	Self-consumption rate	(E4+E5)/(E1+E2+E3)	Ratio of consumed PV energy to total PV production
ISCR	Immediate self-consumption rate	E5/(E1+E2+E3)	Ratio of immediate consumed PV energy to total PV production without storage "natural self-consumption"
SSR	Self-sufficiency ratio	(E4+E5)/(E4+E5+E6)	Ratio of total load demand covered by the PV battery system
PSR	Prosumer ratio	(E1+E2+E3)/(E4+E5+E6)	Ratio of produced PV energy to total energy consumption
SCPVh	Storage capacity PV hours	C/Pn [kWh/kWp], [h]	Ratio of used nominal battery capacity to nominal PV STC power

(ISCR) as given in Table 5.7. The self-sufficiency ratio (SSR) is given by the share of PV consumption to total demanded consumption. If the total amount of annual produced PV electricity equals the total consumption, the prosumer ratio (PR) becomes equal to 1. Even if PR becomes 1, electricity will be fed into the public grid due to the fact that the correlation between PV power generation and energy demand is not equal for each second during a year.

The annual electricity consumption in a single family house of 4500 kWh may be produced by a PV installation of about 4.5 kWp nominal power in most parts of Central Europe. Nevertheless, about two-thirds of the electricity consumption in the house is typically needed at periods with no PV production, for example in the evening. The real example in Fig. 5.28 shows a typical sunny day where nearly all the electricity consumed in the house was produced from PV and/or used after storing in the battery. On this day the self-consumption rate is 98%. Only in the early morning the battery is nearly empty and the needs of the coffee machine have to be powered by the public grid. This house is equipped with an 8 kWp PV system mounted on the rooftop and oriented to the west. The battery has a nominal power of 4.6 kWh (48 V lithium ion system).

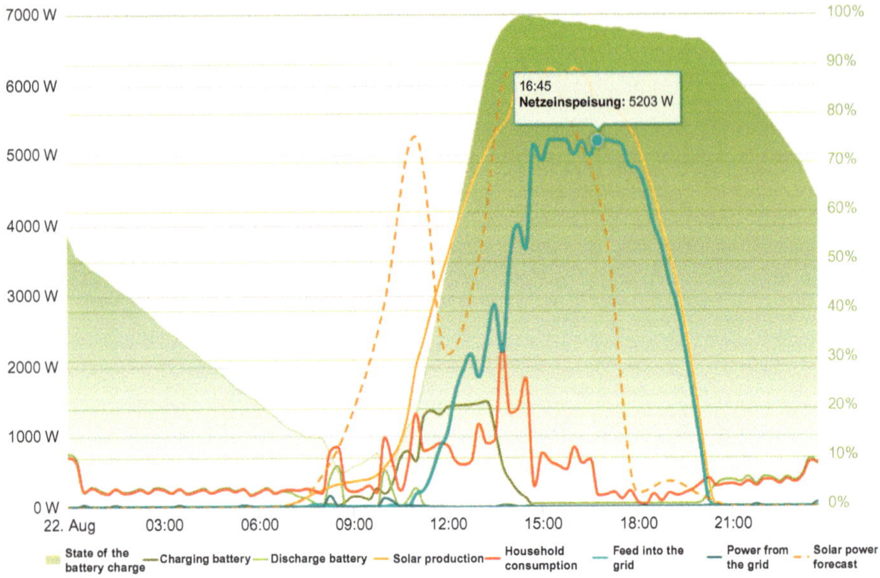

Fig. 5.28 Monitoring data of daily characteristics of 22 August 2015 with an 8 kWp PV rooftop system (PV production on this day E1+E2+E3=45.0 kWh with E1=30.93 kWh feed into the public grid) is shown together with the state of battery charge (green area E3=4.6 kWh loaded and E4=4.3 kWh discharged this day) and red being the total electricity consumption (E4+E5+E6=10.8 kWh this day and 4500 kWh annual demand), for a single family house where two adults and two children are living in the south of Germany. Only 0.2 kWh electricity (E6) comes from the AC grid due to the coffee machine in the morning at low battery capacity [79].

It is obvious that doubling the PV power rating will improve the rate of self-consumption only marginally. The findings of several simulation runs of a standard residential load profile powered by different sizes of PV nominal power and battery capacity are given in Table 5.8. A simple PV battery charging algorithm was used. If PV production exceeds total consumption the difference was charged into the battery. The battery is discharged only to feed the load if PV production is too low.

Starting with a set-up to reach a prosumer ratio of PR = 1, about 56.5% (SSR) of the consumed electricity comes from PV at moderate storage costs of 0.51€/kWh. Thus 43.5% of PV electricity has to be fed into the public grid. By the use of the battery with a capacity to deliver the amount of nominal PV power for 1 h (SCPVh = 1 h), the immediate self-consumption of the PV production (of ISCR = 34.2% without a battery) is increased by 22.3%, to reach a total SCR of 56.5%.

Doubling the battery size increases SSR to 63.1% but also increases storage costs to 0.80€/kWh due to only 163 annual nominal capacity battery cycles during 10 years of operation. Doubling the PV power (PSR = 2) and reducing the battery size to only 0.5 kWh/kWp (SCPVh) leads to even a slightly higher SSR of 65.8% at very low storage costs of 0.46€/kWh. Thus, the PV self-consumption drops to SCR = 32.9%,

Table 5.8 Simulation results of key figures according to Table 5.7 based on measured PV production data of a PV rooftop installation (Germany) [80] in 2012 (annual yield of 1131 kWh/kWp) and a measured typical single-family house in Switzerland with a consumption of 5511 kWh/a, performed at 15 min intervals. The simple cost estimates are calculated on 1300€/kWh installation costs including power electronics and an operating duration of 10 years for the battery system without financing costs or incentives. (Available rather than nominal kWh capacity values are used, with round-trip charging efficiency of 95%.) [81]

PV sizing PSR	Battery sizing SCPVh (h)	Self PV consumption ISCR	From storage SCR-ISCR	Self-sufficiency SSR	Annual usage battery cycles	Costs storage €/kWh
1	1	34.2%	22.3%	56.5%	253	0.51
1	2	34.2%	28.9%	63.1%	163	0.80
1	0.5	34.2%	13.6%	47.8%	307	0.42
2	1	20.3%	16.8%	74.2%	190	0.68
2	0.5	20.3%	12.6%	65.8%	286	0.46
0.8	2	39.8%	31.7%	57.2%	179	0.73
1.2	0.5	30.0%	13.5%	52.2%	305	0.43
1	20	34.2%	34.8%	69.0%	20	6.50

increasing the cost of PV electricity production because two-thirds must be fed into the grid if the feed-in tariffs are low.

This example shows that, depending on the specific battery investment costs, battery cycles and battery lifetime have to be balanced with PV investment costs, the system's SCR and the feed-in tariff to reach an economic optimum at high SSR values. Today, with still relatively high specific storage costs, lower SCPVh design parameters are recommended.

Ref. [82] gives a detailed cost assessment for the German funding and feed-in regime results of a PV battery system on the limit of profitability to the 0.34€/kWh household electricity price for a system design with PSR = 1 and battery capacity of SCPVh = 0.6 h.

However, in the coming years, with higher cost reductions in storage compared to PV production, higher values of SCPVh and lower PSR values will be seen in the markets, and also a higher SSR will be reached. Additional loads like EV charging may change the key figures significantly.

Today PV battery markets are beginning to be developed and some of the first customers are premium option leaders where the costs are not the only issue to place their investments. Thus the battery capacity is designed to reach an SSR close to 1 during the summer period. This is illustrated in Fig. 5.29 by the simulated characteristics in one week in June of two PV battery systems, the second with twice the battery capacity.

Similar to the measured conditions given in Fig. 5.30, a daily simulation run performed (Table 5.8) results in about half of the capacity left in the battery in the early morning after that clear-sky day before. It should be mentioned that half the energy demand is at very low levels, below 0.15 kWh per 15 min interval, as is illustrated in the density plot of Fig. 5.31. Thus, it can be recommended that the most economical solution is first to reduce the load at night after midnight and later on to choose the appropriate battery capacity.

5.6 Reliability and durability

5.6.1 PV module reliability

During the 1970s, the terrestrial PV modules faced a lifetime of only 5 years. Significant advances were made to overcome the relevant failure mechanisms like encapsulant degradation and delamination and cell interconnection fatigue. In 1982 Mon and Ross at the Jet Propulsion Laboratory in Pasadena, California developed a systematic test approach in the form of an accelerated climate chamber test to speed up the development process to increase module lifetime to 20–30 years. For example, they proposed a thermal cycling between −40°C and 90°C lasting only 430 cycles to exhibit the same fatigue effect to the copper-based interconnectors as for a real outdoor field exposure of 30 years at a much lower temperature difference of only 46°C, representing more than 10,000 real cycles [82].

Today, the same principles are applied worldwide during the crystalline silicon terrestrial photovoltaic (PV) module design qualification and type approval test,

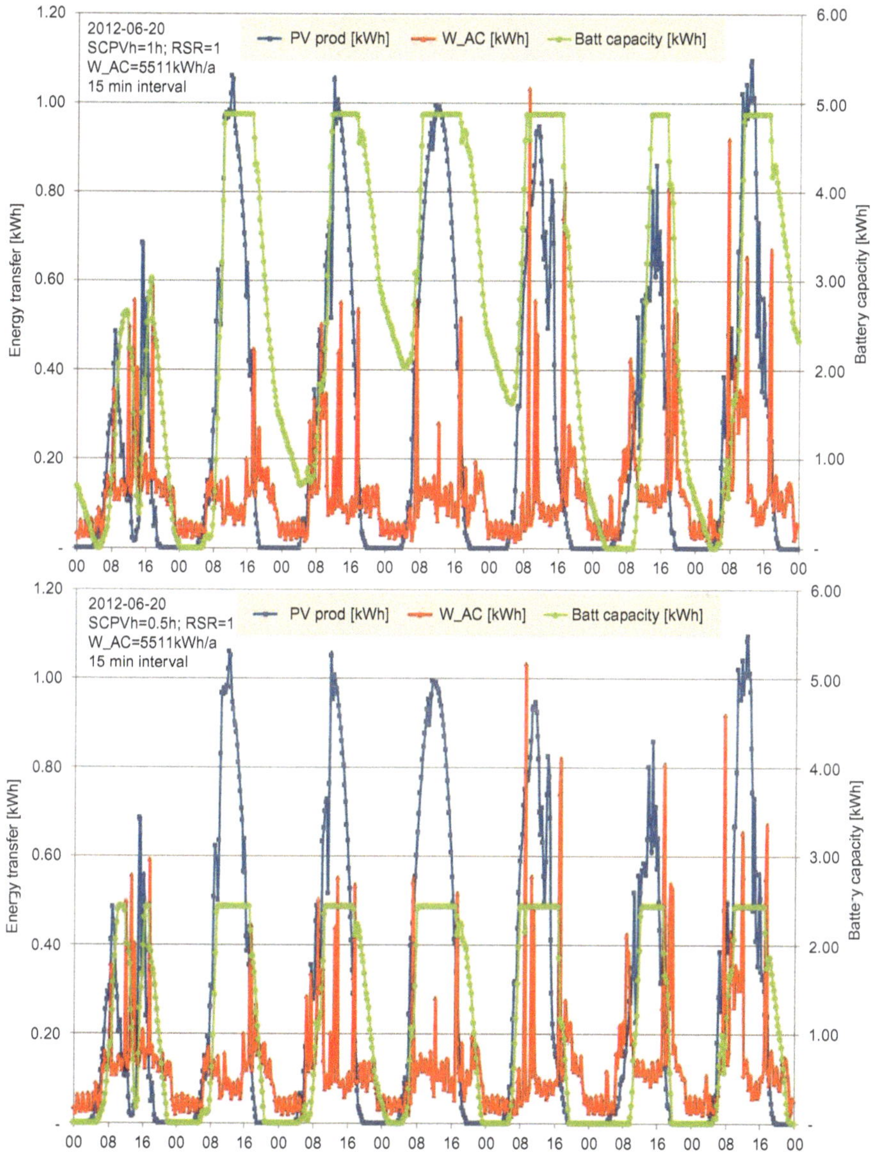

Fig. 5.29 Simulation results of the characteristics of total PV production and total consumption together with the state of charge of the battery storage for two different storage sizings of SCPVh = 1 h and SCPVh = 0.5 h (see Table 5.8 at PSR = 1); the data point markers represent the amount of energy transferred in that 15 min time interval, and the line in between is only indicative.

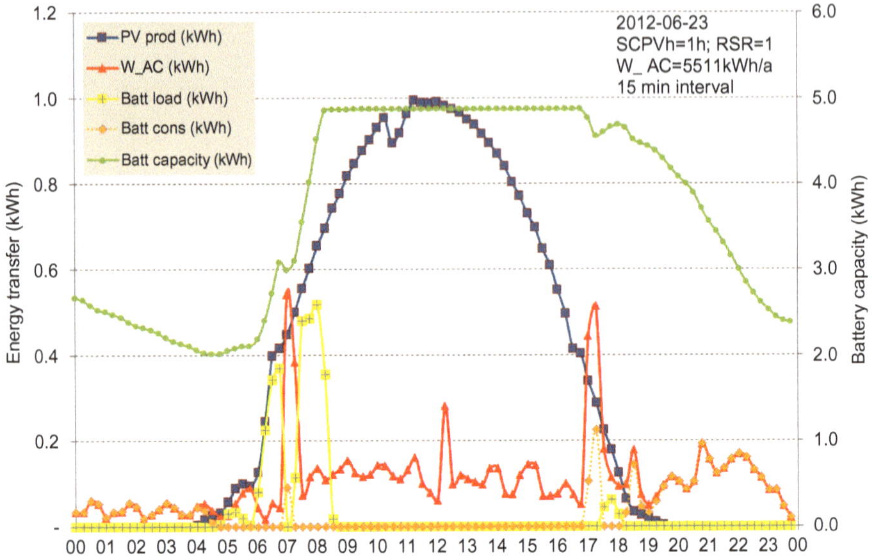

Fig. 5.30 Daily characteristic of energy transfer of the PV battery system including clear-sky PV production total load and charging and discharging of the batteries (storage sizing of SCPVh = 1 h at PSR = 1 corresponding to annual key figures; see Table 5.8).

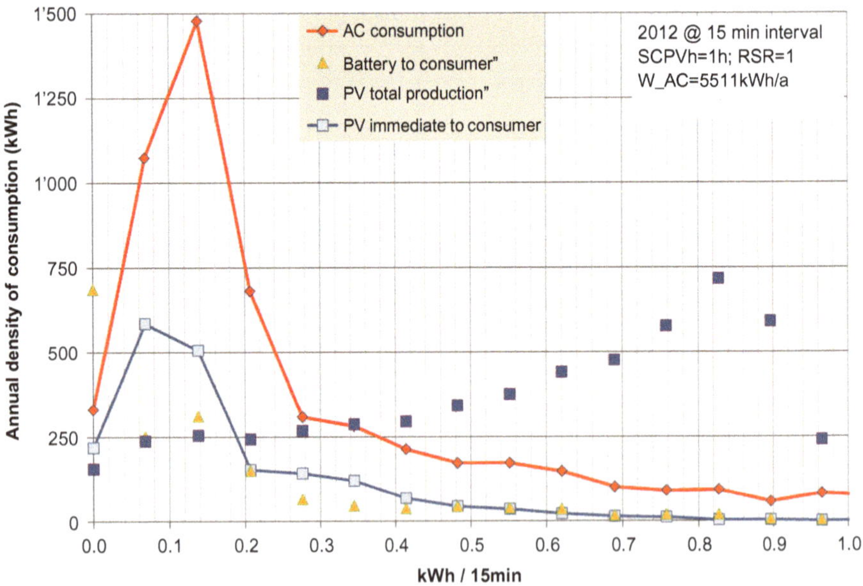

Fig. 5.31 Density plot of the annual total electricity consumption, the PV total production and the energy deliverd by the battery analysed on 15 min intervals (storage sizing of SCPVh = 1 h at PSR = 1 corresponding annual key figures see Table 5.8).

according to IEC 61215. Their test sequence 10.11 was established to determine the ability of the module to withstand thermal mismatch, fatigue and other stresses caused by repeated changes of temperature applied to two modules under test. This test lasts two and a half months by continuously cycling the temperature between −40°C and +85°C over a single period of 9 h, summing up to 200 cycles. The test is successfully passed if the degradation of maximum output power does not exceed 5% of the value measured before the test. These test sequences mainly verify the copper-based intercell connectors.

Upcoming new intercell connectors such as smart-wire technologies have shown much better reliability test results of power degradation below 5% even at 400 thermal cycles [83]. Further outdoor tests have to be applied before a higher operating lifetime of these types of PV modules may be concluded.

Additionally, another test, 10.13, is applied to two more modules of the same type, which are required to show a power degradation less than 5% after 1000 h of damp heat test at 85°C and 85% relative humidity. The focus of this test is the ability of the encapsulation of the module to withstand the effects of long-term penetration of humidity. To test the ability of the module to withstand the effects of high temperature and humidity followed by subzero temperatures, the humidity-freeze cycle was developed with 10 cycles with the same temperature as the 10.11 sequence, each having a period of 20 h at 85% relative humidity. Another mechanical test sequence, 10.16, determines the ability to withstand wind, snow, static or ice loads by applying 3 cycles of 2400 Pa uniform load, applied for 1 h to front and back surfaces in turn. This load corresponds to a wind pressure of 130 km h^{-1} with a safety factor of 3 for gusty winds (see Fig. 5.14). As an option for high snow load conditions, 5400 Pa static test conditions are applied during the last front cycle.

Thin-film modules have to pass another type of approval test, IEC 61646. With this, thermal cycling is not critical because thin-film modules typically do not use similar intercell connectors to standard crystalline cells. But other effects of light-induced degradation and UV testing are more relevant to failure mechanisms.

Some of the PV MW plant developers have set their own test sequences, using a strengthened climate chamber test. They simply increase the number of cycles for the 10.11 test sequence of the IEC 61215 and/or increase the duration of the damp heat test 10.13 by factors of two or four.

Long-term performance tests over 25 years at the European-Joint Research Center in Ispra confirmed that 82% of the 204 analysed modules have a final maximum power greater than 80% of the initial power and meet the manufacturer's warranty criteria [84]. Two-thirds of the modules showed output power values even higher than 90% of nominal power after 25 years in outdoor exposure.

5.6.2 Inverter reliability

Today's PV inverter products are typically offered with a standard 5-year warranty period with an extra charge of about one-third to expand the warranty to 15 years. In most of the operation and maintenance calculations the replacement of today's PV inverter is expected in 13 years, accounting for about 0.5% annual costs relative to the initial plant investment expenses [3].

In a survey performed in 2015 in the United States, the following vulnerabilities or reliability issues of PV inverters were found: electrical transient and grid events, dust and moisture, temperature cooling, components like fans, IGBTs, DC capacitors, boards, interconnects and loss of communication [85].

One critical component limiting the lifetime of the PV inverter is the aluminium electrolytic capacitor used in the decoupling of the AC current ripple of the PWM method from the PV generator. Due to evaporation of that electrolyte investigated in a 2000 h accelerated life test (ALT) at 105°C, the failure rate at lower operation temperatures is smaller by the acceleration factor AF, according to the well-known Arrhenius rule ($\mathrm{AF}(T) = \exp[(E_a/k_b)(1/T_{ref} - 1/T_{used})]$) proportional to the ratio of failure rates with E_a the activation energy, k_b the Boltzmann constant, T_0 the absolute temperature in K of operation and T the accelerated testing temperature [86,87].

The thermic reaction speed is reduced and the evaporation of the electrolyte is roughly halved by reducing the temperature every 10°C. Several inverter manufacturers performed ALT at 50°C of the completed device for two calendar months permanently operated by solar simulators on the DC input [88].

Overvoltage may result in damage to the oxide film of the electrolytic capacitor, which then leads to a failure in the isolation layer and a loss of capacitance due to the anode foil as the dominant failure mechanism, decreasing capacity and increasing thermal losses [89].

To avoid this complex failure mechanism of the cost effective electrolyte capacitors, some manufactures have changed to the technology of foil capacitors, produced without electrolyte but at higher prices per capacity.

A few of the string inverter suppliers are using foil capacitors, such as Schneider Electric with their TL20000 E, and thus higher operating lifetime is expected but not explicitly stated in their advertising data sheets [90].

Some of the manufacturers assume a useful life of 10–15 years using electrolytic capacitors (mean time to failure (MTTF) = 2.7 million h) and 20–30 years for film capacitors (MTTF 2.6 million h) [91].

Concepts to assess the power electronics reliability of AC modules have been discussed, focusing on the same life cycle tests used for PV modules, according to IEC 61215, and damp heat tests at 85°C but for 2000 h [92]. Among the very scanty information about real tests of PV module inverters, TUEV reports of a still-working PV module inverter after 17 years of outdoor exposure in Arizona in the United States [93].

Only three inverters mounted on the back of the PV module have been tested, so that no statistically significant conclusion will be achieved. The type of inverter mounting solution is important, especially the air gap to the PV module back sheet, which is reported to be close to a maximum temperature of 95°C on this rooftop installation. The lack of information about the results of such reliability tests is expected to be replaced in the coming years by a broader techno-economic discussion driven by the increasing market share of microinverters and power conditioners.

5.7 Trends and Outlook in BOS

In 2009 the research group of ISE Freiburg exceeded the 99% maximum efficiency level by implementing a SiC power transistor in a laboratory scale PV inverter

[94]. The remaining margin in improving Euro inverter efficiency on industrial products above 98–99% is limited by the costs of low-loss devices based on wide bandgap (WBG) semiconductors such as SiC or GaN. Reducing power losses also diminishes space, packaging material and thus the overall system costs of the inverter. Lower losses and lower maximum temperature of the critical power electronic components on the other side improves the service life of the system and thus the final generation costs of PV electricity.

In 2014 Google and the Institute of Electrical and Electronic Engineers (IEEE) launched the Little Box Challenge. It is an open competition with a US$1 million first prize to demonstrate a 2 kW DC/AC inverter and to shrink space down to less than the size of a small laptop (655 cm^3) at a maximum surface temperature of the housing below 60°C [95].

Residential PV battery systems will follow a steady cost reduction as has happened in the last decades in PV module manufacturing, due to the industrial economy-of-scale effects. Starting from 0.35€/kWh storage costs in 2015 and assuming an annual cost reduction of 8%, it is expected that cost parity to the household grid cost, without energy, will be reached about 2025. This will only work for battery systems on the daily scale. The connection to the public grid will also remain with that connection cost, which will be oriented to the maximum of demand power and not to energy demand below a certain level. Later costs may diminish if the user disconnects from the public grid in a totally self-sufficient energy system, assuming an economic solution of seasonal storage.

The integration of the PV battery system into a residential system will be continuously improved by the most economical solutions such as controlling thermal loads via heat pumps or the charging of electrical vehicles, to reach the highest self-sufficiency ratios.

In these complex residential solar energy systems, reduction of the operation and maintenance (O&M) costs will be a key challenge in the coming years. It is expected that PV modules and inverters will not be the main O&M cost drivers compared to the batteries after 10 years of operation.

To fulfil these requirements, it is indispensable to implement design for reliability (DfR) in the development process of power electronics. In detail, attention must be paid to modularity and repair design, derating components, temperature management at reduced operating temperatures, different types of accelerated testing and applying predictive methods such as MTTF and DFMEA (design failure modes and effects analysis), fault tree assessments, and learning from reliability in allied fields, such as automotive and drives.

References

[1] F. Baumgartner, in: Talk 4CO.12.2, 28th European Photovoltaic Solar Energy Conference and Exhibition (28th EUPVSEC), Paris, October, 2013.

[2] A. Jäger Waldau, in: Closing Session of 29th European Photovoltaic Solar Energy Conf. (EUPVSEC) 2014, Amsterdam, 2014. http://www.eupvsec-proceedings.com.

[3] F.P. Baumgartner, O. Maierl, D. Schär, D. Sanchez, P. Toggweiler, Operation and maintenance costs (O&M) of PV plants in Switzerland, in: 31th EUPVSEC, Proceedings; Talk: 5AO.8.2; Hamburg; Germany; 14–18 September 2015, 2015.

[4] www.eex.com.
[5] N. Allet, F. Baumgartner, et al., EUPVSEC, 2012.
[6] F. Baumgartner, A. Büchel, R. Bartholet, in: 24th EUPVSEC, 2009.
[7] F. Carigiet, F.P. Baumgartner, J. Sutterlueti, N. Allet, M. Pezzotti, J. Haller, 29th EUPVSEC, 2014.
[8] International Technology Roadmap for Photovoltaic Results 2014_Rev.1, July 2015, http://www.itrpv.net/.cm4all/iproc.php/Reports%20downloads/ITRPV_Roadmap_2015_Rev1_July_150722.pdf?cdp=a.
[9] SMA Workshop MW Inverter Proceedings, 15 Jan 2008, Neu Ulm, 2008.
[10] Market Data PV System price in Germany: BSW-Solar, Graph: PSE AG 2015, http://www.pse.de/.
[11] V. Cassagne, in: Total, 26th EUPVSEC, 2012, p. 4584.
[12] S. Ringbeck, J. Sutterlueti, BoS costs: status and optimization to reach industrial grid parity, Prog. Photovolt. Res. Appl. 21 (2013) 1411–1428.
[13] I. Clover, PV Magazine Deutschland, 03, December 2013.
[14] D. Brearley, et al., Racking Manufacturer Survey, SolarPro Magazine (4.3 April/May) (2011). D. Brearley, et al., Ground-Mount PV Racking Systems Survey, SolarPro Magazine (6.2 February/March) (2013).
[15] 2015 ISEP™ International Solar Energy Provisions™ Commentary, ISBN: 978-1-60983-624-5.
[16] Polysun PV Constructor, polysun.solarprotool.com.
[17] D. Schär, Master thesis 2015, ZHAW Zurich University of Applied Sciences, Winterthur.
[18] J. Libal, V.D. Mihailetchi, R. Kopecek, Photovoltaics International, www.pv-tech.org, 23rd March 2014, pp. 35–45.
[19] F. Baumgartner, in: WCPEC 6 Kyoto, 2014. session 8WePo.9.33LN.
[20] M. Šúri, T.A. Huld, E.D. Dunlop, H.A. Ossenbrink, Potential of solar electricity generation in the European Union member states and candidate countries, Solar Energy 81 (2007) 1295.
[21] Haselhuhn, DGS PV Planning Guide Book, 2007. Leitfaden Photovoltaische Anlagen, DGS Berlin, 2005, ISBN 3-9-9805738-3-4.
[22] PVSYST, www.pvsyst.com/.
[23] F. Carigiet, F. Baumgartner, et al., in: EUPVSEC, 2013.
[24] R.C. Neville, Solar energy collector orientation and tracking mode, Solar Energy 20 (1978) 7–11.
[25] T. Huld, http://re.jrc.ec.europa.eu/pvgis/apps4/pvest.php.
[26] T. Huld, M. Šúri, E. Dunlop, Comparison of potential solar electricity output from fixed-inclined and two-axis tracking photovoltaic modules in Europe, Prog. Photovolt. Res. Appl. 16 (2008) 47–59.
[27] F. Baumgartner, A. Buechel, R. Bartholet, in: Proceedings of the 27th European Photovoltaic Solar Energy Conference, Frankfurt, 25 September, 2015. 4BV.1.60.
[28] F. Baumgartner, A. Buechel, et al., EUPVSEC 2010, Valencia.F. Baumgartner, A. Buechel, et al., 27th EUPVSEC 2012, Frankfurt.
[29] F. Baumgartner, A. Buechel, R. Bartholet, in: 3rd_EU_PVSEC, Valencia 04th September, 2008. invited talk 4DO.9.5.
[30] M. Osborne, 23 March 2015, PV News according to market research firm IHS expects report. http://www.pv-tech.org/news/.
[31] P.V. Fakten, ISE, www.pv-fakten.de, May 2015.
[32] DuraTrack HZ v3 Solar Tracker from Array Technology Inc, Albuquerque, US, http://arraytechinc.com/utility/duratrack-hz-v3-solar-tracker/.

[33] W. Shockley, The theory of *p-n* junctions in semiconductors and *p-n* junction transistors, Bell Syst. Tech 28 (1949) 435–489.
[34] F. Baumgartner, Energieertrag unterschiedlicher Solarmodultechnologien, Electro suisse Bulletin (7) (2011) 17.
[35] D. Schäer, F. Baumgartner, in: EUPVSEC, 2014.
[36] F. Baumgartner, R. Grabherr, in: 21st EUPVSEC, 2006.
[37] N. Allet, F. Baumgartner, et al., in: EUPVSEC 26th EUPVSEC, 5-9 September, 2011. Hamburg, Session 4DO.6.2.
[38] N. Allet, F. Baumgartner, et al., in: 27th EUPVSEC, 2012.
[39] R. Bründlinger, et al., On the characterisation of PV inverters' efficiency—introduction to the concept of achievable efficiency, Prog. Photovolt. Res. Appl. 19 (4) (2011) 423.
[40] F. Baumgartner, A. Büchel, et al., EUPVSEC, 2009.
[41] D.P. Hohm et al., 2003.
[42] N. Onat, Recent developments in maximum power point tracking technologies for photovoltaic systems, Int. J. Photoenergy 2010 (2010) 11. Article ID 245316.
[43] L. Kazmerski, shown in his talk at World PV Conf Nov 2014, Kyoto, the Bell Telephone System News article from 1954, "Bell System Solar Battery Converts Sun's Ray into Electricity!".
[44] C. Gilligan, in: PV Inverter Market Tracker, IHS, 06 July, 2015. technology.ihs.com.
[45] J. Kolar, et al., ETHZ, EEPE; Conf March, Integrated Power Electronic Systems CIPS, Nürnberg, 2012.
[46] N. Mohan, Power Electronics, 1995, ISBN 0-471 58408-8.
[47] Photon Inverter database, www.photon.com, Status 2015.
[48] PLECS (Piecewise Linear Electrical Circuit Simulation) software tool, www.plexim.com.
[49] M. Calais, in: 17th European Photovoltaic Solar Energy Conference, Munich, Germany, 22-26 October, 2001, p. 437.
[50] H. Schmidt, Patent EP1369985B1, priority date 15.05.2002.
[51] R. Teodorescu, M. Liserre, P. Rodríguez, Grid Converters for Photovoltaic and Wind Power Systems; Text Book, John Wiley & Sons, Ltd., USA, ISBN: 978-0-470-05751-3, 2011.
[52] Photon International Magazine, June 2014, 64.
[53] Products see Heric topology with additional DC/DC booster from Sunways AT series in 2006 and Bosch, Voltwerk in 2012 with the V3 series.
[54] F. Carigiet, F. Baumgartner, et al., Case study of a low-voltage distribution grid with high PV penetration in Germany and simulation analyses of cost-effective measures, Prog. Photovolt. Res. Appl. (2015), http://dx.doi.org/10.1002/pip.2352. EUPVSEC 2015.
[55] F. Baumgartner, Euro Realo Inverter Efficiency: DC-Voltage Dependency, in: 20th European Photovoltaic Solar Energy Conference, 6-10 June, Barcelona, Spain, 2005.
[56] R. Hotopp, Private PV-Stromerzeugungsanlagen im Netzparallelbetrieb, RWE Essen 1991, pp. 28.
[57] EN 50530:2010+A1:2013; Overall efficiency of grid connected photovoltaic inverters; table D1 shows the Euro weighting factors; DIN EN 50530 VDE 0126–12:2011–04.
[58] F. Baumgartner, S. Roth, H. Scholz, MPP Voltage Monitoring to Optimise Grid Connected System Design Rules, EU PVSEC, Paris, 2004.
[59] A. Kreutzmann, P. Welter, PV Inverter Market Survey, PHOTON Magazine, March 2005, 98.
[60] F. Baumgartner, H. Schmidt, B. Burger, R. Bründlinger, H. Häberlin, M. Zehner, *Status and relevance of the DC voltage dependency of the inverter efficiency*, in: EUPVSEC, 2007. Milano, Session 4DO.4.6.
[61] EN 50530, ÖVE/ÖNORM, Overall efficiency of grid connected photovoltaic inverters, 2014–02-01, see also IEC62116.

[62] R. Bründlinger, N. Henze, H. Häberlin, B. Burger, A. Bergmann, F. Baumgartner, prEN 50530 – the New European standard for performance characterisation of PV Inverters, in: EUPVSEC, 2009.
[63] H. Häberlin, Photovoltaics: System Design and Practice, 2012, ISBN-10: 380073351X.
[64] F. Baumgartner, in: EUPVSEC Paris, 2004.
[65] B. Bletterie, R. Bründlinger, G. Lauss, On the characterisation of PV inverters' efficiency—introduction to the concept of achievable efficiency, Prog. Photovolt. Res. Appl. 19 (4) (2011) 423–435.
[66] Test report of the Schneider Electric TL20000 E, 20 kW string inverter, Photon Profi, issue Aug 2013, pp. 53–59.
[67] Test report of the Suntellite ZDNY-TL 17000, 17 kW string inverter; Photon Profi, issue Sept 2013, pp. 44–52.
[68] L. Kazmerski, World PV Conference, Kyoto Nov 2014, a citation of D. Chapin inventor of the Bell Solar Battery is shown; photograph see also http://www.beatriceco.com/bti/porticus/bell/belllabs_photovoltaics.html.
[69] S. Enkhardt in, http://www.pv-magazine.de; 18 January 2016.
[70] M. Boust, citation in PV-magazine, I. TSAGAS, 22. JANUARY 2016, http://www.pv-magazine.com/.
[71] C. Stegner, P. Luchscheider, J. Bogenrieder, R. German, C. Brabec, in: EUPVSEC, 2015.
[72] Patent EP2293410A2, Schmiegl, Firma Voltwerk heute Bosch, angemeldet Sept 2009.
[73] M. Braun, et al., Photovoltaic self-consumption in germany using lithium-ion storage to increase self-consumed photovoltaic energy, in: EUPVSEC, 2009.
[74] Patent EP2293410A2, Schmiegel, Firma Voltwerk heute Bosch, September 2009.
[75] M. Bortolini, et al., Technical and economic design of photovoltaic and battery energy storage system, Energy Convers. Manage. 86 (2014) 81–92.
[76] A. Colthorpe in http://www.pv-tech.org/, Jan 18, 2016; products see www.sma.com product sunny boy storage as well as www.fronius.at product Fronius Symo Hybrid; see also.
[77] V.J. Weniger, T. Tjaden, V. Quaschning, http://www.photovoltaik.eu/; Photovoltaik 10/2012, S.50-54.
[78] J. Weniger, J. Bergner, T. Tjaden, V. Quaschning, EUPVSEC, 2014.
[79] Data from the residential PV battery installation of Antje Boll in Allensbach, Germany, provided by Sunny Solartechnik GmbH; www.sunny-solartechnik.de/.
[80] 5.4kWp multi-crystalline Silicon Modules, inclination 20°, azimuth 155°; location Constance, Germany http://www.pvlog.de/solar.phtml?nav=data&lid=V58tK2P4Q2r0If1G&profil=7.
[81] F. Baumgartner, T. Baumann, PV2050 an NFP70 project, www.nrp70.ch, Midterm Project Report 2016.
[82] R.G. Ross Jr., in: PVSEC-1, Japan, Kobe, 1984, p. 337.
[83] T. Söderström et al. 2014, with paper SWCT_140519, Meyer Burger, www.meyerburger.com.
[84] A. Skoczek, T. Sample, E.D. Dunlop, The results of performance measurements of field-aged crystalline silicon photovoltaic modules, Prog. Photovolt. Res. Appl. 17 (2009) 227–240.
[85] Survey of the attendants of the NREL-Workshop PV Performance and Reliability R&D, 2015 Inverter Workshop; February 25, 2015; www.nrel.gov/pv.
[86] This Arrhenius rule also applied for silicon devices in a slight modification, see JEDEC Standard JEP122, Failure Mechanisms and Models for Silicon Devices, www.jedec.org.
[87] T. Ashburn, D. Skamser, in: SMTA Medical Electronics Symposium – Anaheim, California, 2008. www.kemet.com.

[88] R. Vidano, NREL-Workshop PV Performance and Reliability R&D, 2015 Inverter Workshop, February 25, 2015, www.nrel.gov/pv.
[89] Manufacturers technical report, http://www.elna.co.jp/en/capacitor/alumi/trust/#anc02.
[90] http://cdn.solar.schneider-electric.com/wp-content/uploads/2014/10/conext-tl-15-20-datasheet-20141002_eng.pdf.
[91] M. Ursino, NREL-Workshop PV Performance and Reliability R&D, 2015 Inverter Workshop, February 25, 2015, www.nrel.gov/pv..
[92] C. Verhoeve, W. Sinke, Int. PVSEC – 9 Japan 1994, pp. 187, see also IEC 61215-2:2016 IEC 61215-1:2016 IEC 61215-1-1:2016.
[93] M.G. TamizhMani, NREL-Workshop PV Performance and Reliability R&D, 2015 Inverter Workshop, February 25, 2015, http://www.nrel.gov/pv/performance_reliability/pdfs/2015_pvmrw_126_tamizhmani.pdf.
[94] D. Kranz, B. Burger, H. Schmidt, in: EUPVSEC, 2009.
[95] https://www.littleboxchallenge.com/.

Prediction and measurement of photovoltaic system energy yield

6

N.M. Pearsall
Northumbria University, Newcastle upon Tyne, United Kingdom

6.1 Introduction

In previous chapters of this book, the performance of different components of the photovoltaic (PV) system has been considered, but, for the PV system users, interest mainly lies in the performance of the system as a whole and whether the output will meet their needs, technically and financially. The first step is to achieve a prediction of system output, which may consider only the overall output in its simplest form but is more likely to also encompass the variations in performance arising from variations in operating conditions. Once the system is operational, there must also be a way of assessing whether the performance matches the prediction and interpreting the comparison between measured and expected performance. In some cases, this assessment will be linked to a legal agreement in relation to a guarantee of output or service or to a financial payment in regard to the energy produced. In this chapter, we will consider how both the predictions and the measurements are undertaken, together with some discussion of how system losses can be identified. We will also consider developments in relation to output prediction, particularly in relation to solar forecasting, which may impact on the penetration level of PV systems into the electricity distribution network.

6.2 Predicting PV system performance

The instantaneous output of the PV system will depend directly on the operating conditions experienced by the system, together with some aspects relating to the detailed system design. The most influential of the operating conditions are the solar irradiance received and the operating temperature of the modules, noting that the latter is a function of the solar irradiance, ambient temperature, other thermally related conditions such as wind speed and direction, array mounting and thermal behaviour of the module. Additional parameters of influence include the spectrum of the incident light, the angle of incidence of the irradiance and system level aspects such as the amount of soiling, any shading conditions and losses within the power conditioning system. The ambient temperature may also affect the operation of the power conditioning equipment and battery storage, if extremes of heat or cold are experienced.

The Performance of Photovoltaic (PV) Systems. http://dx.doi.org/10.1016/B978-1-78242-336-2.00006-9

However, instantaneous output is only of limited use and the PV system user is generally more interested in the cumulative energy output over a specific period. There are many uses for this information, such as:

- Determining the technical and/or financial viability of a particular system design
- Optimizing system design by comparison of output for different options
- Providing a benchmark against which the actual system performance can be assessed, leading to the possibility of fault/loss identification
- Providing a benchmark against which a minimum annual output can be agreed, as part of a legal contract for the supply of a PV system

The accuracy with which the output should be determined and the flexibility of the simulation employed will vary depending on the purpose of the output prediction.

The following sections consider some of the most important aspects of the output simulation, starting with the solar resource and then going on to consider the operating temperature and the simulation of the system performance.

6.2.1 Solar data sources

It may appear to be an obvious statement but it is nevertheless true that the accuracy of the prediction of PV system output can never be better than the accuracy with which the solar input is known. In order to determine the solar input for a PV system at a given location, a suitable source of historical solar data must be selected, with the main criterion being that it should be as representative as possible of what the system is expected to receive over its operating lifetime. Because of the variability of solar irradiation from year to year, output predictions generally use observations over several years (usually at least a decade, although longer periods are now available and generally preferred), since the uncertainty in the solar data due to random variations is reduced by having a large data set. The solar data are generally provided as an annual average or as a data set which represents a typical year for the period (eg, typical meteorological year (TMY)—a set of 12 monthly sets of real data selected to be representative of long-term distributions of selected climatic conditions [1]) and are presented as hourly, daily or monthly values depending on the data set. It is worth noting that neither long-term average nor TMY data sets represent the extreme variations of solar irradiation, by definition, and so are not suitable for considering worst-case conditions in system design. They are used for determining an average or typical output for the PV system based on long-term historical measurements.

Solar data for a specific location are obtained by ground measurement using suitable sensors, by interpolation of ground measurements from other locations, preferably nearby, or by extraction from satellite observations. Most ground measurements of global solar irradiation for long-term data sets are made using a pyranometer, which is a thermopile device whereby a change in temperature caused by the absorption of sunlight provides a voltage signal proportional to the incident irradiation. Advantages of the pyranometer are that its response is spectrally insensitive and that it measures across a wide spectral range from 280 to 2800 nm [2]. This makes the irradiation measurements suitable for a wide range of applications. The thermal effects mean that it has a relatively slow response to changes in irradiance (typically a few seconds) compared to a PV device, but this is not important for long-term measurements of global irradiation and is outweighed by the stability of the measurement device. For measurement of the direct irradiance, which requires tracking of the sun position, a pyrheliometer may be used. This also operates by measuring the change in temperature due to absorption of solar irradiation, but the instrument's large field of view (typically 5°) reduces the effect of small tracking errors [1]. In both cases, the instruments need to be regularly calibrated and properly maintained.

Different grades of instrument are defined by their calibration method and precision. Stoffel has considered expanded measurement uncertainties for commercial instruments and estimates the overall uncertainty for a thermopile pyranometer as ±4.0% (for a 95% confidence interval) and for a thermopile pyrheliometer as ±2.8% [3]. This is in general agreement with Sengupta et al. [1], who suggest that ±5% is an achievable uncertainty for a good quality pyranometer for daily irradiation totals (again with a 95% confidence interval), whilst noting that this would rise to ±8% for hourly irradiation totals using the same instrument.

Earth observation satellites use imaging channels in the visible and infrared regions of the electromagnetic spectrum to obtain information on cloud cover, surface temperature, albedo (reflections from the earth's surface and clouds), water vapour content, snow cover and vegetation. Some satellites can also measure ozone and CO_2 concentrations in the atmosphere. A more comprehensive discussion of satellite measurements can be found in Perez et al. [4]. The determination of solar irradiance at the Earth's surface is based on the principle that the visible radiance of the Earth, as measured by the satellite, is inversely proportional to the global horizontal irradiance (GHI) at the surface, for a given solar zenith angle. This is because the more opaque the cloud coverage, the more light is reflected back to the satellite. The model to determine GHI will include a calculation of clear sky irradiance, based on the extraterrestrial irradiance (which varies with Earth–Sun distance), the sun position (as given by the solar zenith angle), atmospheric composition (water vapour, ozone, etc.), aerosol content and elevation above sea level. The combination of water vapour and aerosol content is usually termed turbidity and, together with the path length through the atmosphere, which is a function of the zenith angle and the elevation of the ground surface, determines the attenuation of light as it passes through the atmosphere. The model must also account for challenging conditions at the Earth's surface, which can modify the visible radiance, such as snow cover, deserts and dense forests. Turbidity conditions also vary rapidly at coastal sites, due to the proximity of land and water, in mountainous regions, due to changing altitude, and for urban/industrialized sites, due to higher aerosol content [4].

The uncertainty in regard to satellite measurements is dependent on the spatial resolution of the satellite image, the time interval of the measurement and the integration period. Again, as for the ground-based measurement, integration over longer periods reduces the uncertainty. Thus, lower uncertainty values are obtained for high spatial resolution (eg, 1 km), frequent readings (eg, 1–30 min intervals) and long integration times (eg, daily or longer) [4]. The method of determining irradiation values from the satellite images is also influential. However, for monthly or yearly integration periods, satellite data can provide a similar level of uncertainty to ground measurements in some cases [5]. An advantage of satellite measurements is the spatial coverage, since this only relies on the view of the satellite, whereas ground measurements are limited by the number of measurement stations.

The measurement uncertainty can be reduced by combining satellite data with ground measurements, using the latter to validate the satellite-derived values and quantify any systematic errors. This is the approach used by many of the currently used solar databases, for example, that provided by Photovoltaic Geographical Information System (PVGIS) [6] for Africa, where data from Meteosat Eastern geostationary satellites was used with validation from ground measurement sites [7]. This has allowed a full coverage of Africa that would not have been possible with ground measurements alone, but with lower uncertainty than would have been the case without the ground validation.

In most cases, the measurements are of global irradiation on a horizontal surface (GHI) and the values must be converted to global irradiation in the plane of the PV array in order to determine PV system output. The first step is to split the GHI value into its components of direct irradiation and diffuse irradiation. The diffuse irradiation is that which is scattered in the atmosphere and so is incident on the array at a range of incidence angles. In some cases, one or both of these quantities will be measured alongside the global measurement and this simplifies the translation. In most cases, however, a model is used to predict the two components (eg, Perez [8]) taking account of the clear sky irradiance and the turbidity of the atmosphere to determine the direct/diffuse ratio. A more comprehensive discussion of solar data models can be found in the report by Sengupta et al. [1] and a comparison of the performance of selected models in predicting global irradiance on inclined surfaces has been carried out by Gracia and Huld [9]. It should be noted that the atmospheric turbidity is of particular importance to concentrator PV systems, due to attenuation of the direct irradiation.

The direct irradiation is corrected to the array plane using a trigonometric function to account for the angle of incidence, whereas the correction of the diffuse is more complex depending on whether an isotropic distribution is assumed. The ground reflected irradiation must also be considered for nonhorizontal array planes, with the importance of this component depending on the nature of the surrounding ground and the tilt angle of the array.

Long-term averages of solar irradiation smooth out the annual variations, but these variations may need to be considered in terms of understanding the year on year variability of system output, for example in terms of predicting yearly cash flow variations. The long-term variability depends on the climate, generally being higher for mountain and coastal regions and lower for arid regions [5]. Richter et al. have reported values of variability, defined as the standard deviation over the measurement period, of between 4.0% and 7.1% for a range of European cities [5]. Long-term studies of solar irradiation data have also shown drift of the long-term averages for Europe with a reduction of 4–6% in the period from 1950 to 1990, often referred to by the term 'global dimming'. However, Wild et al. have shown that later data, covering the period from 1985 to 2000, exhibited an increase in the annual average [10]. They term this 'global brightening' and attribute this to reductions in cloud cover and the aerosol burden in the atmosphere leading to higher transmission of sunlight. They also note that increases have also been observed in some other parts of the world.

Huld and Trentmann [11] have considered the variation in satellite measured irradiation values for Europe, Africa and Asia by comparing sequential decades over the periods 1983–2013 (for Europe and Africa) and 1999–2014 (for Asia). They concluded that the temporal variability is dependent on location, with temperate climates showing higher values than for desert areas (3–5%) and the highest values obtained for mountainous regions (7–8%). The analysis demonstrates a slight negative trend for West Africa (−1% to −2% per decade) and small positive trends for Eastern Europe (+2% to +3% per decade) and Central Asia (up to +5% per decade). They also noted that major events, such as volcanic eruptions, could be seen in the data sets and may influence the long-term averages derived.

It is not clear how climate change will affect the long-term averages of measured irradiation in different locations across the world, but this will need to be monitored to enable it to be taken into account in PV output prediction. Nevertheless, it is only possible to use historic data for the prediction and so this introduces some unavoidable uncertainty regarding the expected irradiation received by the PV array due to the potential future trends in long-term average irradiation, whether increasing or decreasing.

6.2.2 Operating temperature

After the irradiation received, the operating temperature is the next most influential parameter in terms of predicting the PV system output. The operating temperature of a PV module is a function of:

- The irradiation received
- The ambient temperature
- The module design, in terms of how it rejects heat
- The module efficiency, since this determines the percentage of the solar energy converted to electricity and hence the percentage converted to heat
- The mounting system, which determines the ventilation around the module
- Other ambient conditions such as wind speed and direction that may affect heat transfer from the module

As discussed in Chapter 1, the module temperature is generally found to be directly proportional to the irradiation but the exact relationship is determined by the detail of the system design. In terms of system performance, an increase in module temperature results in a reduction in power from the PV array for the same irradiance condition. This is mainly because of a reduction in the voltage output and the magnitude of that reduction is dependent on the PV technology used. In general, we can consider that the larger the bandgap of the absorber material within the PV cell, the smaller the temperature coefficient and therefore the smaller the power loss for a given temperature rise. However, the temperature coefficient is also dependent on cell and module design. Values for the temperature coefficient for all the parameters of open circuit voltage, short circuit current and power are provided on the module data sheet and can be used to predict the change in output as result of an operating temperature above the standard test condition (STC) output. Since the STC operating temperature of 25°C is consistent with factory or laboratory ambient conditions, rather than typical operating conditions, the power output of the module in operation is often lower than the rated value.

Thus, to predict the output of the PV system taking into account the operating temperature, we must:

i. Determine the operating temperature of the PV module, taking into account ambient temperature, array mounting conditions and other climatic conditions (eg, wind).
ii. Correct the power output of the module for operating temperature using the temperature coefficient for the module in question (for example, using Eq. (6.1) following).

$$\text{Power at temperature } T = \text{Power at STC} + [\text{temperature coefficient of power} \times (T - 25)] \quad (6.1)$$

Eq. (6.1) assumes that the temperature coefficient is given in W/°C. In some data sheets, it is given as a percentage of the rated power and should first be converted before using in the calculation. Also, the equation considers only temperature and assumes that the irradiance remains at the STC level. Procedures for the simultaneous correction for both irradiance and temperature can be found in IEC standard 60891 [12].

In order to provide a closer estimate to the power output in operation, it is usual to measure the nominal operating cell temperature (NOCT) of a module and also provide this value on the module data sheet. NOCT is measured for a specific set of operating conditions, these being 800 W/m^2 irradiance, open rack mounting, open circuit condition, ambient temperature of 20°C and wind speed of 1 m/s [13]. Whilst, like STC, this relates to only one set of operating conditions, the combination of STC values and NOCT values can be used in the correction procedures given in Ref. [12] and NOCT is often used as an input to performance simulation.

For some technologies, most notably amorphous silicon, the behaviour with temperature is more complex, since heating the module can reverse some of the light-induced degradation that occurs with this type of module. Thus, the module output can show a seasonal variation, with higher outputs in summer months, in contrast to other types of module. This should be taken into account in performance simulations.

Finally, it should be noted that high operating temperatures and temperature cycling are stress factors for the PV modules and may result in output degradation and reduction of lifetime for some modules. This has been considered in more detail in Chapter 3 of this volume.

6.2.3 PV system output simulation

Detailed assessment of system design options and prediction of output are usually carried out using system simulation software. Several commercial products are available (eg, PVSyst [14], PV*Sol [15]) along with some free to use software, generally developed by research organizations (eg, PVGIS system tool combined with the solar database [6]). PV system supply companies may use their own in-house software or one of the commercial products. There is also some software available on balance of system (BOS) supplier web sites that will allow some system design and assessment, in order to facilitate selection of product [16]. This section provides an overview of the approaches used and the required steps to predict energy performance, but it is not the intention to provide a comprehensive review of the simulation programmes available or to make any recommendations as to which programmes should be used. The field of PV system simulation programmes is one of continuous change and development and new features are regularly introduced in order to refine system design and reduce uncertainty in output predictions. The simulation programmes also differ in regard to purpose, with some being intended for commercial system design and others for research and development. This leads to differences in emphasis and complexity in terms of the simulation process and the subsequent analysis of results.

Most simulation programmes will include the following key elements:

(a) Solar irradiation data for a range of locations and representing a measurement period of several years (either embedded or able to be imported from designated sources).
(b) A method to translate irradiation data to the location of interest (if not included in the original data set).
(c) A method to translate the irradiance from a horizontal surface (as measured) to the surface of interest for the array (as defined by the user), taking into account reflection from the ground (albedo) and climate variations.
(d) A database of operating characteristics of commercial PV modules and BOS components (inverters, batteries, charge regulators, etc.) in order to allow the performance of different choices to be assessed.
(e) The ability for the user to input system details, including components used, connection options, location, array orientation and, where necessary, the load(s) to be powered.
(f) A model of the operation of all components, so as to be able to determine periodic output values according to the defined frequency (often hourly simulations over a period of one year).
(g) The ability to define additional performance aspects such as shading, complex connection options (eg, definition of sharing between multiple maximum power point tracking (MPPT) inputs on an inverter), modification of default values to reflect specific environments or applications, definition of own components, definition of cabling parameters to determine resistive losses and variations in parameters for different months of the year.
(h) A range of options for the output of results, including tabular, graphical and, in some cases, system visualization—simulations will usually provide total annual electrical energy output, efficiency and performance ratio values, together with monthly values in most cases, and may also give indications of losses of different types.
(i) Some simulation packages provide the option to determine system cost and/or environmental impact (eg, equivalent CO_2 emissions).

Increasingly, simulation software includes more complex performance aspects such as, but not limited to, light-induced degradation of modules, mismatch losses, detailed battery performance, detailed electrical losses due to cabling, soiling, snow accumulation and angle of incidence losses. Of course, the resulting simulation is only as accurate as the ability of the user to specify these aspects numerically for their design and location.

Most simulation models return annual output values based on the initial parameters of the system design. Whilst this is of interest to the user, the viability of the system is based on the lifetime output and so some models now consider the reduction in output over the system lifetime by introducing degradation factors relating to the PV modules and/or BOS components. The determination of component lifetimes and the long-term performance of system components are discussed in other chapters in this volume. The lifetime output may be of interest in two ways. Firstly, the overall output of the system can be used to determine the overall financial return on investment and contribution to the reduction of greenhouse gas emissions resulting from the replacement of other sources of electricity production. Secondly, the change in performance across the lifetime can be considered on an annual basis as an input to cash flow considerations. In this second case, it is important to consider the profile of the change in output. Most studies assume a linear reduction in output across the system lifetime, with typical values being in the range of 0.5–1% per year, but the actual output reduction is likely to be a more complex function of time than this. System performance may also depend strongly on the maintenance procedures adopted for the system.

Nevertheless, no matter how complex the system simulation, there will be some level of uncertainty in the output prediction, based on the accuracy with which the input data are known and the accuracy of the system simulation algorithm. A summary of the main issues relating to the output prediction uncertainty will be given here. A comprehensive review of system modelling uncertainty can be found in the report by Richter et al., resulting from the Performance Plus project [5].

Since the assumption of the solar resource is the most important factor in determining the output of a PV system, it is also the most important source of uncertainty. This arises from two aspects, the first being the uncertainty in the measurement of the solar data used in the simulation and its translation to the array plane. The second source of uncertainty is in the year to year variation of the resource, which has been discussed in Section 6.2.1. The measurement precision for various sensors has also been discussed in Section 6.2.1, and it can be seen that the uncertainty in irradiation measurement will be dependent on the type of sensor used for ground measurement and the measurement procedures, especially in regard to cleaning and recalibration. For satellite measurements, it depends on the spatial resolution of the satellite image and the method for extraction of the irradiance value.

Simulation programmes will use a sequence of procedures to convert the measured solar data set to values on the array plane of interest in the location of interest. As an example, the sequence used by the programme PVSyst is summarized here [17]. The first step is the synthetic generation of hourly values of irradiance from the monthly averages used in the solar data base, using stochastic models. An ambient temperature profile is also produced, related to the daily irradiation and incorporating a phase shift to account for thermal lag. The process produces a random set of hourly values, within the overall monthly values of the original measurements, and each simulation will produce a different random set. It is noted that this may lead to 0.5–1% difference in simulated annual output for the same input details.

A detailed analysis of output simulation programmes also shows that there are some limitations due to the assumptions made for the output modelling itself.

Ransome et al. [18,19] have identified the following aspects that can lead to a difference in the energy yield between the modelled performance and the actual performance of a PV system:

- Fitting the module *I–V* characteristics to the usual one-diode model may lead to the derivation of a different temperature coefficient of power and efficiency change at low light levels, in comparison with measured values.
- The use of hourly weather data for the simulation averages any weather transients and, in general, will overestimate the fraction of irradiation received at low irradiance levels, potentially leading to increases in yield prediction.
- Seasonal annealing, degradation and module spectral response are not usually modelled well—these factors tend to affect thin-film modules to a greater extent than crystalline silicon modules, thus leading to a larger uncertainty level in the prediction.
- The variation of shunt resistance varies between module types and may not be known and/or modelled.
- The models do not usually account for variability between modules of the same nominal rating.

Whilst it is clear that the simulation of PV systems has some inherent uncertainty due to both the solar input and the algorithms used, the simulation programmes are becoming more inclusive of the known behaviour patterns of all the components. The result of a simulation should be treated as an indication of the probable average output over the system lifetime, but variability should be considered when using the value to compare with the measured output over any specific period.

6.3 Monitoring PV system performance

A PV system is one of the simplest electricity generating technologies in terms of operation and maintenance, since it has no moving parts (unless solar tracking is incorporated) and its solar fuel is delivered by nature. However, since it does not move and there are no emissions, including noise, it is not always straightforward to judge its performance against expectation. The only way to do this is to measure and record the electrical output, together with any other parameters that are either desired or required, and inspect the results at suitable intervals. For stand-alone systems performing a specific function, eg, pumping water, it is also possible to judge the effectiveness of the service, eg, the amount of water provided per day, to assess whether the system is performing as expected. Nevertheless, it may also be necessary to investigate the electrical output of this type of system in the event that the system effectiveness is below expectation, so as to determine the reasons. Therefore, it is a general recommendation that all PV systems be monitored to an extent consistent with their function. In this section, we will consider the methods for monitoring and assessing the performance of the PV system and comparing this with the prediction of output at the design stage.

As with the output prediction, there can be various reasons why the system owner, or another interested party, would wish to monitor the operation of their system, including one or more of the following:

- Assessment of the quality of the design, feeding into decisions about modification, replacement of components and/or future designs.
- Compliance with guarantees of output (or service) provided by the system installer.
- Loss and fault identification, to allow action to be taken to reduce output and revenue losses to the minimum possible.

For grid-connected systems, the guarantee of overall system performance often relates to either minimum energy yield or performance ratio, with both of these usually being annual values with the inclusion of a suitable degradation rate across the period of guarantee. In considering energy yield, there must be some allowance for the variation in weather conditions and this can be addressed by either setting a value estimated to allow for poor conditions or by comparing the measured output with a simulation of the output under the weather conditions experienced, rather than the average conditions considered in the design simulation. If this latter approach is used, then the irradiation received by the array and the ambient temperature must be monitored and a suitably robust simulation package must be used. If the former approach is used, then the tendency would be to set the value rather low so as to protect the installer in the case of abnormally poor weather conditions and, in this case, it is questionable as to whether the guarantee is of use. The consideration of the performance ratio is a conceptually simpler option, since this parameter expresses the losses in the system and takes into account the irradiation received by the array in the calculation. The installer is therefore guaranteeing that the system losses will be below a prescribed limit but does not need to consider how the irradiation will vary. However, it is also necessary to obtain a measurement of the irradiation in this case.

One of the most important considerations in assessing the system performance is the actual installed module power. In the prediction of output, the nominal power capacity of the PV array is considered, this being the sum of the nameplate power for each module. In a real system, we must account for the variation in power output between modules as a result of manufacturing variations and the uncertainty in the power measurement under STC in the factory. Initially, this variation can be estimated from the limits set by the manufacturer, who will provide an uncertainty on the module power (typically $<5\%$). Manufacturers try to minimize this uncertainty by sorting the module output into a range of power batches to ensure the power level quoted. Nevertheless, some variation remains and this can lead to a perceived performance that is somewhat higher or lower than prediction due to the difference in nominal and actual installed power. Variation in output also leads to a mismatch loss for series connected modules, due to differing maximum power point values under the same operating conditions. In some cases, the measured output values for each module are provided and the uncertainty can be estimated. The installed module power is used in the calculation of performance ratio, so also affects the assessment of this parameter.

6.3.1 Monitoring parameters and procedures

The level of monitoring, in terms of parameters measured and the precision and frequency of those measurements, depends on the size and complexity of the PV system and the speed required for the identification of any faults. The rapid identification of system faults allows the minimization of downtime and reduces the loss of output and hence revenue (or value in the case of using the electricity on site). The monitoring of system performance can be seen as an insurance against loss output and, hence, revenue (or benefit in the case that the PV system is providing electricity that would normally have been purchased from another source). The potential loss of revenue must be balanced against the expenditure on the system monitoring, remembering that includes not only the required hardware (including any recalibration and servicing required) but also the analysis of the measured data to identify any issues.

In general, monitoring can be classified into two main groups, usually termed global and analytical monitoring. Global monitoring includes the measurement of a limited set of parameters, sometimes only the energy output of the system, and is designed to ensure that the system is operational at a reasonable level. It is most often implemented for those small systems where the costs of monitoring need to be minimized. However, this approach usually requires additional information to determine the cause of any losses observed. Analytical monitoring provides a more comprehensive dataset of operating parameters and allows investigation of performance trends and problems. In the remainder of this section, analytical monitoring is considered. Table 6.1 provides a list of the commonly measured parameters for grid-connected systems, together with some comments on the measurement process and considerations.

The measurements discussed in Table 6.1 can be multiplied as necessary for large PV systems. For example, for a PV system consisting of several subsystems, the measurements of array output power, voltage and current would typically be made for each subsystem (or a selection of subsystems if comprehensive monitoring is deemed to be too costly) as well as at the overall system level. This is one of the most effective ways of identifying faults, since a subsystem with a significantly lower output than its neighbouring subsystems is likely to have some operating problem. For a large site, it may also be useful to measure irradiance at different locations, especially where there are any variations in array orientation or the likelihood of variable conditions across the site.

Additional measurements may be made where these would be helpful for characterizing the system performance and where they can be financially justified. These may include wind speed, utility grid voltage, current to the utility grid and durations of system outage, as specified in Woyte et al. [2]. For a concentrator system, direct normal irradiance measurements are required, along with some assessment of the energy used by the tracking system and, if possible, the tracking accuracy. Information on the determination of the performance of concentrator systems can be found in IEC standard 62670-2 [21]. In the case of building integrated systems, it may also be useful to measure the effectiveness of any other functions that the PV system has been designed to perform, for example, the measurement of internal building temperature to determine the level of shading provided by a solar shading system.

The discussion so far has considered grid-connected PV systems. For stand-alone systems, the performance assessment must take into account the service provided, ie, the amount of time for which the load can be met. The monitoring system will require the provision of a power supply, usually from the PV system itself. In this case, care must be taken to include the power requirements of the monitoring in the load estimation for the system design and to allow for the energy used by the monitoring

Table 6.1 **Monitoring parameters and measurement conditions for grid-connected PV systems**[a]

Measurement parameter	Units	Measurement instrument (suggested)	Comments on measurement
In-plane irradiance	W/m^2	Pyranometer or reference cell[b] for on-site measurements, mounted in-plane with the solar array and in an unshaded location, with regular calibration[c]	This parameter determines the energy input to the system and so is essential for the calculation of efficiency and performance ratio. It is also an input to most fault finding procedures and the frequency of measurement should be sufficient to identify faults. Typical frequencies would be 1 min intervals, but may be lower for detailed monitoring of, for example, cloud shading. Irradiance measurements would be averaged over a suitable period to determine solar irradiation received. The sensor should be coplanar with the PV array, with a maximum error of $\pm 2^o$ If on-site measurements are not practical, satellite data may be used if available, although this will require translation to the array plane and so introduce uncertainty. Horizontal irradiance may also be measured if desired
Ambient temperature	°C	Shielded temperature sensor	The operating temperature of the array is directly influenced by the ambient temperature. The measurement instrument should be shielded to ensure that it is not affected by climate conditions such as direct sunlight or wind
Module temperature	°C	Thermocouple on rear of module	For given irradiance and ambient temperature conditions, the module operating temperature depends on the module construction and the mounting system, including the ventilation of the module. These are both influential in determining the heat transfer from the module. Therefore, it is useful to measure the module operating temperature using a thermocouple, which should be placed on the rear of the module behind a cell (so as to avoid exposure to sunlight) and away from the frame and junction box. This gives an approximation to the cell temperature, since it includes transfer of heat to the rear surface

PV array output power	kW[d]	DC power meter with appropriate range	It is possible to measure the energy output over a defined period, as an alternative. However, the measurement of the power output allows direct comparison with the irradiance level for fault diagnosis. The power readings are then integrated over a suitable period to determine the energy output of the array
PV array output voltage	V	Voltage sensor with appropriate range	In some cases, only the output DC power is measured. However, measurement of the constituent voltage and current can assist in the diagnosis of operational problems
PV array output current		Current sensor with appropriate range	See the comments for PV array output voltage above
PV system output power[e]	kW	AC power meter with appropriate range	See the comments for PV array output power above
Power to utility grid	kW	AC power meter with appropriate range	Where there are local loads, this quantity can be used to determine the excess generation from the PV system which is exported to the utility grid and, hence, the effect on the grid can be analysed
Power from the utility grid	kW	AC power meter with appropriate range	This is only relevant for systems with local loads (eg, building integrated/ mounted system). This value, together with the measurement of the output of the PV system and the export to the grid, can determine the solar fraction, ie, that portion of the local loads that are able to be met from the PV system

[a] A more comprehensive treatment of monitoring requirements, including sensor accuracy and usage, can be found in IEC standard 61724 [20].
[b] For long-term solar datasets, a pyranometer is generally used, since this has a wide wavelength response making the measurements suitable for a range of purposes. For on-site measurement for PV systems, a lower cost option is to use a reference cell, usually crystalline silicon to ensure suitable stability. The reference cell will respond across a similar wavelength range to the PV array. However, it should be noted that a pyranometer and a reference cell will give different readings for the same incoming irradiance. due to this difference in spectral response and in their angle of incidence response. The difference between the readings differs with location, but is suggested to be typically 2–4% [2] with the reference cell returning the lower reading. This may need to be taken into account in performance assessment.
[c] A description of solar sensor calibration methods can be found in Chapter 3 of Sengupta et al. [1].
[d] The unit of kW is typical for a small PV system, but would be scaled for larger systems. This applies to all the power categories within the table.
[e] All AC power measurements should measure active (real) power, with due consideration of power factor and harmonic distortion.

system in the performance assessment. For some small systems, it is only appropriate to use simple monitoring of energy output, but it may be useful to record the periods when the battery is fully charged (since this gives a measure of the safety margins for poor weather conditions) and of any periods when the load is not met.

In general, the monitored data can be used to determine energy yields, system efficiency and performance ratio over suitable periods. For long-term trends, monthly or annual values are often used, but it is also useful to record and store data at higher frequency to enable the fault identification discussed in the next section.

6.3.2 Approaches to fault identification

Over time, with the exception of complete failure of components, any faults or additional losses that have developed are observed with reference to an expected performance level. Indeed, in this section, we are defining a fault as a reduction in performance and will use this term for all system issues resulting in this reduction, whether or not they result from faulty equipment or operating changes. The expected performance level can be established at the time of installation, and over a short period thereafter, by measurement of initial parameters and system behaviour. Of course, the performance should be related to the original prediction at the time of designing the system, but now with details of the actual conditions on site. Measurements would include:

- Verification of the installed power of the PV array (from manufacturer measurements, the measurement of a sample of the modules at a suitably recognized test facility or by field measurements, according to the preference of the system purchaser).
- Verification of the correct installation of the system, including electrical connections and array orientation.
- Testing of the correct operation of the monitoring system and the verification of the correct installation of all monitoring sensors.
- Recording of all parameters to be measured over a suitable initial period to ensure values are within the expected range.

The recorded values would also be used to calculate the usual system figures of merit, such as energy yield, performance ratio and system efficiency, taking into account the period over which the values have been measured.

It should be noted that PV systems are, in general, very reliable and may operate for long periods with no evident faults or problems. This sometimes makes it difficult to identify problems, because there is a large amount of performance data to consider in order to identify them. The failure of a major component, such as an inverter, is rather straightforward to identify since it results in no output from that part of the system. However, less severe faults that simply lead to a reduction in output below the usual value are more difficult to observe, not least because that reduction in output is a symptom of most types of fault.

The first step in fault identification is to establish that there is a fault of sufficient severity to reduce the system performance by an amount that is greater than the uncertainty of the measurement. One of the most useful methods to do this is by comparing the system output with that of one or more other systems operating under similar conditions. For a large power plant, it is usual to measure and compare the different subsystems, making it easy to identify any subsystem that is performing in a different manner. For single, small PV systems this is more difficult, but some monitoring services will use comparison with other systems in the same area. Alternatively, faults may be identified by comparison with previous performance of the same system, with the performance ratio most often used as the considered parameter.

Once the occurrence of reduced performance has been identified, the cause of that reduction can then be investigated by consideration of how the performance varies with the operating conditions, particularly irradiance and temperature. Because there is a variation in operating conditions, it is difficult to determine the cause of a fault simply on the basis of a single measured value. Many approaches use a footprint method, which compares a certain measured parameter with a single influencing factor to determine whether the dependence is as expected. Table 6.2 summarizes some potential loss mechanisms and the general method by which they can be identified from the performance data using graphical methods.

Clearly, there is a wide range of faults that can be investigated in this manner. Figs. 6.1 and 6.2 present examples of performance interpretation by graphical methods, one indicating good performance and the other indicating a performance problem. Both of the examples are taken from projects on the monitoring of domestic PV systems undertaken by the author.

Similar approaches have been taken by other researchers. The experts from the IEA Photovoltaic Power Systems Programme Task 13 have developed what they refer to as a 'stamp collection' as a monitoring tool [2]. This is a series of graphs that characterize the system performance and consists of:

- Graphs of irradiation on the array plane, ambient temperature, module temperature, DC voltage, DC current, AC power and performance ratio against time.
- Graphs of DC power, AC power, DC voltage, DC current, performance ratio, inverter efficiency, ambient temperature and module temperature against irradiation in the array plane.
- Graph of AC power against DC power.
- Graph of performance ratio, DC voltage against module temperature.
- Graph of DC current against DC voltage.

Drews et al. [22] also describe a footprint method for the detection of faults based on the comparison of measured output with simulated output based on satellite solar data. The energy loss is characterized in terms of amount, duration, changes in output and how this compares with measurements of neighbouring systems. This provides a profile of the loss that can be compared with predefined profiles for different types of loss. Table 6.3 shows the failure sources that they can detect using this approach.

It is clear that the detection of faults is dependent on the quality of the measurement of system parameters, the availability of information on irradiation and temperature and the quality of the simulation of the expected performance of the system. Of course, the larger the fault, the easier it is to detect, so rapid identification of small losses requires both high-quality measurement and analysis. The expenditure on system monitoring varies with the importance of identifying losses at a particular level, determined by the expected energy and financial loss if the fault is not detected.

Table 6.2 **Summary of graphical investigations for common loss mechanisms**

Loss mechanism	Performance parameter	Description of graph	Description of evidence
Inverter threshold	AC output	Plot of AC output and in-plane irradiance (*y*-axis) against time of day (*x*-axis) for a clear day	The AC output of the system should follow the in-plane irradiance over the day, but will not commence until a defined level has been reached and will cut off once the irradiance falls below a defined value. Inspection of the output curve at the beginning and end of the day will indicate the thresholds for inverter operation
System outages	AC output	**(a)** Plot of AC output and in-plane irradiance (*y*-axis) against time of day (*x*-axis) **(b)** Plot of daily AC output (*y*-axis) against daily in-plane irradiation (*x*-axis) **(c)** Plot of daily AC output and daily irradiation (*y*-axis) against day (*x*-axis) for the required period	The AC output falls to zero at times when the in-plane irradiance is not zero (or below the inverter threshold—see above) or the daily AC output is zero or substantially below the expected value based on the irradiation received. Note that system outages only lasting a short period but occurring regularly (eg, inverter shutdown due to grid voltage fluctuation) can reduce output by a large cumulative amount The detailed data should be investigated to determine the cause of the outage (see later entries for some examples) and to eliminate other causes for loss of output
Shading	AC output	Plot of AC output and in-plane irradiance (*y*-axis) against time of day (*x*-axis), preferably for a clear day	The AC output falls below the expected value, based on the irradiance profile, for a portion of the day, lasting a few minutes to a few hours. Generally, the output does not fall to zero. The diagnosis of shading can be checked by considering output profiles for several adjacent days, since the effect should be seen at approximately the same time over a period, and by considering the seasonal variation of the effect. On-site checks should also be made to determine the shading source

Poor MPPT behaviour	AC output	Plot of AC output and in-plane irradiance (*y*-axis) against time of day (*x*-axis) for a day with irradiance fluctuations	For good MPPT behaviour, the AC output should follow fluctuations in the irradiance closely and without too much delay. If the output fails to follow or drops to zero when there are rapid changes (usually a decrease in irradiance), the tracking algorithm may be inappropriate for the array type
Grid voltage fluctuations	AC output	Plot of AC output and in-plane irradiance (*y*-axis) against time of day (*x*-axis), preferably for a clear day	This behaviour is characterized by short to medium-term system outages, lasting from a few minutes to a few hours. Since it is caused by a fluctuation in the external grid, it is not generally correlated with changes in irradiance but may be seen when irradiance levels are sustained at a high level. This is because the presence of the PV system on the grid increases the grid voltage and, when loads are low, can cause the grid to be out of specification
Poor inverter efficiency at low light levels	AC output or performance ratio	Plot of AC output or performance ratio or inverter efficiency (*y*-axis) against irradiance (*x*-axis)	**(a)** AC output should vary linearly with irradiance, but falls well below expected values at low light levels (note that some reduction in efficiency would be expected and it is anomalous reductions that are being considered here) **(b)** Performance ratio should be roughly constant with irradiance (apart from temperature considerations) but would reduce at low light levels due to inverter efficiency changes—the profile should be compared with expectation **(c)** If DC power measurements are available, the conversion efficiency of the inverter can be derived and plotted directly, again looking for any variation that is greater than expectation

Continued

Table 6.2 Continued

Loss mechanism	Performance parameter	Description of graph	Description of evidence
Inverter output plateau (incorrect inverter sizing)	AC output	**(a)** Plot of AC output and in-plane irradiance (y-axis) against time of day (x-axis) for a clear day with high sunlight levels **(b)** Plot of AC output (y-axis) against irradiance (x-axis) for a clear day with high sunlight levels or over a period of several days (to provide more data)	**(a)** The AC output follows the in-plane irradiance most of the time, but is restricted to a level below expectation when irradiance is high—often leading to a horizontal section of the plot across the middle of the day **(b)** The AC output varies in a linear fashion with the irradiance until high irradiance levels are reached when the AC output then levels off In both cases, this demonstrates that the output is limiting. It is sometimes also possible to observe this from a rise in DC voltage when the inverter limits the power level from the array by moving off the maximum power point
Temperature effects	AC output or performance ratio	**(a)** Plot of AC output (y-axis) against irradiance (x-axis) for one or more days **(b)** Plot of performance ratio against module temperature	**(a)** If the plot shows two separated lines, then this may indicate a temperature induced hysteresis—check whether the lower values occur for afternoon periods when the module temperature should be higher for a given irradiance condition **(b)** Whilst performance ratio should reduce with increasing temperature, a large effect may indicate that the modules are insufficiently cooled

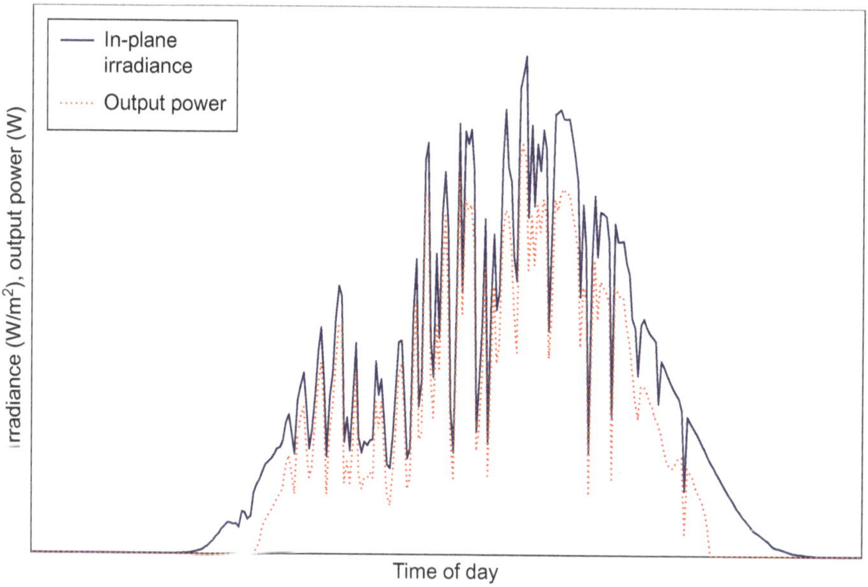

Fig. 6.1 Plot of measured in-plane irradiance and PV system output power as a function of time of day for a UK domestic PV system. As can be observed, this was a very variable day in terms of irradiance, due to cloud cover, and the output power follows the changes in irradiance very closely. This shows good maximum power point tracking. The turn-on and turn-off thresholds for the inverter at each end of the day can also be observed for this plot and it may be worthwhile adjusting these to reduce threshold losses.

6.4 Forecasting of PV system output

In Section 6.2, the prediction of system output was discussed, with specific reference to understanding how the PV system would perform in terms of average annual energy output and, often, what this means in terms of expected financial viability. The technique is also used as a benchmark to assist in fault finding and assessment of the optimum system design. However, as the capacity and number of PV systems increases, especially with regard to those which are connected to the electricity distribution network, it becomes important to consider not only the average output we might expect over a period of, say, a month or a year, but a forecast of the actual output of the system in a range of timeframes from a few minutes to a few days. This information feeds into decisions in regard to control of the grid and financial remuneration for electricity supplied from PV systems. The difference between the prediction of output considered previously and the forecast of output considered here can be likened to the difference between the average weather conditions for the time of year and the forecast of the actual weather conditions to be experienced today or over the next few days. The development of robust forecasting techniques is receiving a high degree of research interest as an important aspect of the move to a high contribution of solar electricity to the overall electricity sector.

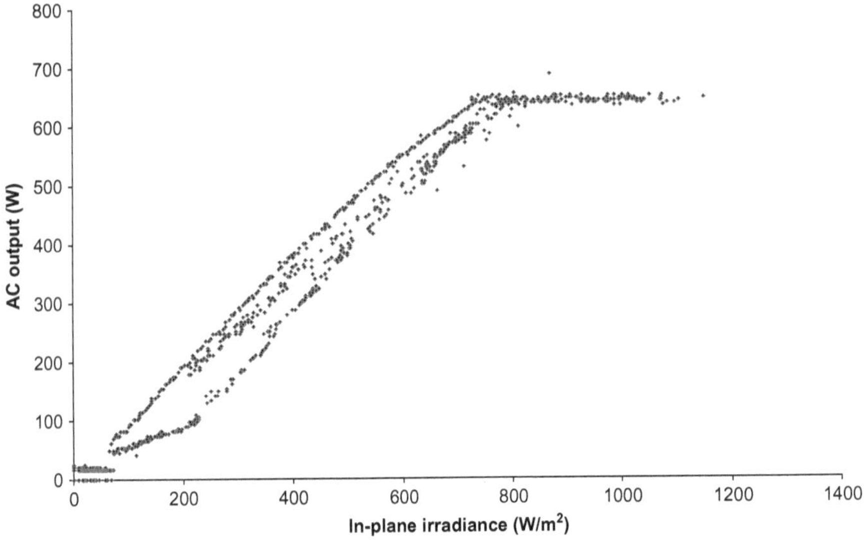

Fig. 6.2 Plot of AC output against in-plane irradiance for a small domestic PV system in the UK. The expected linear relationship can be seen for the irradiance range of around 100–750 W/m^2, but at higher irradiance levels the output limits to around 650 W. This indicates that the selected inverter is too small for the PV array and so is unable to allow the AC output to increase above this value. Whilst it is common to undersize the inverter for locations with significant amounts of energy received at low irradiance levels, such as in the UK, this plot indicates that too much output is being lost at high irradiance levels. It is also possible to observe the effects of module temperature in the spreading of the data points in the linear area of the graph, this being due to higher temperatures in the afternoon compared to the morning for the same irradiance level.

Table 6.3 Detectable failure sources as identified by Drews et al. [22]

General failure type	Failure
Constant energy loss	Degradation
	Soiling
	Module defect
	String defect
Changing energy loss	Shading
	Grid outage
	High losses at low power
	Power limitation
	MPP tracking
	Inverter temperature
	High module temperature
Snow cover	Snow cover
Total blackout	Defective inverter
	Defective control devices

The ability to accurately forecast the output of any energy system depends directly on the ability to forecast the availability of the incoming energy (fuel, solar energy, wind energy, etc.) over the period being considered. Therefore, for a PV system, we need to be able to forecast the irradiation falling on the solar array in order to be able to make a good forecast of system output. The solar irradiation received at ground level is a function of the position of the sun in the sky, resulting in a specific directionality of the light and path length, and the prevailing climatic conditions, including cloud cover. Whilst the sun position varies, it can be determined precisely and so there is a part of the solar forecast that is known. The effect of the clouds is the unknown part and this may have significant short- and medium-term impacts on the PV system output. For grid-connected PV systems, rapid variations in output may cause voltage flicker and longer-term variations in output require balancing with other electricity generation responding on a similar timescale.

The variability of the solar output is regularly cited as one of the main challenges of having a large percentage contribution from PV systems in the electricity network and, although in some cases the problem is overstated, it is important to address this challenge. One approach to address the variability is to include storage capacity in the system (either locally or centrally on the electricity grid) so as to modify the availability profile of the PV electricity and this is being widely investigated. A major challenge here is the reduction in cost of the storage technology. A second approach is to be able to forecast changes in PV output so as to ensure sufficient alternative generation is available but without excessive provision that would not be cost effective. In practice, a mixture of the two approaches is likely to be the best option.

Different purposes require forecasting over different time periods and spatial regions. In relation to interaction with the electricity grid, network operators are interested in both short-term forecasting (between a few minutes and a few hours) and medium-term forecasting (eg, day ahead, 48 h ahead). The former is useful for despatch purposes, which often works on a 30 min to 1 h horizon. Slightly longer timescales allow for the activation of reserve generation to meet any predicted shortfall in electricity supply. An accurate forecast can minimize the amount of reserve generation being held, without jeopardizing the ability of the network to meet all required loads. As the option of storage for grid-connected PV systems becomes more widely used, forecasting can also help with the decisions on the control of that storage.

The most suitable forecasting technique depends on the required resolution and the time period. For short-term forecasts, the simplest technique is persistence, ie, assuming that the climatic conditions persist and allowing only for the change in path length due to the movement of the sun. This gives a good result for clear days or those that are heavily overcast. When the conditions are changeable due to the movement of clouds, sky imaging or stochastic techniques may be used. For longer time periods, for example day ahead or longer, numerical weather prediction models can be used to predict the likely irradiation values, using physical parameters such as radiative transfer of solar radiation, air temperature, wind, air pressure and water vapour evaporation and condensation in a physical model.

Short-term forecasts up to 30 min can be based on observation of current conditions together with a measure of cloud fluctuation. This requires characterization of the cloud pattern together with its movement, based on wind speed and direction at cloud level, and some knowledge of cloud height. Typically, cloud patterns can be determined with a sky camera for a specific location, although there are limitations to this approach. The method of prediction can be summarized as follows [23]:

i. Conversion of the camera image from a spherical to a rectangular grid.
ii. Determination of the apparent velocity field of the cloud motion using pairs of images and an algorithm to express the correlation of velocity with the changes in position of the clouds.
iii. Choice of a suitable velocity vector for the cloud or cloud pattern.
iv. Classification of each pixel of the sky image as either clear sky or cloud.
v. Determination of cloud fractions for a set of grid elements defined by the velocity vector in a 'ladder' extending away from the sun position.
vi. Use of the cloud fractions as input to a forecasting algorithm to determine future obscuration of the sun by the moving clouds.

Depending on wind speed, clouds that may cause reduction of solar irradiance at ground level may not appear in the camera's image for moderate time horizons and so there is a limit to the forecast period. Marquez and Coimbra suggest a typical useful forecast period of 12–18 min. [24], although this depends on the cloud velocity. Note that there are also some problems with very short forecast timescales, since the sky camera uses a shadow band to obscure the sun and clouds very close to the sun cannot be resolved [24]. Therefore, it is becoming more common to extract the cloud patterns from satellite images that cover a wider area and therefore allow predictions over a longer timescale. The cloud-motion vector approach is used, where both the cloud speed and direction are extracted from two successive images and used to predict where the cloud will be in the next image. Regardless of the approach, forecasting of cloud patterns usually makes the assumption of persistence, in terms of cloud size and shape, as the cloud moves and this introduces an uncertainty that is dependent on cloud type and weather conditions. For dynamic cloud systems, where clouds are forming and dissipating, this can lead to substantial errors, but there remains considerable research to be done on cloud modelling in order to improve on the persistence assumption. (Note that the assumption of persistence here is only in relation to the cloud motion and only for the period to the next observation, in comparison to the persistence model as a forecasting technique which assumes no change in all the climatic conditions over the period considered except for the known movement of sun position.) A more comprehensive discussion of solar forecasting methods can be found in Coimbra et al. [25] and Pelland et al. [26].

So far, we have considered the forecast of system output for a specific location, but in practice the total PV capacity on the grid would be comprised of a large number of systems with a wide geographical area. The aggregation of output from distributed systems tends to reduce the variability and the ramp rates of specific variable events, depending on the number of systems and their spatial distribution. This has been shown empirically by a range of researchers. An example of this reduction in variability can be found in the paper by Lave and Kleissl [27], where they studied the correlation in GHI for four sites in Colorado with spacings up to 197 km. Their findings included a strong reduction of correlation for time periods less than 3 h and a decrease in ramp rate of up to 51% for the averaged data compared to a single site. They suggest that this would lead to a reduced requirement for spinning reserve in regard to grid operation. Hoff and Perez [28] have derived a general model for the determination of variability for a fleet of PV systems. They have shown that the maximum standard deviation of the change in output of the fleet in the time period Δt, $\sigma_{\Delta t}^{\text{Fleet}}$, can be given by Eq. (6.2), for a fleet of N similar systems with a total capacity of C^{Fleet} where the outputs of all the systems are uncorrelated (ie, there is no link between the solar conditions at different system locations). It should be noted that this allows for a change in output from full power to zero power, which implies a fully opaque cloud, so this gives a maximum value and the actual value is likely to be lower than this.

$$\sigma_{\Delta t}^{\text{Fleet}} = \frac{C^{\text{Fleet}}}{\sqrt{2N}} \tag{6.2}$$

Hoff and Perez also considered the variation of correlation between sites with distance, time period and relative speed of cloud movement. They showed, for a large database of solar irradiance measurements for a range of sites in the United States, that [28]:

- The correlation factor reduces with the distance between sites, although the rate of decrease varies with the time period considered.
- The decrease occurs at a similar rate when considered as a function of distance divided by time interval.
- The speed of the cloud movement is an additional factor in determining correlation between sites.

The topic of correlation of solar irradiance (and hence of PV system output) has received considerable research attention over the last few years, leading to the refining of models and consideration of many parameters and timescales (see, for example, Lauret et al. [29] and Perez and Fthenakis [30], who both consider the correlation of changes in clearness index with distance between sites, but for intraday and interday, respectively).

The forecasting methods discussed here determine the future irradiance level (usually on a horizontal surface) and this must then be translated into power output from the PV systems, using a knowledge of the PV technology, array orientations, system sizes and system distribution in the area of interest. In this translation, the principles already discussed in relation to output prediction of a single system are applied. Lorenz et al. [31] have considered forecasting methods to arrive at a prediction of the regional PV output for control areas of two transmission system operators in Germany and have shown that this approach can contribute to improved control of an electricity grid system with embedded PV systems.

As with weather forecasts, the PV output estimates from solar forecasting have an uncertainty level that varies with both the technique adopted and the climatic conditions. Changes in solar irradiance can be rapid and, if the duration of the change (commonly referred to as ramp rate) is shorter than the temporal resolution of the forecasting technique, then these fluctuations may not be identified in the specific forecast. In this case, it is usual to consider the statistical probability of such an event occurring, based on historical data. The uncertainty of the forecast has to be included in the decisions on the provision of balancing power and the costs assigned to the electricity from different sources. How the user of the forecast deals with the uncertainty is too complex a subject to be discussed here, but a more comprehensive treatment, together with examples, can be found in Mathieson et al. [32].

6.5 Summary

In order to achieve the highest level of performance from a PV system, it is necessary to understand the dependence of the output on the operating conditions and the components chosen and to regularly check the performance obtained against expected values. In this chapter, we have considered how to predict the system performance from knowledge of the design parameters and how to measure and analyse the real performance of the system. We have noted that there are both inherent and measurement uncertainties that should be taken into account when judging the outcomes of both activities. We have also considered how to forecast the PV system output in the short and medium term, by using solar forecasting techniques based on observation of changing sky conditions. This is expected to become more important as the penetration level of PV systems within the electricity grid increases. The topic of system performance, both predicted and measured, is very detailed and it has only been possible to provide an overview here, but there are references to more detailed reports provided for the interested reader. Finally, it should be acknowledged that the correct monitoring and assessment of system performance over the system lifetime is at least as important in regard to overall electricity output as increasing the performance of the individual PV modules and should be given due attention.

References

[1] M. Sengupta, A. Habte, S. Kurtz, A. Dobos, S. Wilbert, E. Lorenz, T. Stoffel, D. Renné, C. Gueymard, D. Myers, S. Wilcox, P. Blanc, R. Perez, Best practices handbook for the collection and use of solar resource data for solar energy applications, National Renewable Energy Laboratory technical report NREL/TP-5D00-63112, 2015.

[2] A. Woyte, M. Richter, D. Moser, N. Reich, M. Green, S. Mau, H. Georg Beyer, Analytical monitoring of grid-connected photovoltaic systems, Report IEA-PVPS T13-03, 2014.

[3] T. Stoffel, Terms and definitions, in: J. Kleissl (Ed.), Solar Energy Forecasting and Resource Assessment, Academic Press, 2013, p. 12 (Chapter 1).

[4] R. Perez, T. Cebecauer, M. Ŝúri, Semi-empirical satellite models, in: J. Kleissl (Ed.), Solar Energy Forecasting and Resource Assessment, Academic Press, 2013 (Chapter 2).

[5] M. Richter, K. de Brabandere, J. Kalisch, T. Schmidt, E. Lorenz, Best Practice Guide on Uncertainty in PV Modelling, Deliverable 2.4, Performance Plus Project, European Commission Seventh Framework Programme No. 308991, 2015.

[6] Photovoltaic Geographical Information System (PVGIS), European Commission Joint Research Centre, Institute for Energy and Transport, web based solar data source. http://re.jrc.ec.europa.eu/pvgis/.

[7] A. Gracia Amillo, T. Huld, R. Müller, A new database of global and direct solar radiation using the eastern meteosat satellite, models and validation, Remote Sens. 6 (2014) 8165–8189.
[8] R. Perez, P. Ineichen, R. Seals, J. Michalsky, R. Stewart, Modeling daylight availability and irradiance components from direct and global irradiance, Sol. Energy 44 (5) (1990) 271–289.
[9] A. Maria Gracia, T. Huld, Performance comparison of different models for the estimation of global irradiance on inclined surfaces, JRC technical report, EUR 26075 EN, 2013.
[10] M. Wild, H. Gilgen, A. Roesch, A. Ohmura, C.N. Long, E.G. Dutton, B. Forgan, A. Kallis, V. Russak, A. Tsvetkov, From dimming to brightening: decadal changes in solar radiation at Earth's surface, Science 398 (2005) 847–850.
[11] T. Huld, J. Trentmann, Variability and trend in the annual solar irradiation determined from Meteosat satellite data, in: Proc. of the 31st European PV Solar Energy Conference (EU PVSEC 2015), 2015, pp. 2073–2076.
[12] IEC 60891:2009, Photovoltaic devices—procedures for temperature and irradiance corrections to measured *I–V* characteristics, International Electrotechnical Commission, 2009.
[13] IEC 61215:2005, Crystalline silicon terrestrial photovoltaic (PV) modules—design qualification and type approval, International Electrotechnical Commission, 2005.
[14] PVSyst, commercial photovoltaic system simulation software, PVsyst SA. http://www.pvsyst.com/en/ (accessed on February 2016).
[15] PV*Sol, commercial photovoltaic system simulation software, The Solar Design Company. http://www.solardesign.co.uk/index.php (accessed on February 2016).
[16] See, for example, the Fronius Configuration Tool. http://www.fronius.com/ (accessed on February 2016); other suppliers also have similar tools, although specifications change regularly.
[17] A. Mermoud, B. Wittmer, PVSyst6 User's Manual, January 2014, *http://www.pvsyst.com/images/pdf/PVsyst_Tutorials.pdf*.
[18] S. Ransome, J. Sutterlueti, S. Sellner, PV technology differences and discrepancies in modelling between simulation programs and measurements, in: Proc. of the 38th IEEE Photovoltaic Specialists Conference, 2012, pp. 3061–3066.
[19] S. Ransome, J. Sutterlueti, Improving and understanding kWh/kWp simulations, in: Proc. of the 26th European Photovoltaic Solar Energy Conference, 2011, pp. 3255–3260.
[20] IEC 61724-1, Photovoltaic System Performance—Part 1: Monitoring, International Electrotechnical Commission. At the time of writing, this standard was undergoing revision to update the monitoring specification and the new version is expected to be published in 2016.
[21] IEC 62670-2:2015, Photovoltaic Concentrators (CPV)—Performance Testing: Part 2—Energy Measurement, International Electrotechnical Commission, 2015.
[22] A. Drews, A.C. de Keizer, H.G. Beyer, E. Lorenz, J. Betcke, W.G.J.H.M. van Sark, W. Heydenreich, E. Wiemken, S. Stettler, P. Toggweiler, S. Bofinger, M. Schneider, G. Heilscher, D. Heinemann, Monitoring and remote failure detection of grid-connected PV systems based on satellite observations, Sol. Energy 81 (2007) 548–564.
[23] C.F.M. Coimbra, H.T.C. Pedro, Stochastic-learning methods, in: J. Kleissl (Ed.), Solar Energy Forecasting and Resource Assessment, first ed., Academic Press, 2013 (Chapter 15).
[24] R. Marquez, C.F.M. Coimbra, Intra-hour DNI forecasting based on cloud tracking image analysis, Sol. Energy 91 (2013) 327–336.
[25] C.F.M. Coimbra, J. Kleissl, R. Marquez, Overview of solar-forecasting methods and a metric for accuracy evaluation, in: J. Kleissl (Ed.), Solar Energy Forecasting and Resource Assessment, first ed., Academic Press, 2013 (Chapter 8).

[26] S. Pelland, J. Remund, J. Kleissl, T. Oozeki, K. De Brabandere, Photovoltaic and Solar Forecasting: State of the Art, IEA PVPS Task 14, Subtask 3.1, Report IEA-PVPS T14-01: 2013, 2013.
[27] M. Lave, J. Kleissl, Solar variability of four sites across the state of Colorado, Renew. Energy 35 (2010) 2867–2873.
[28] T.E. Hoff, R. Perez, Modeling PV fleet output variability, Sol. Energy 86 (2012) 2177–2189.
[29] P. Lauret, R. Perez, L. Mazorra Aguiar, E. Tapache's, H. Maimouna Diagne, M. David, Characterization of the intraday variability regime of solar irradiation of climatically distinct locations, Sol. Energy 125 (2016) 99–110.
[30] M.J.R. Perez, M. Vasilis, Fthenakis, on the spatial decorrelation of stochastic solar resource variability at long timescales, Sol. Energy 117 (2015) 46–58.
[31] E. Lorenz, T. Scheidsteger, J. Hurka, D. Heinemann, C. Kurz, Regional PV power prediction for improved grid integration, Prog. Photovolt. Res. Appl. 19 (2011) 757–771.
[32] P. Mathieson, J. Kleissl, C. Collier, Case studies of solar forecasting with the weather research and forecasting model at GL-Garrad Hassan, in: J. Kleissl (Ed.), Solar Energy Forecasting and Resource Assessment, first ed., Academic Press, 2013 (Chapter 14).

Environmental life-cycle assessment of photovoltaic systems

7

*V. Fthenakis**,†, *M. Raugei*‡
*Columbia University, New York, NY, United States, †Photovoltaic Environmental Research Center, Upton, NY, United States, ‡Oxford Brookes University, Oxford, United Kingdom

7.1 Introduction

Assessments of the environmental impacts of energy-generation technologies are essential in evaluating their sustainability, and life-cycle assessment (LCA) is arguably the most comprehensive framework to assess such impacts. LCA aims at quantifying the environmental impacts caused by material and energy flows in each and all the stages of the 'life cycle' of a product or process, from 'cradle to grave'. The cycle typically starts from the mining of materials from the ground and continues with the processing of the materials, manufacturing of the compounds and chemicals used in processing, manufacturing of the product, transport, installation if applicable, use, maintenance and eventual decommissioning and disposal and/or recycling. Common metrics for life-cycle evaluations of solar energy systems include energy pay-back time (EPBT), energy return on investment (EROI or EROEI), greenhouse gas (GHG) emissions and toxic emissions along their cradle-to-grave life cycles. A literature review reveals a great divergence in estimates reflecting different assumptions regarding module efficiencies, mounting structures, solar irradiation, performance ratio (PR) and lifetime. However, all recent LCA studies concur with the fact that photovoltaic (PV) technologies have very low environmental impacts compared to those of conventional electricity generation [1] and the range of the mentioned diversity is drastically reduced when one dismisses LCAs based on outdated information or on worst-case conditions.

7.2 Methodology

The methodology employed in PV LCA studies should conform to the ISO standards 14040 and 14044 [2,3], which prescribe the following four steps for conducting an LCA: (a) goal and scope definition; (b) life-cycle inventory (LCI); (c) life-cycle impact assessment (LCIA) and (d) interpretation (Fig. 7.1).

The first step calls for the unambiguous definition of the research objective and the system boundaries used for the analysis. The inventory step quantifies the flows of materials, energy and emissions in each stage of the life cycle of a PV system. In LCIA

The Performance of Photovoltaic (PV) Systems. http://dx.doi.org/10.1016/B978-1-78242-336-2.00007-0

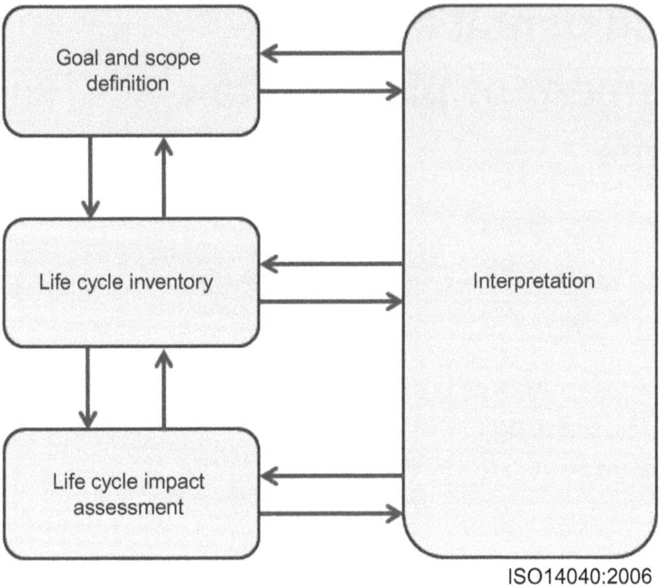

Fig. 7.1 The four-stage structure of a life-cycle assessment.

the impact on the environment of energy consumption, resource consumption, pollutant emissions and GHG emissions are quantified and their cumulative impact on the environment is estimated (separately for a number of different impact categories such as depletion of nonrenewable primary energy resources, global warming, potential toxicity, etc.). The fourth step of the LCA as defined by the ISO is interpretation of the results; this is extremely important as different assumptions can result in very divergent results. To this end the International Energy Agency (IEA) Task 12 add specificity regarding what major assumptions are valid in stationary PV LCAs and regarding the interpretation of the results. The IEA Task 12 recommendations on such assumptions are summarized below [4]:

Life expectancy

- Modules: 30 years.
- Inverters: 15 years for small plants (residential PV); 30 years with 10% of replacement of parts every 10 years for large plants.
- Structure: 30 years for rooftop and façade installations and 60 years for ground-mounted ones.
- Cabling: 30 years.

PR

- The PR accounts for the losses from DC to AC conversion, thermal losses and wire losses and depends on the kind of installation. In general, the ratio increases with a decline in temperature build-up and monitoring the PV systems for early detection of defects; this means that well-ventilated and large-scale systems have a higher PR. A PR of 75% and 80% may be generally assumed respectively for rooftop and for ground-mount utility-scale installations.

Degradation

- The degradation of the modules lowers their efficiency over the lifetime. For mature module technologies, a linear degradation is assumed leading to 80% of the initial efficiency at the end of the 30-year life.

Regarding the interpretation of the results, transparency in reporting is of the utmost importance, as parameters vary with geographical zones, and a system's boundary conditions and modelling approach can affect the findings significantly. The IEA Task 12 LCA Guidelines [5] recommend that the following parameters should be reported: (1) On-plane irradiation level and location; (2) module-rated efficiency; (3) system's PR; (4) time-frame of data; (5) type of system (eg, rooftop, ground-mount fixed tilt or tracker); (6) expected lifetime for PV and balance of system (BOS); (7) system study boundaries [whether capital goods, installation, maintenance, disposal, the transportation and recycling stages are included for both PV modules and BOS (frame, mounting, cabling and inverter; for utility applications the transformer, site preparation and maintenance)]; (8) the place/country/region of production modelled [eg, average grid, site-specific power use (eg, hydro, coal)] and (9) explicit goal of the study (eg, static or prospective LCA, prototype or commercial production and current performance or expected future development). These parameters should either be listed directly in the captions of figures showing the results of the LCA, or otherwise clearly listed in a separate box or table and referred to in all figure captions. In addition, the report should identify the following: the LCA method used, especially if is not process-based; the LCA software tool [eg, SimaPro (http://simapro.com/), GaBi (http://www.gabi-software.com), other]; databases used [eg, Ecoinvent (http://ecoinvent.org), ELCD (http://eplca.jrc.ec.europa.eu/ELCD3/), Franklin (http://www.fal.com/lifecycle-services.html), NREL (http://www.nrel.gov/lci/about.html), other]; the adopted EPBT and/or EROI calculation methods, if applicable (cf. Sections 7.2.2 and 7.2.3 following); commercial representativeness of the study (required if the data are from a pilot-scale production), and assumptions for production of major input materials, eg, solar grade silicon, aluminium (primary and/or secondary production), etc.

7.2.1 Life-cycle stages of a PV system and cumulative energy demand

The life-cycle stages of PVs involve: (1) the production of raw materials, (2) their processing and purification, (3) the manufacture of solar cells, modules and BOS components, (4) the installation and operation of the systems and (5) their decommissioning, disposal or recycling (Fig. 7.2). Typically, separate LCAs are undertaken for the modules and the BOS components (inverters, transformers, mounting, supports and wiring), as the module technologies evolve more rapidly and entail more options than do the BOS structures. Processes in the life-cycle stages of PV systems include, but are not limited to, the following:

(A) Upstream processes

- Raw-material acquisition (eg, mining and concentrating the ores, extracting the fossil fuel, growing trees if wood is used in the system).

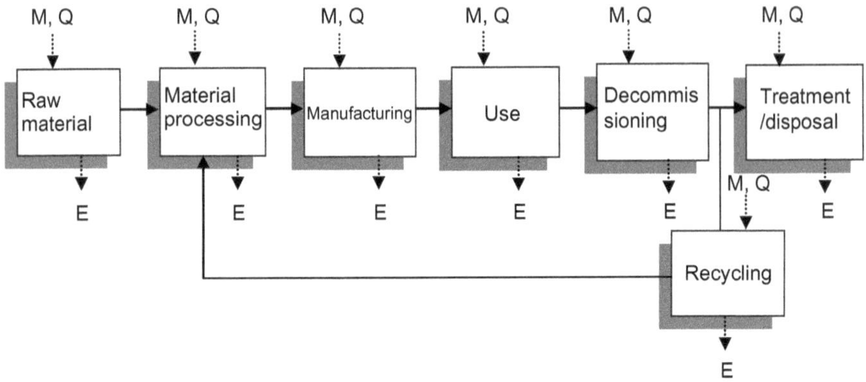

Fig. 7.2 Flow of the life-cycle stages, energy, materials and wastes for PV systems. M and Q—material and energy inputs.

- Materials production (eg, metal smelting, purification, alloying, polymerization, solar silicon production and wafer production).
- Solar-cell production or thin-film deposition (eg, wafer cutting, sputtering, chemical-vapour deposition and vapour-transport deposition).
- Module production (eg, contact formation, encapsulation, wiring and assembly).
- Module and BOS installation (eg, installing module, inverter and support structures).

(B) Operation and maintenance (eg, tracker operation if applicable, cleaning if applicable, other scheduled maintenance and office use for utility scale plant).

(C) Downstream processes

- Decommissioning and disposal; demolition and transportation.
- Recycling; collection, disassembly, shredding and material separation.

The life-cycle cumulative energy demand (CED) [6] of a PV system is the sum total of the (renewable and nonrenewable) primary energy harvested from the geo-biosphere in order to supply the direct energy (eg, fuels and electricity) and material (eg, Si, metals and glass) inputs used in all its life-cycle stages (excluding the solar energy directly harvested by the system during its operation). Thus,

$$\mathrm{CED}\left[\mathrm{MJ}_{\text{PE-eq}}\right] = E_{\text{mat}} + E_{\text{manuf}} + E_{\text{trans}} + E_{\text{inst}} + E_{\text{EOL}}$$

where E_{mat} [$\mathrm{MJ}_{\text{PE-eq}}$] is the primary energy demand to produce materials comprising the PV system; E_{manuf} [$\mathrm{MJ}_{\text{PE-eq}}$] is the primary energy demand to manufacture the PV system; E_{trans} [$\mathrm{MJ}_{\text{PE-eq}}$] is the primary energy demand to transport materials used during the life cycle; E_{inst} [$\mathrm{MJ}_{\text{PE-eq}}$] is the primary energy demand to instal the system and E_{EOL} [$\mathrm{MJ}_{\text{PE-eq}}$] is the primary energy demand for end-of-life management.

The life-cycle nonrenewable CED is a similar metric in which only the nonrenewable primary energy harvested is accounted for.

7.2.2 Energy pay-back time

EPBT is defined as the period required for a renewable energy system to generate the same amount of energy (*in terms of equivalent primary energy*) that was used to produce (and manage at end-of-life) the system itself [4].

$$\text{EPBT[years]} = (E_{\text{mat}} + E_{\text{manuf}} + E_{\text{trans}} + E_{\text{inst}} + E_{\text{EOL}}) / \left((E_{\text{agen}} / \eta_{\text{G}}) - E_{\text{O\&M}} \right)$$

where E_{mat}, E_{manuf}, E_{trans}, E_{inst} and E_{EOL} are defined as previously; additionally, E_{agen} [MJ_{el}/year] is the annual electricity generation; $E_{\text{O\&M}}$ [$\text{MJ}_{\text{PE-eq}}$] is the annual primary energy demand for operation and maintenance and η_{G} [$\text{MJ}_{\text{el}}/\text{MJ}_{\text{PE-eq}}$] is the grid efficiency, ie, the average life-cycle primary energy to electricity conversion efficiency at the demand side.

The annual electricity generation (E_{agen}) is converted into its equivalent primary energy, based on the efficiency of electricity conversion at the demand side, using the grid mix where the PV plant is being installed. Thus, calculating the primary-energy equivalent of the annual electricity generation ($E_{\text{agen}}/\eta_{\text{G}}$) requires knowledge of the life-cycle energy conversion efficiency (η_{G}) of the country-specific energy mixture used to generate electricity and produce materials. The average η_{G} for the United States and Western Europe are, respectively, ~0.30 and 0.31 [7,8].

7.2.3 Energy return on investment

The CED of a PV system may be regarded as the energy 'investment' (I) that is required in order to be able to obtain an energy 'return' (R) in the form of PV electricity.

When looking at the performance of a PV system over its entire lifetime (T), and expressing the electricity it produces *in terms of equivalent primary energy*, we have: $R = T \times ((E_{\text{agen}}/\eta_{\text{G}}) - E_{\text{O\&M}})$.

The overall energy return on (energy) investment (EROI) may thus be calculated as:

$$\text{EROI}_{\text{PE-eq}} \left[\text{MJ}_{\text{PE-eq}} / \text{MJ}_{\text{PE-eq}} \right] = R/I = T \times \left((E_{\text{agen}} / \eta_{\text{G}}) - E_{\text{O\&M}} \right) / (E_{\text{mat}} + E_{\text{manuf}} + E_{\text{trans}} + E_{\text{inst}} + E_{\text{EOL}}) = T/\text{EPBT}$$

The fundamental idea underpinning EROI is that, in order for an energy production system to provide a positive net energy 'return' to the end user ($R_{\text{net}} = R - I$), the gross energy return (R) must be larger than the energy 'investment' (I), ie, $\text{EROI}_{\text{PE-eq}}$ must be >1.

$\text{EROI}_{\text{PE-eq}}$ and EPBT provide complementary information.

In fact, $\text{EROI}_{\text{PE-eq}}$ looks at the overall energy performance of the PV system as a whole over its entire lifetime. In reality, however, the largest part of the energy 'investment' for PV (ie, $E_{\text{mat}} + E_{\text{manuf}} + E_{\text{trans}} + E_{\text{inst}}$) is required up front before the system even starts to produce any electricity, while the energy 'return' is spread over the years of operating phase. Thus, from a dynamic perspective, EPBT measures the point in time (t) after which the system is able to provide a net energy 'return', that is:

$$R_{\text{net}}(t) = R(t) - I > 0 \text{ when } t > \text{EPBT}$$

Finally, it is also noteworthy that in the literature the EROI of PV has also sometimes been calculated by defining the energy 'return' as the direct electricity output produced over the system's lifetime (R_{el}) and divide it by the energy invested (as primary energy) [9]:

$$\begin{aligned}\text{EROI}_{\text{el}}\left[\text{MJ}_{\text{el}}/\text{MJ}_{\text{PE-eq}}\right] &= R_{\text{el}}/I \\ &= \left(E_{\text{agen}} - E_{\text{O\&M}}\right)/\left(E_{\text{mat}} + E_{\text{manuf}} + E_{\text{trans}} + E_{\text{inst}} + E_{\text{EOL}}\right)\end{aligned}$$

It must be noted, however, that in so doing the numerator and denominator are no longer measured in the same units, and hence R_{net} is no longer strictly defined, and the 'intuitively appealing interpretation that EROI > 1 is the absolute minimum requirement a resource must meet in order to constitute a net energy source' is lost [10,11].

Further discussion of the EROI methodology can be found in an IEA PVPS Task 12 report [12].

7.2.4 GHG emissions and global warming potential

The overall global warming potential (GWP) due to the emission of a number of GHGs along the various stages of the PV life cycle is typically estimated using an integrated time-horizon of 100 years (GWP_{100}), whereby the following CO_2-equivalent factors are used: 1 kg CH_4 = 23 kg CO_2-eq, 1 kg N_2O = 296 kg CO_2-eq and 1 kg chlorofluorocarbons = 4600–10,600 kg CO_2-eq [55]. It is noted that the CO_2-equivalent factor of CH_4 has recently been updated to 32; the effect of this update on the reported GHG has not been assessed yet. Electricity and fuel use during the production of the PV materials and modules are the main sources of the GHG emissions for PV cycles, and specifically the technologies and processes for generating the upstream electricity play an important role in determining the total GWP of PV systems, since the higher the mixture of fossil fuels is in the grid, the higher are the GHG (and toxic) emissions. For instance, the GHG emission factor of the average US electricity grid is higher than that of the average Western European (UCTE) grid, and that of China is higher still. During 1999–2002, the US average GHG emissions inventory from electricity production and transmission was 676 g CO_2-eq/kWh, while, in Germany it was 539 g CO_2-eq/kWh, and in China it was 839 g CO_2-eq/kWh [14]. The emissions in the US and Europe are lower today, but assuming that the differences remain the same, would result to 20% higher GHG emissions for producing modules and BOS components in China than in the United States. Furthermore, transporting the modules and BOS components from China to either the United States or Europe could add an additional 15% to the GHG emissions [15].

7.3 Commercial PV technologies considered in current analyses

7.3.1 Crystalline Si PV

An LCA of crystalline-Si PV starts with the assessment of the mining of the raw materials (ie, the quartz sand for silicon PV and the metals used in the BOS components), and continues with their processing and purification (Fig. 7.3). The silica in the quartz sand is reduced in an arc furnace to metallurgical-grade silicon, which is purified into solar-grade polysilicon by either a silane (SiH_4)- or a trichlorosilane ($SiHCl_3$)-based process. The energy requirement for purification is the largest expense for making crystalline Si PV modules, accounting for 45% of the primary energy used in fabricating multi-Si modules. Two technologies are employed for producing polysilicon from silicon gases: the Siemens reactor method, and the fluidized bed reactor (FBR) method. In the former, which accounts for the majority (~90% in 2004) of solar-grade silicon production in the United States, silane or trichlorosilane is introduced into a thermal-decomposition reactor with high-temperature (~1100–1200°C) polysilicon rods [16,17]. The silicon rods grow as silicon atoms from the gas phase

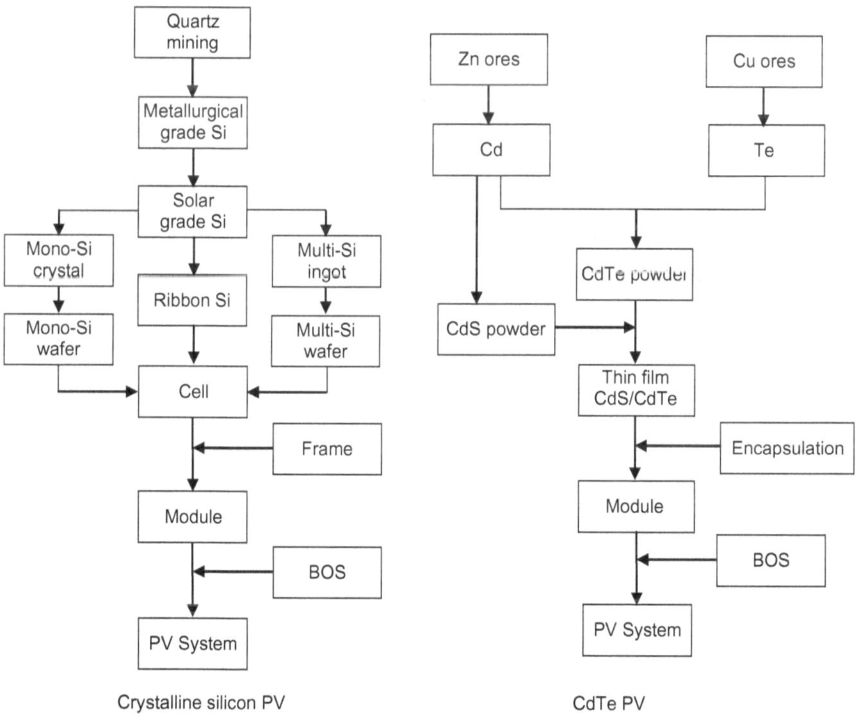

Fig. 7.3 Detailed flow diagram for PV from the acquisition of raw material to its manufacturing stage.

deposit on them [16]. In 1987, MEMC Electronic Materials commercialized the FBR method on a small scale; recent developments show promise for scaling-up production to meet the needs of the emerging solar industry. In this approach, granular polysilicon is produced by decomposing silane around seed particles in a mixture of silane/hydrogen fed into the reactor. This continuous production process offers inherent advantages over the Siemens batch process, requiring lower energy consumption while enabling higher throughput. However, such advantages have not been quantified in LCAs [18]. The LCI data used in the studies reviewed here correspond to polysilicon production via the Siemens method, and were gathered in the European 'Crystal Clear' project. Detailed LCIs can be found in IEA PVPS Task 12 reports [19,20]. The average module efficiencies for monocrystalline and polycrystalline Si PV described in these inventories are respectively ~15% and ~14%. It is noted that as of 2016, these module efficiencies have considerably increased, and the current EPBT and EROI estimates are expected to be better than the ones reported herein.

7.3.2 Cadmium telluride PV

Fthenakis et al. [21–23] reported the LCI data for cadmium telluride (CdTe) thin-film technology gleaned from the 2005 production data from First Solar's plants, Perrysburg, United States in 2005, and from Frankfurt-Oder, Germany data in 2008. Held and coworkers [24] and Raugei et al. [25] used the same data in their independent investigations.

The life-cycle inventories of the minor metals used in thin-film CdTe PVs (ie, Cd and Te) are related closely to the production cycle of the respective base metals, Zn and Cu [26]. Cadmium is obtained from the waste streams in zinc smelting, mainly in the slimes from the Zn-electrolyte purification stages, and the particulates collected in bag-houses, electrostatic precipitators (ESPs), and cyclones in the thermal-oxidation units; thereafter, it is processed further and purified to the 99.999% purity required for synthesizing CdTe. Te is recovered after treating with dilute sulfuric acid the slimes produced during electrolytic copper refining; these slimes also contain Cu and other metals. After its cementation with copper, CuTe is leached out with caustic soda to produce a sodium telluride solution that is used as the feed for Te and TeO_2. Additional leaching and vacuum-distillation produces Te- and TeO_2-powders of 99.999% purity required for synthesizing pure CdTe. This same route is followed in generating solar-grade CdS, used as a window layer in CdTe PV. Both the CdTe and CdS layers are placed by vapour-transport deposition, based on subliming the powders and condensing the vapours on glass substrates. A stream of inert carrier gas guides the cloud of sublimed dense vapour to deposit the films on glass substrates at 500–600°C at a line speed of up to 2.5 m/s [27], followed by depositing layers of metals. Thereafter, a series of scribing and heat treatments form interconnections and back contacts. In 2015 First Solar produced CdTe PV modules with an average efficiency of 15.5% [28].

Glass is the heaviest part of the PV module, particularly in frameless CdTe modules where two panes of glass ensure their structural toughness. The double-glass design eliminates the need for an aluminium frame that accounts for a significant fraction of emissions from making silicon modules. The use of thin-film layers per unit area

Table 7.1 Materials and energy inputs in the production of Si and CdTe PV modules—per m^2 of module (excluding frames, commonly used for Si modules, http://ecoinvent.org)

Category	Inputs	Multi-Si	Mono-Si	CdTe
Components (kg)	Cell materials	1.6	1.5	0.065
	Glass	9.1	9.1	19.2
	Ethylene vinyl acetate	1.0	1.0	0.6
	Other	1.8	1.8	2.0
Consumables (kg)	Gases	2.2	7.8	0.001
	Liquid	6.8	6.6	0.67
	Other	4.3	4.3	0.4
Energy	Electricity (kWh)	248	282	59
	Oil (L)	0.05	0.05	0.05
	Natural gas (MJ/)	308	361	–

of thin-film cadmium telluride (CdTe), copper indium gallium selenide (CIGS), and amorphous silicon–germanium (a-SiGe) is minimal compared with silicon modules since the thickness of cell materials in the former is 1–3 μm compared with 150–200 μm for the latter. Thin-film modules also require lesser amounts of consumables than do silicon modules (Table 7.1), as fewer process steps are involved (Fig. 7.3).

7.3.3 Copper indium gallium selenide PV

Currently, CIGS PV modules are available in several major-layer configurations depending on the substrate and encapsulation materials, including glass/glass, glass/polymer and polymer/polymer. Reported module efficiencies range from 9.4% to 13.1%. In double glass configurations manufacturers offer frameless modules, as the two glass sheets provide sufficient structural integrity.

The European Commission's project ECLIPSE (Environmental and Ecological Life Cycle Inventories for present and future Power Systems in Europe) was the first major research on the LCI of commercial CIGS PV. This project compiled life-cycle inventories of emerging options for power generation, including PV, wind and biomass. The PV technologies investigated were single- and multicrystalline silicon modules, and thin-film modules, such as a-Si, copper indium gallium diselenide and CdTe. This study reported the usage of a mass of metals, glass, encapsulating polymer and a frame, along with electricity requirements during manufacturing based on data from the literature [13].

Raugei et al. (2007) [25] collected the first LCI from an actual production line of CIGS PV, from a pilot-scale manufacturing process of Würth Solar, Germany. The SENSE report [29] gave information on the production of CIGS solar modules by Würth Solar; Lozanovski and Held [30] summarize it. The sputtering of molybdenum aims at coating the 'back contact'. The substrates are structured by laser techniques. Targets of copper, indium, gallium and selenium are co-evaporated, and substrates carry the vapours under an inert-gas flow process. A buffering layer of CdS (cadmium sulphide) is put onto the substrates by dip coating. The front electrode is deposited by the sputtering of zinc oxide, doped with Al_2O_3. Laser scribing follows the deposition of each layer. A functional test identifies the defective products. The addition of interconnects and front glass completes the module production. Section 7.4 shows the results of the SENSE report; however, the corresponding LCI data are not available. The average module efficiency of CIGS PV modules described in these studies is ~12% [20].

7.3.4 Concentrated photovoltaics

We report LCA results for the most common design of concentrated photovoltaics (CPV) that is point focus high concentration employing III–V cells and Fresnel lenses in megamodules, mounted on two-axis trackers. A representative of this technology is the Amonix 7700 CPV system for which a complete LCA was conducted. The rated capacity of the system is 60 kWp and its total area is 328 m^2; its aperture area is 267 m^2. Sunlight is concentrated on each of 7560 multijunction GaInP/GaInAs/Ge solar cells at a concentration ratio of 500 suns. The cells were rated at 37% under 500 kW/m^2, 25°C, and AM 1.5D spectrum.

Fthenakis et al. [56] reported an LCA of the Amonix 7700 system. Table 7.2 shows the materials used in the production of the cells and Table 7.3 shows the aggregate of the materials in the whole system; while the measurements of the mass of

Table 7.2 Materials and energy inputs for processing 1000 wafers, corresponding to 915 kWp[a] of GaInP/GaInAs/Ge terrestrial concentrator solar cell [2]

Inputs	Amount
Materials for components (kg)	
Wafer/precursors	21.7
Contact metals	3.4
Antireflection coating	0.02
Materials use for process (kg)	
Hydrogen	57.2
Nitrogen	0.2
Photoresist	2.3
Solvents	1073.4
Acids	255.8
Bases	161.3
Electricity (kWh)	
MOVPE	2365
Other (gas scrubbing and cell processing)	470

[a] At 37% rated efficiency.

Table 7.3 **Material breakdown of the Amonix 7700**

Submodule	Components	Materials	Mass (kg)	Fraction (%)
MegaModules	Cell	Semiconductor	0.2	0.001
	Frame	Steel	6566	23.0
	Fresnel lenses	Acrylic	1143	4.0
	Heat Sink	Aluminium	3086	10.8
Tracker	Foundation	Concrete	3126	10.9
	Hydraulic drive	Steel	2724	9.5
	Pedestal and torque tube	Steel	11,260	39.4
	Motor	Various	16	0.1
Electrical	Inverter	Various	500	1.7
	Transformer	Various	100	0.3
	Cables	Copper/PVC	35	0.1
Other	Controller	Various	18	0.1
	Sensor	Various	1.4	0.005
	Anemometer	Various	0.1	0.0003

manufactured parts were taken directly from the assembly line, the quantity of concrete used was calculated from the foundation's dimensions: 5.5 m deep and 1.1 m diameter. The detailed material compositions of electrical parts, ie, motor, transformer and inverter, were estimated from Mason et al. [31]. The LCI includes the materials used in scheduled maintenance over an expected lifetime of 30 years; they include changing the hydraulic and bearing oils, cleaning the lens and changing the air and oil filters. The MegaModules (36%) and tracker (58%) account for most of the components, while steel (75%), concrete (11%) and aluminium (11%) dominate the material usages.

7.3.5 Balance of system

PV modules are either rooftop or ground mounted. Silicon modules need an aluminium frame for structural robustness, while a second glass back-sheet performs the same function for thin-film CdTe PV. For a rooftop PV application, the BOS typically includes inverters, mounting structures, cables and connectors. Large-scale ground-mounted PV installations require additional equipment and facilities, such as transformers for grid connections, office facilities, and, in some cases, concrete for installations.

De Wild-Scholten [32] report the material inventories for two types of rooftop-mounting systems: on-roof mounting wherein the system is built on existing roofing material, and in-roof mounting where the modules replace the roof tiles. Table 7.4 shows the material LCI of common rooftop mounting systems, cabling, inverters and transformers, including two sizes (500 and 2500 W) of small inverters adequate for rooftop PV designs.

Table 7.4 LCI of balance of system [14]

	On-roof		In-roof	
	Phönix, TectoSun	**Schletter Eco05+EcoG**	**Schletter, Plandach 5**	**Schweizer, Solrif**
(a) Mounting system (kg/m² module)				
Low alloy steel	0	0	0	0
Stainless steel	0.49	0.72	0.28	0.08
Aluminium	0.54	0.97	1.21	1.71
Concrete	0	0	0	0
Frame	3.04	0	0	0

	Helukabel, Solarflex 101, 4 mm², DC	Helukabel, NYM-J, 6 mm², AC
(b) Cabling (g/m²)		
Copper	83.0	19.9
Thermoplastic elastomer (TPE)	64.0	0.0
PVC	0.0	16.9

	Philips PSI 500 (500 W)	Mastervolt SunMaster 2500 (2500 W)
(c) Inverters (g)		
Steel	78	9800
Aluminium	682	1400
Copper	2	
Polycarbonate	68	
ABS	148	
Other plastics	5.4	
Printed circuit board	100	1800[a]
Connector	50	
Transformers, wire-wound	310	5500
Coils	74	
Transistor diode	10	
Capacitor, film	72	
Capacitor, electrolytic	54	
Other electric components	20	

[a] Including electric components (eg, connectors and transistors).

Table 7.5 Mass balance of major components for the 3.5 MW Tucson Electric Power Generating plant in Springerville, AZ, based on 30 years of operation

	Balance of system	Mass (kg/MW$_p$)	Total (%)
BOS	PV support structure	16,821	10.3
	Module interconnections	453	0.3
	Junction boxes	1385	0.8
	Conduits and fittings	6561	4.0
	Wire and grounding devices	5648	3.4
	Inverters[a] and transformers	28,320	17.3
	Grid connections	1726	1.1
	Office facilities	20,697	12.6
	Concrete	76,417	46.6
	Miscellaneous	5806	3.5
	Total	163,834	100.0
Frame[a]		18,141	

Based on 12.2% rated efficiency for monocrystalline-Si module.
[a] Xantrec PV-150 inverter type.

Data on materials and energy use for a ground-mounted BOS were obtained from the 5 MW$_p$ PV plant at the Springerville, Arizona facility operated by Tucson Electric Power (TEP) (Table 7.5) [31]. The inverters are utility-scale models with a wide-open frame, allowing the facile replacement of failed parts. Thus, inverters and transformers are expected to last 30 years, with parts amounting to 10% of their total mass, being replaced every 10 years. The LCI includes the office facility's materials and energy use for administrative, maintenance and security staff, as well for running the maintenance vehicles. Detailed LCIs are shown in an IEA PVPS Task 12 report [19].

7.4 Results

7.4.1 Energy pay-back time and energy return on investment

Fthenakis et al. [23] showed that the primary energy used in the life cycle of a complete rooftop Si PV system is 2700 and 2900 MJ/m^2, respectively, for multi-, and monocrystalline Si modules, down from 5000 and 2700 in 2006. The EPBT of these systems was estimated to be 1.8 years for module efficiencies of 13.2% and 14% correspondingly, in rooftop installations under Southern European insolation of 1700 kWh/(m^2 year), with a grid efficiency (η_G) of 0.31 and a PR of 0.75 (Fig. 7.4). The corresponding EROI$_{PE\text{-}eq}$ ratio, assuming a 30-year lifetime, is 17. In these estimates, the BOS for rooftop applications accounts for 0.3 years of EPBT, but there are different types of rooftop mounting systems with different energy burdens.

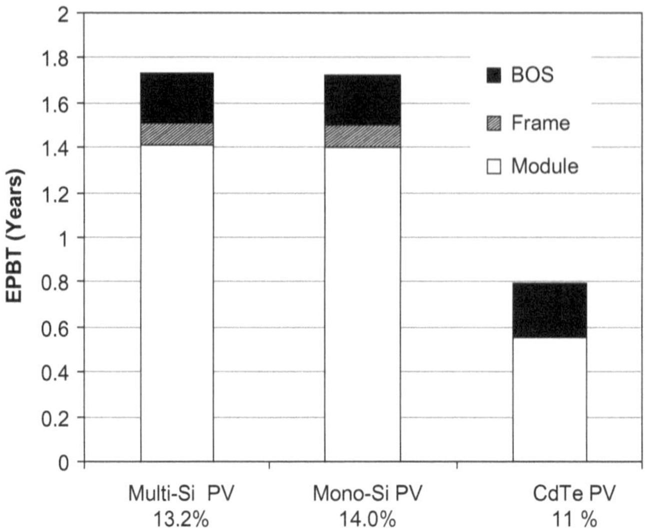

Fig. 7.4 EPBT of PV systems: rooftop installed module with insolation = 1700 kWh/(m^2 year) and performance ratio = 0.75; European production.

For CdTe PV, the primary energy consumption was estimated to be 850 and 970 MJ/m^2, correspondingly, based on actual production from First Solar's plant in Frankfurt-Oder, Germany and Perrysburg, OH, United States. The energy burden at the US plant is higher than that in Germany because the former includes R&D and additional administrative functions. For insolation levels of 1700 kWh/(m^2 year) and a grid efficiency (η_G) of 0.31, the EPBT and $EROI_{PE\text{-}eq}$ values for the European production are: EPBT = 0.8 years, $EROI_{PE\text{-}eq}$ = 37.5; and for the US production: EPBT = 0.9 years, $EROI_{PE\text{-}eq}$ = 33.

EPBT decreases (and EROI increases) as the solar irradiation levels increase; for example, in Phoenix, AZ, US Southwest [latitude optimal irradiation of 2370 kWh/(m^2 year)], the EPBTs of crystalline silicon and cadmium telluride PV in fixed-tilt ground mount utility installations, respectively, are 1.3 and 0.6 years. At the same location, the two-axis Amonix HCPV would have an EPBT of 0.9 years (Table 7.6).

The EROI of PV as calculated by different research teams has recently been the object of controversy in the scientific literature. A critical overview of this issue can be found in Ref. [33].

7.4.2 GHG emissions

Brookhaven National Laboratory (BNL) in collaboration with the National Renewable Energy Laboratory (NREL) recently undertook a harmonization study of greenhouse estimates from the hundreds of reported LCAs [34]. It consisted of

Table 7.6 Direct, atmospheric Cd emissions during the life cycle of the CdTe PV module (allocation of emissions to coproduction of Zn, Cd, Ge and In)

	Air emissions (g Cd/ tonne Cd[a])	Allocation (%)[b]	Air emissions (g Cd/ tonne Cd[a])	mg Cd/GWh[c]
Mining of Zn ores	2.7	0.58	0.016	0.02
Zn smelting/ refining	40	0.58	0.23	0.3
Cd purification	6	100	5.97	7.79
CdTe production	6	100	5.97	7.79
PV manufacturing	3	100	0.40	0.52
Operation	0.3	100	0.05	0.06
Disposal/recycling	0	100	0.10	0.13
Total			12.74	16.61

[a] Tonnes of Cd produced.
[b] Emissions are allocated between coproducts (Zn, Cd and other) based on mass and cost proportioning, according to ISO 14040 guidelines.
[c] Energy produced assuming average Southern European insolation [ie, 1700 kWh/(m^2 year)], 9% electrical conversion efficiency, and a 30-year life for the modules.

systematically gathering nearly all publicly available LCAs that referred to PVs, screening them to ensure minimum standards of quality and relevance, and then reconciling several key performance characteristics (eg, irradiation, system lifetime, module efficiency and PR) with standard values following the IEA Task 12 LCA Guidelines [5].

For crystalline-Si PV, 400 studies were screened; before matching them, the median life-cycle GHG emissions reported in the c-Si LCAs passing screens was 57 g CO_2-eq/kWh, with an interquartile range (IQR) of 44–73. After harmonization, the median estimate declined to 32–57 g CO_2-eq/kWh and IQR tightened to 27–35 [34]. The median GHG decreased because of higher irradiation and a longer system lifetime. Standardizing irradiation and system lifetime were the most effective parameters in reducing variability. The most valid LCAs for crystalline-silicon modules are those based on data obtained from the CrystalClear Project. The GHG emissions of Si modules are ~29 g CO_2-eq/kWh range, for a rooftop application under Southern European insolation of 1700 kWh/(m^2 year), and a (PR) of 0.75 (Fig. 7.5). Notably, under these estimates, the BOS for rooftop applications accounts for ~5 g CO_2-eq/kWh of GHG emissions. These calculations were based on the electricity mixture for the 2009 production of Si that is powered by hydroelectricity.

The harmonization process of thin-film PVs involved 109 LCA studies [34]. As with the work on c-Si activity, the initial screening eliminated abstracts, short conference papers, presentations without supporting documentation, and unrelated analyses; 90 studies passed this screening. In the primary screening, rigorous criteria were

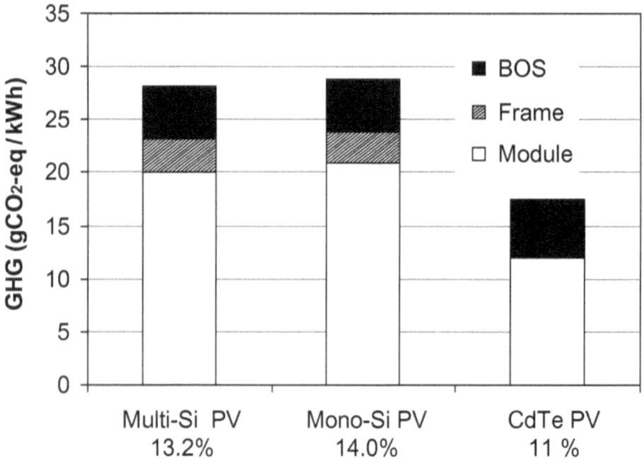

Fig. 7.5 GHG emissions of PV systems: rooftop-installed module with insolation = 1700 kWh/(m^2 year) and performance ratio = 0.75; European production.

applied to assure the completeness of reporting, the validity of LCA methods, and the relevance to present times of the PV system studied. Additionally, it was determined whether the product is a commercial one, whether the production line still exists, and whether the study's core data is original or secondary. These screenings identified four studies as the best representations of the carbon footprint of the modern thin-film PV technologies: Pacca et al. [35] for a-Si; Fthenakis et al. [23,36] for CdTe; Raugei and Fthenakis [13,37] and SENSE for CIGS [29,30]. The corresponding GHG emissions are 28, 17, 55 and 35 g CO_2-eq/kWh for irradiation levels of 1700 kWh/(m^2 year). The wide difference in the CIGS estimates highlights the need for more complete analyses of these technologies.

For assessing BOS, estimates were used of GHG emissions from rooftop systems, based on the Crystal Clear project [38,39]; they correspond to 5 g CO_2-eq/kWh under 1700 kWh of insolation, with 14% module efficiency, and PR (ie, the ratio of the DC rated and the actual AC electricity output) of 0.75. For ground-mounted BOS, data were employed corresponding to the TEP power plant in Springerville, Arizona [31]. They are 5.4 g CO_2-eq/kWh, harmonized for the same conditions as the rooftop application. Emissions from the structural part of the BOS were adjusted according to the conversion efficiency of PV, whereas emissions from the inverter portion of the BOS that depend on power output were unchanged. Thin-film CdTe PV modules have no frames, and the same was assumed for CIGS modules. A triple-junction a-Si module deposited on a stainless-steel substrate, manufactured by Uni-Solar, uses a frame with a very thin aluminium profile, specifically, 15 g of anodized extruded aluminium per m^2 of module, except for building-integrated applications [35]. Crystalline silicon modules employ a single glass sheet, and have an aluminium frame for structural stability, weighing $\sim$3 kg/m^2 of module.

Both the as-published and harmonized LCA results suggest that the carbon footprint of thin-film PV technologies fall significantly as production capacity increases, reflecting the technological advances in process and device designs. For example, between 2005 and 2010, First Solar's annual production capacity of CdTe PV jumped from 25 to 1100 MWp, and, during the same period, the module efficiency of CdTe PV increased from 9% to 11.6% while the life-cycle GHG estimates fell by ~30% [23,36,39]. Scaling up a CIGS PV prototype to a large commercial line also likely will result in significant, but yet to be accurately determined, reductions in GHG emissions [25,30].

Fig. 7.6 compares GHG emissions from the life cycle of PV with those of conventional fuel-burning power plants, revealing the environmental advantage of using PV technologies. The majority of GHG emissions come from the operational stage for the coal, natural gas and oil fuel cycles, while the material and device production accounts for nearly all the emissions for the PV cycles. With over 50% contributions, the GHG emissions from the electricity demand in the life cycle of PV are the most impactful input. Therefore, the LCA results strongly depend on the available electricity mix. The GHG emissions from the nuclear fuel cycle mainly are related to the fuel production, ie, mining, milling, fabrication, conversion and the enrichment of uranium fuel. The details of the US nuclear fuel cycle are described elsewhere [41].

Other comparisons between the life cycles of PVs and conventional power generation cover land use [42] and water use [43]. Accounting for the land occupation in coal mining, it is shown that the life cycles of PV in the US-SW occupy about the same or less land, and orders of magnitude less water, on an electricity produced (GWh) basis, than the average coal-based power generation.

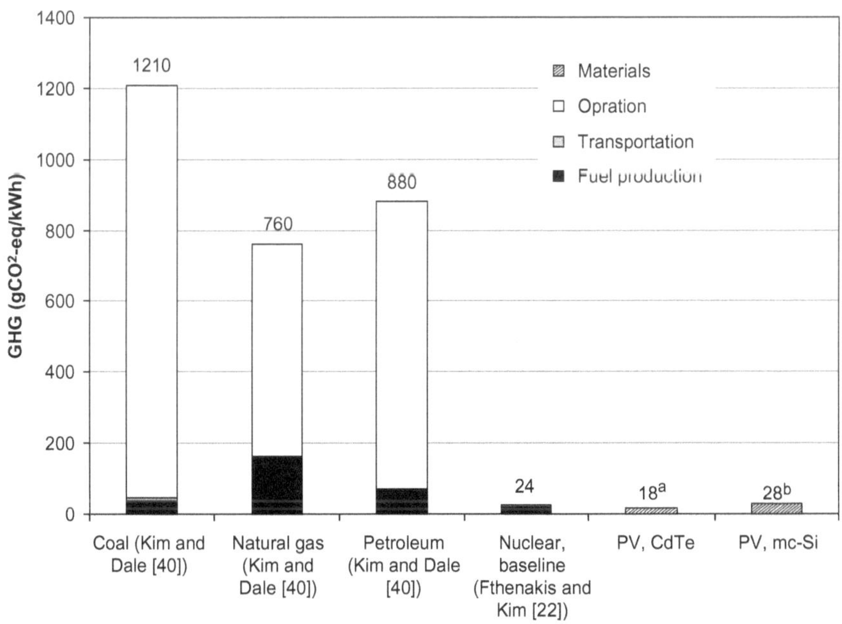

Fig. 7.6 Comparison of GHG emissions from PV with those from conventional power plants [insolation of 1700 kWh/(m^2 year)], performance ratio of 0.75.

7.4.3 Toxic gas emissions

The emissions of toxic gases (eg, SO_2, NO_x) and heavy metals (eg, As, Cd, Hg, Cr, Ni, Pb) during the life cycle of a PV system are largely proportional to the amount of fossil fuel consumed during its various phases, in particular, processing and manufacturing PV materials. Fig. 7.7 shows estimates of SO_2 and NO_x emissions. Heavy metals may be emitted directly from material processing and PV manufacturing, and indirectly from generating the energy used at both stages. For the most part, they originate as trace metals in the coal used.

Direct emissions of cadmium in the life cycle of CdTe PV were assessed in detail [44]. They total 0.016 g/GWh of PV-produced energy under an irradiation of 1700 kWh/(m^2 year) (Table 7.7); this includes emissions during fires on rooftop residential systems, quantified in experiments at BNL that simulated actual fires [21]. These experiments were designed to replicate average conditions, and the estimated emissions were calculated by accounting for US fire statistics pointing to 1/10,000 houses catching fire over the course of a year in the United States where most houses have wood frames, by assuming

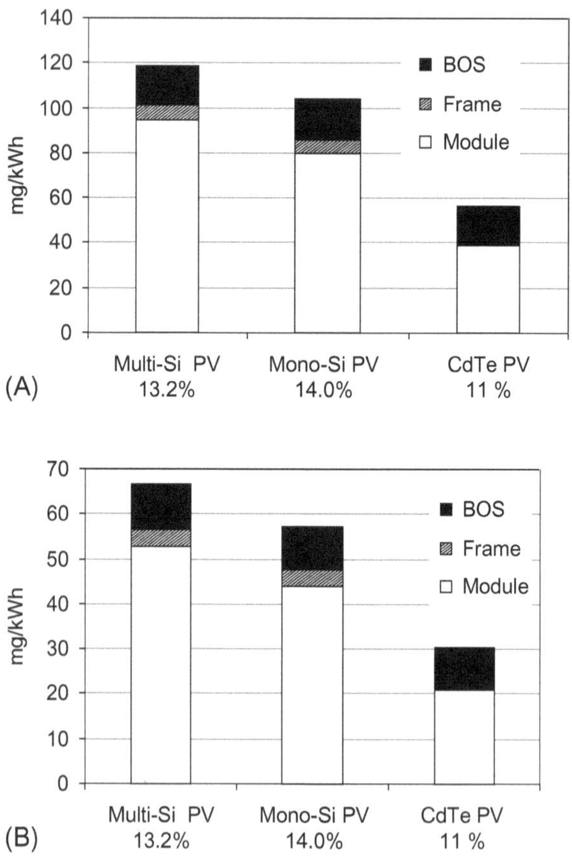

Fig. 7.7 Life-cycle emissions of (A) SO_2, and (B) NO_x emissions from silicon and CdTe PV modules. BOS includes module supports, cabling and power conditioning. The estimates are based on rooftop-mount installation, insolation of 1700 kWh/(m^2 year), performance ratio of 0.75, lifetime of 30 years, and European production with electricity supply from the UCTE grid.

Table 7.7 Comparison of life-cycle parameters and performances across PV technologies (normalized for Phoenix insolation)

PV system	Amonix HCPV, 7700 current	Multi crystalline-Si ground-mount	Mono crystalline-Si ground-mount [2,6]	CdTe ground-mount [7]
Module DC efficiency (%)	37 (cell)	13.2	14	10.9
Insolation [kWh/(m^2 year)]	2480[a]	2370[b]	2370[b]	2370[b]
EPBT (years)	0.9		1.8	0.7

100 years of integrated time-horizon.
[a] Direct normal insolation with two-axis tracker.
[b] Longitude optimal insolation for flat-plate collectors.

that all fires involve the roof. The indirect Cd emissions from electricity usage during the life cycle of CdTe PV modules (ie, 0.24 g/GWh) are an order of magnitude greater than the direct ones (routine and accidental) (ie, 0.016 g/GWh) [22].[1] Cadmium emissions from the electricity demand for each module were assigned, assuming that the life-cycle electricity for the silicon and CdTe PV modules was supplied by the Union for the Coordination of Transmission of Energy's (European) grid. The complete life-cycle atmospheric Cd emissions, estimated by adding those from the usage of electricity and fuel in manufacturing and producing materials for various PV modules and BOS, were compared with the emissions from other electricity-generating technologies (Fig. 7.8) [36]. Undoubtedly, displacing the others with CdTe PV markedly lowers the amount of Cd released into the air. Thus, every gigawatt-hour of electricity generated by CdTe PV modules can prevent around 5 g of Cd air emissions if they are used instead of, or as a supplement to, the UCTE electricity grid. In addition, the direct emissions of Cd during the life cycle of CdTe PV are 10 times lower than the indirect ones due to use of electricity and fuel in the same life cycle, and about 30 times less than those indirect emissions from crystalline PVs. Furthermore, examining the indirect heavy metal emissions in the life cycle of the three silicon technologies discussed earlier revealed that, among the PV technologies, CdTe PV had the lowest energy burden and, consequently, the fewest heavy-metal emissions (Fig. 7.9) [36]. Regardless of the particular PV technology, these emissions are extremely small compared to the emissions from the fossil-fuel based plants

[1] Indirect emissions of heavy metals result mainly from the trace elements in coal and oil. According to the US Electric Power Research Institute's (EPRI's) data, under the best/optimized operational- and maintenance-conditions, burning coal for electricity releases into the air between 2 and 7 g of Cd/GWh [45]. In addition, 140 g/GWh of Cd inevitably collects as fine dust in boilers, baghouses and ESPs. Furthermore, a typical US coal-powered plant emits per GWh about 1000 tons of CO_2, 8 tons of SO_2, 3 tons of NO_x and 0.4 tons of particulates. The emissions of Cd from heavy-oil burning power plants are 12–14 times higher than those from coal plants, even though heavy oil contains much less Cd than coal ($\sim$0.1 ppm); this is because these plants do not have particulate-control equipment [44].

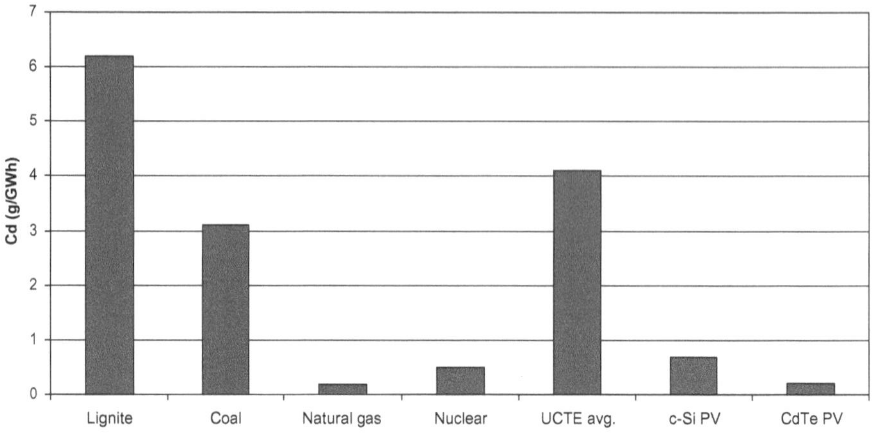

Fig. 7.8 Life-cycle atmospheric Cd emissions for PV systems from electricity and fuel consumption, normalized for a Southern Europe average insolation of 1700 kWh/(m^2 year), performance ratio of 0.75, and lifetime of 30 years.

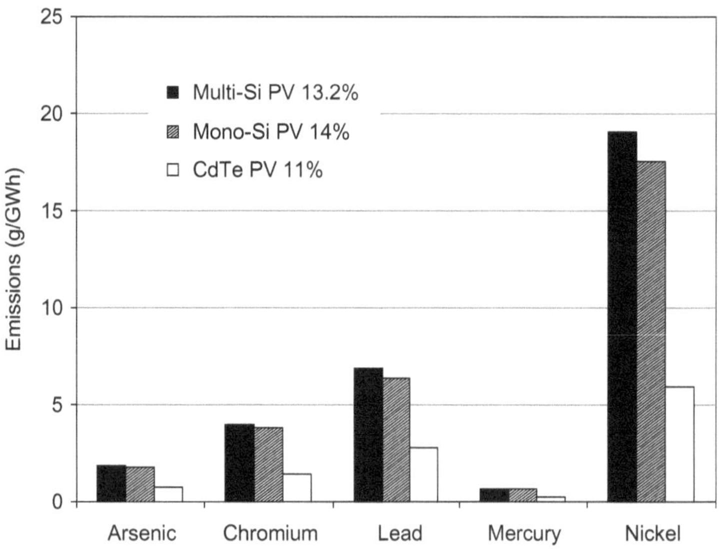

Fig. 7.9 Emissions of heavy metals due to electricity use, based on European UTCE averages (Ecoinvent database). Emissions are normalized for Southern European average insolation of 1700 kWh/(m^2 year), performance ratio of 0.75, and lifetime of 30 years.

that PV will replace. Furthermore, the external environmental costs of PVs are negligible in comparison to the external costs of fossil fuel life cycles [9,46,47].

Ongoing LCA studies include evaluating replacing building materials with PV in building-integrated PVs [57], and recycling of PV modules at the end of their useful life [48–54].

7.5 Conclusion

This chapter gives an overview of the life-cycle environmental performance of PV technologies. EPBT is a basic metric of this performance; it measures the time it takes for a PV system to generate as much electricity as could be generated by the current electric mix, when using the same amount of primary energy that is used for the production of the PV system. The lower the EPBT, the lower will also be the emissions to the environment, because emissions mainly occur from using fossil-fuel-based energy in producing materials, solar cells, modules and systems. These emissions differ in different countries, depending on that country's mixture in the electricity grid and the varying methods of material/fuel processing. Under average US and Southern European conditions (eg, 1700 $kWh/m^2/year$), the EPBT of crystalline Si-, and CdTe rooftop PV systems was estimated respectively to be 1.8 and 0.8 years. Ground-mount installations entail a greater amount of BOS and, therefore, have a longer EPBT. Under US SW irradiation [eg, Phoenix, AZ, 2370 $kWh/(m^2$ year) at a fixed latitude tilt], the EPBT of ground-mount installations is estimated to be 1.3 and 0.6 years, respectively, for crystalline silicon and cadmium telluride PV. At the same location, the EPBT of the Amonix 2-axis HCPV is assessed as 0.9 years.

The environmental impacts of the life cycle of PVs as assessed by the common metrics of GHG emissions, toxic emissions and heavy metal emissions are very small in comparison to those of the power generation technologies they replace. In addition the PV industry continues to evolve with new module efficiency records and their EPBT and GHG emissions continue to decrease.

References

[1] V.M. Fthenakis, H.C. Kim, Photovoltaics: life-cycle analyses, Sol. Energy 85 (2011) 1609–1628.

[2] ISO, 14040—Environmental Management. Life Cycle Assessment: Principles and Framework, International Organization for Standardization, Geneva, 2006.

[3] ISO, 14044—Environmental Management. Life Cycle Assessment: Requirements and Guidelines, International Organization for Standardization, Geneva, 2006.

[4] E. Alsema, D. Fraile, R. Frischknecht, V. Fthenakis, M. Held, H.C. Kim, W. Polz, M. Raugei, M. Wild Scholten, Methodology Guidelines on Life Cycle Assessment of Photovoltaic Electricity, International Energy Agency, Paris, 2009. Report IEA-PVPS T12-01:2009.

[5] V. Fthenakis, R. Frischknecht, M. Raugei, H.C. Kim, E. Alsema, M. Held, M. De Wild-Scholten, Methodology guidelines on life cycle assessment of photovoltaic electricity, in: International Energy Agency Photovoltaic Power Systems Programme, IEA PVPS Task 12, second ed., 2011. Report IEA-PVPS T12-03:2011.

[6] R. Frischknecht, N. Jungbluth, H.-J. Althaus, C. Bauer, G. Doka, R. Dones, S. Hellweg, R. Hischier, S. Humbert, M. Margni, T. Nemecek, Implementation of life cycle impact assessment methods, Swiss Centre for Life Cycle Inventories, Dübendorf, 2007. Ecoinvent report No. 3, v2.0.

[7] R. Dones, et al., Sachbilanzen von Energiesystemen. Final Report Ecoinvent 2000, vol. 6, Swiss Centre for LCI, PSI, Duebendorf, 2003.
[8] Franklin Associates, USA LCI Database Documentation, Prairie Village, Kansas, 1998.
[9] C. Sener, V. Fthenakis, Energy policy and financing options to achieve solar energy grid penetration targets: accounting for external costs, Renew. Sust. Energ. Rev. 32 (2014) 854–868.
[10] A. Arvesen, E.G. Hertwich, More caution is needed when using life cycle assessment to determine energy return on investment (EROI), Energ Policy 76 (2015) 1–6.
[11] M. Raugei, P. Fullana-I-Palmer, V. Fthenakis, The energy return on energy investment (EROI) of photovoltaics: methodology and comparisons with fossil fuel life cycles, Energ Policy 45 (2012) 576–587.
[12] M. Raugei, R. Frischknecht, C. Olson, P. Sinha, G. Heath, Methodological guidelines on Net Energy Analysis of Photovoltaic Electricity, 2016, IEA-PVPS Task 12, Report T12-07:2016, Available at: http://www.iea-pvps.org.
[13] M. Raugei, Material intensity analysis of CdTe PV vs. conventional electricity, in: 25th EU PV-SEC, Valencia, September 6–10, 2010.
[14] US Department of Energy, Energy Information Administration, Voluntary Reporting of Greenhouse Gases, Form EIA-1605, OMB No. 1905-0194. Appendix F: Electricity Emission Factors, EIA, Washington, DC, 2007.
[15] M. De Wild-Scholten, Energy payback time and carbon footprint of commercial photovoltaic systems, Sol. Energy Mater. Sol. Cells 119 (2013) 296–305.
[16] H.A. Aulich, F.-W. Schulze, Crystalline silicon feedstock for solar cells, Prog. Photovolt. Res. Appl. 10 (2002) 141–147.
[17] P. Woditsch, W. Koch, Solar grade silicon feedstock supply for PV industry, Sol. Energy Mater. Sol. Cells 72 (2002) 11–26.
[18] E. Williams, Global Production Chains and Sustainability: The Case of High-Purity Silicon and its Applications in IT and Renewable Energy, Institute of Advanced Studies, The United Nations University, Tokyo, 2000.
[19] V. Fthenakis, H.C. Kim, R. Frischknecht, M. Raugei, P. Sinha, M. Stucki, M. de Wild Scholten, Life Cycle Inventories and Life Cycle Assessments of Photovoltaic Systems, International Energy Agency, Paris, 2011. Report IEA-PVPS T12-02:2011, ISBN: 978-3-906042-01-5.
[20] R. Frischknecht, R. Itten, P. Sinha, M. De Wild-Scholten, J. Zhang, V. Fthenakis, H.C. Kim, M. Raugei, M. Stucki, Life Cycle Inventories and Life Cycle Assessment of Photovoltaic Systems, International Energy Agency, Paris, 2015. PVPS Task 12, Report T12-04:2015, http://www.iea-pvps.org.
[21] V.M. Fthenakis, et al., Emissions and encapsulation of cadmium in CdTe PV modules during fires, Prog. Photovolt. Res. Appl. 13 (2005) 713–723.
[22] V.M. Fthenakis, H.C. Kim, CdTe photovoltaics: life cycle environmental profile and comparisons, Thin Solid Films 515 (2007) 5961–5963.
[23] V. Fthenakis, H.C. Kim, M. Held, M. Raugei, J. Krones, Update of PV energy payback times and life-cycle greenhouse gas emissions, in: 24th European Photovoltaic Solar Energy Conference, Hamburg, Germany, September 21–25, 2009, 2009.
[24] M. Held, R. Ilg, Update of environmental indicators and energy payback time of CdTe PV systems in Europe, Prog. Photovolt. Res. Appl. (2011), http://dx.doi.org/10.1002/pip.1068.
[25] M. Raugei, S. Bargigli, S. Ulgiati, Life cycle assessment and energy pay-back time of advanced photovoltaic modules: CdTe and CIS compared to poly-S, Energy 32 (8) (2007) 1310–1318.

[26] V. Fthenakis, W. Wang, H.C. Kim, Life cycle inventory analysis of the production of metals used in photovoltaics, Renew. Sust. Energ. Rev. 13 (2009) 493–517.
[27] R.C. Powell, U. Jayamaha, G.L. Dorer, H. McMaster, Scaling and quantifying CdTe/CdS module production, in: AIP Conf. Proc.—March 5, 1999—vol. 462, 1999, pp. 31–36.
[28] First Solar. Guidance Call. http://investor.firstsolar.com, 2016.
[29] SENSE, Sustainability Evaluation of Solar Energy Systems, 2006. Summary Report on LCA, http://www.sense-eu.net.
[30] A. Lozanovski, M. Held, Update of the environmental indicators and energy payback time (EPBT) of CIS modules in Europe and scenario analysis of solar cell printing, in: Proceedings 25th European PVSC, Valencia, Spain, September 6–10, 2010, 2010.
[31] J.E. Mason, et al., Energy payback and life-cycle CO_2 emissions of the BOS in an optimized 3.5 MW PV installation, Prog. Photovolt. Res. Appl. 14 (2006) 179–190.
[32] M.J. De Wild-Scholten, Environmental Life Cycle Inventory of Crystalline Silicon Photovoltaic System Production—Status 2005/2006, ECN Solar Energy, Petten, 2007.
[33] M. Carbajales-Dale, M. Raugei, C.J. Barnhart, V. Fthenakis, Energy return on investment (EROI) of solar PV: an attempt at reconciliation, Proc. IEEE 103 (7) (2015) 995–999.
[34] V. Fthenakis, H.C. Kim, Life cycle GHG emissions of PV electricity generation: harmonization of published estimates, 4AV.3.59, in: 26th European Photovoltaic Solar Energy Conference, GER, Hamburg, 2011.
[35] S. Pacca, D. Sivaraman, G. Keoleian, Life Cycle Assessment of the 33 kW Photovoltaic System on the Dana Building at the University of Michigan: Thin Film Laminates, Multi-crystalline Modules, and Balance of System Components, University of Michigan, Ann Arbor, MI, 2006. CSS05-09.
[36] V.M. Fthenakis, H.C. Kim, E. Alsema, Emissions from photovoltaic life cycles, Environ. Sci. Technol. 42 (2008) 2168–2174.
[37] M. Raugei, V. Fthenakis, Cadmium flows and emissions from CdTe PV: future expectations, Energ Policy 38 (9) (2010) 5223–5228.
[38] M.J. De Wild-Scholten, E. Alsema, E.W. ter Horst, M. Bachler, V.M. Fthenakis, A cost and environmental impact comparison of grid-connected rooftop and ground-based PV systems, in: Proceedings 21st European Photovoltaic Solar Energy Conference, Dresden, Germany, September 4–8, 2006, 2006, pp. 3167–3172.
[39] M.J. De Wild-Scholten, et al., A. Cost, Environmental impact comparison of grid-connected rooftop and ground-based PV systems, in: 21st European Photovoltaic Solar Energy Conference, Dresden, Germany, 2006.
[40] S. Kim, B.E. Dale, Life cycle inventory information of the United States electricity system, Int. J. LCA 10 (2005) 294–310.
[41] V.M. Fthenakis, H.C. Kim, Greenhouse-gas emissions from solar electric- and nuclear power: a life-cycle study, Energ Policy 35 (2007) 2549–2557.
[42] V. Fthenakis, H.C. Kim, Land use and electricity generation: a life-cycle analysis, Renew. Sustain. Energy Rev. 13 (2009) 1465–1474.
[43] V.M. Fthenakis, H.C. Kim, Life cycle uses of water in U.S. electricity generation, Renew. Sustain. Energy Rev. 14 (2010) 2039–2048.
[44] V.M. Fthenakis, Life cycle impact analysis of cadmium in CdTe PV production, Renew. Sust. Energ. Rev. 8 (4) (2004) 303–334.
[45] Electric Power Research Institute (EPRI), PISCES Data Base for US Power Plants and US Coal, EPRI, Palo Alto, CA, 2002.
[46] V. Fthenakis, E. Alsema, Photovoltaics energy payback times, greenhouse gas emissions and external costs: 2004-early 2005 status, Prog. Photovolt. Res. Appl. 14 (2006) 275–280.

[47] V. Fthenakis, Considering the total cost of electricity from sunlight and the alternatives, Proc. IEEE 103 (3) (2015) 283–286.

[48] V.M. Fthenakis, End-of-life management and recycling of PV modules, Energ Policy 28 (14) (2000) 1051–1058.

[49] V.M. Fthenakis, W. Wang, Extraction and separation of Cd and Te from cadmium telluride photovoltaic manufacturing scrap, Prog. Photovolt. Res. Appl. 14 (2006) 363–371.

[50] V.M. Fthenakis, Sustainability metrics for extending thin-film photovoltaics to terawatt levels, MRS Bull. 37 (4) (2012) 425–430.

[51] J.K. Choi, V.M. Fthenakis, Design and optimization of photovoltaics recycling infrastructure, Environ. Sci. Technol. 44 (22) (2010) 8678–8683.

[52] J.-K. Choi, V. Fthenakis, Crystalline silicon photovoltaic recycling planning: macro and micro perspectives, J. Clean. Prod. 66 (1) (2014) 443–449.

[53] H. Kim, K. Cha, V. Fthenakis, P. Sinha, T. Hur, Life cycle assessment of cadmium telluride photovoltaic (CdTe PV) systems, Sol. Energy 103 (2014) 78–88.

[54] K. Burrows, V. Fthenakis, Glass needs in a growing PV industry, Sol. Mater. Sol. Cells 132 (2015) 455–459.

[55] Intergovernmental Panel on Climate Change, Climate Change 2001: The Scientific Basis, Cambridge University Press, Cambridge, 2001.

[56] V.M. Fthenakis, H.-C. Kim, Life cycle assessment of high-concentration PV systems, Prog. Photovolt. Res. Appl. 21 (3) (2013) 379–388.

[57] M. Perez, V. Fthenakis, H.-C. Kim, A. Pereira, Façade-integrated photovoltaics: a life-cycle and performance assessment case study, Progr. Photovolt. Res. Appl. 20 (8) (2012), 975–990.

Part Three

Maximising photovoltaic (PV) system performance

Building-integrated photovoltaics (BIPV)

8

C. Ferrara, H.R. Wilson, W. Sprenger
Fraunhofer Institute for Solar Energy Systems, Freiburg, Germany

8.1 Introduction

8.1.1 Future energy system and building-integrated photovoltaics

In the near to mid-term future, our energy demand will be met by an energy system based on 100% renewable energy sources such as wind, hydroelectricity, biomass and solar energy [solar thermal and photovoltaic (PV)]. PV, including building-integrated PV (BIPV), will be one part of this future energy system.

Worldwide energy systems are currently undergoing a transition from highly centralized, large fossil and nuclear power plants to decentralized, small and medium-sized renewable power plants. For example, in Germany in 2014, 25.8% of the total electricity supply was provided by renewable sources (wind 3.4%, biomass 7.0%, solar 5.8%, hydroelectric power 8.6% and domestic waste 1.0%) [1]. This 5.8% of 610 TWh is generated by an installed PV power capacity of 38 GWp.[1] Recent studies by Fraunhofer ISE [2] show that for the reduction of carbon dioxide (CO_2) emissions by 80–95% of the complete energy system, we need among other measures up to 122–290 GWp installed PV power in Germany, corresponding to ~1200–2900 km^2 (with a total area of 360,000 km^2 equal to 0.78%, in comparison to 13.4% settlement and transport-related areas [3]). With roughly 80 million inhabitants in Germany, the area needed for PV installations results in 25–31 m^2/capita, compared to ~300 m^2/capita for settlement. To minimize further sealing of land, it seems obvious to use existing built-up areas for the installation of PV systems. In Germany, between 1500 and 3000 km^2 of roof and façade area is available on existing buildings. An obvious option is to install PV systems on buildings.

8.1.2 BIPV – requirements and definition

When a PV module is designed as a BIPV module and integrated in to a building envelope, it is converted from a purely electrical device to a construction product that generates electricity, among other features. Thus, BIPV modules are subject to all local conditions (relating to climate, building code, construction, etc.), and national and international regulations regarding construction products. In this function, they replace traditional construction elements like roof and façade cladding, windows,

[1] Watt peak (Wp): Nominal power mentioned on the name plate of PV modules measured under standard test conditions (STC) defined by standards (IEC 61215 Ed. 3.0, UL 1703).

The Performance of Photovoltaic (PV) Systems. http://dx.doi.org/10.1016/B978-1-78242-336-2.00008-2

blinds and static or dynamic shading elements. When a BIPV module replaces traditional construction elements, it assumes the function of the replaced building element (definition of BIPV, source: EN 50583 [4]). Taking the European Construction Product Regulation [5] as a guide, the building element's functions in the context of BIPV are one or more of the following:

- mechanical rigidity or structural integrity
- primary weather impact protection: rain, snow, wind and hail
- energy economy, such as shading, daylighting and thermal insulation
- fire protection
- noise protection
- separation between indoor and outdoor environments
- security, shelter or safety.

8.1.3 Costs of BIPV

According to Renken [6], the costs per square metre of BIPV modules replacing façade elements are similar to common cladding materials. Frontini et al. [7] published similar results for façades but found a difference of roughly 200€/m^2 compared to conventional roofing materials (Table 8.1).

8.1.4 Operation and maintenance

PV technology provides on-site renewable energy, it is silent, there are no hazardous emissions during operation and as it intrinsically has no moving parts, it needs little maintenance during its expected lifetime of 25–30 years, qualities that recommend it for building-integrated use. The need to clean BIPV systems is highly dependent on their geographical place of installation and the meteorological conditions (rain, humidity, wind, dust, etc.), the tilt angle of the system and the surface morphology.

Table 8.1 Comparison of prices of cladding materials for facades in Switzerland 2015

Example	Cladding materials	Price per m^2 [CHF]/[Euro][a]
1	Wood	220.00/198.00
2	Fibre cement	310.00/279.00
3	Natural stone	360.00/324.00
4	Coloured glass	145.00/130.50
5	Aluminium	130.00/117.00
6	Acrylic glass	155.00/139.50
7	PV panel crystalline—glass/back sheet	200.00/180.00
8	PV panel crystalline—glass/glass translucent	500.00/450.00
9	Thin film PV panel—fixed dimensions[b]	80.00/72.00

[a]Original prices in CHF, conversion factor 0.90 (03 Feb. 2016) [6].
[b]Not suitable for BIPV.

Fig. 8.1 Light and shadow pattern in the atrium of Fraunhofer ISE generated by BIPV as roof.
Courtesy of Ch. Erban, Herzogenrath DE.

Studies on the impact of soiling and cleaning have been performed on field-mounted PV systems installed [8,9] comparably to roof-mounted BIPV systems. They find a maximum degradation in efficiency of approximately 6%. Depending on all these factors mentioned, cleaning can improve the overall efficiency of BIPV systems (Fig. 8.1).

8.1.5 Current obstacles

Although the mentioned facts are obvious, there is still no consistent market for BIPV. Several obstacles are causing this status quo. The planning and construction process is currently greatly complicated by the use of BIPV, because the recently approved BIPV standard EN 50583 is not yet widely known, there are only a few certified BIPV construction products and no easy installation methods (plug and play, plug and function) for BIPV. BIPV construction products are not yet integrated into widely available construction product catalogues and planning tools. As a result, the planning and installation effort is high and for many projects, specialists are needed for the electrical design and wiring of the BIPV systems. Only a small number of architects, engineers and planners are aware of the different technologies, their potential and advantages when installed on a building. During the design and construction process, BIPV is not part of the early design phase, because of the reasons mentioned here, and because of a lack of awareness by architects and planners. This finally leads to higher costs and limited application of BIPV.

8.2 Best practice for design

PV is, in a diluted form, already ubiquitous, starting from mW on watches, calculators, bags, etc. and extending to GW-scale PV power plants (eg, Solar Star Project, Capacity: 579 MW, USA [15]). BIPV, starting with some kWp on free-standing houses

(≥5 m^2) and reaching MWp scale on industrial and administrative buildings (≥5000 m^2), is in the middle of this range. As buildings are very heterogeneous regarding size, usage, construction, etc., there will be a variety of BIPV modules for all buildings. In the medium term, we will see mainly customized BIPV in many locations. Alternatively, it can be camouflaged by new materials and still generate electricity on roofs and façades.

As we have seen in the previous subsection, there are several requirements coming from the construction of the building itself and from building regulations addressing mechanical stability, residual load capacity, fire safety, etc., as well as design issues. These requirements lead to the definition of the following construction classification.

8.2.1 Mechanical properties

Rigid BIPV modules

Rigid BIPV modules can be constructed using all PV technologies available on the market by including a rigid substructure or back sheet, such as sheets of glass or metal plates. They differ in module electrical efficiency, depending on the technology used (see Table 8.2). Based on their mechanical properties, rigid BIPV modules can easily replace conventional cladding materials for façades and roofs. It is important to note here that they have to fulfil the same requirements defined by national building codes (see the next section) and constructional requirements (Fig. 8.2).

Flexible BIPV modules

Flexible BIPV modules can be constructed with all so-called new technologies such as organic PV (OPV), dye-sensitized solar cells (DSC), perovskite solar cells (PSC) and all thin-film technologies, including amorphous silicon (a-Si), microcrystalline silicon (μ-Si), a combination of a-Si/μ-Si, copper indium gallium selenide (CIGS) and

Table 8.2 Maximum efficiency of photovoltaic technologies achieved in lab dimensions [10] and efficiency of PV modules from production lines [11]

Technology	Max. efficiency (%)	Module efficiency (%)	Required area (m^2/kWp)
c-Si	25.0	22.9	4.4
CIGS	22.3	17.5	5.7
CdTe	21.5	17.5	5.7
a-Si	13.6	10.9	9.2
DSC	11.9	n.a.	n.a.
OPV	11.5	n.a.	n.a.

The required area is the active area of a module on the building envelope plus 10% for construction and safety issues.
n.a., data not available.

Fig. 8.2 Rigid building-integrated photovoltaic modules on a laboratory building of Fraunhofer ISE. The BIPV modules were developed and produced in a pilot production facility at the institute.
© Fraunhofer ISE.

Fig. 8.3 Demonstration sample of a flexible organic solar cell module.
© Fraunhofer ISE.

cadmium telluride (CdTe). The substructure can be based on polymer films or metal sheeting (Fig. 8.3).

8.2.2 Optical properties

Coloured BIPV modules

All so-called new technologies (DSC, OPV and PSC) have intrinsic potential for different colours. The materials of thin-film and crystalline technologies can be coloured to a limited extent or the appearance changed by the use of coloured front covers,

Fig. 8.4 Dye Solar Panel comprising three module prototypes (60 cm × 100 cm) with internal series connection and glass-frit edge seal.

either polymer sheets or glass. Adding coloured layers in front of the PV-active components of a module will reduce efficiency of the module (Fig. 8.4).

Transparent BIPV modules

Physical principles determine that it is not possible to make a 100% transparent BIPV module, because the PV active layer has to absorb some of the incident irradiation to convert it into electricity. There are examples of thin-film based BIPV modules [a-Si (ASI THRU by SCHOTT), CIGS] that have a semitransparent appearance, because they apply transparent conductive oxides instead of metals for the electrodes and are coloured in a neutral grey tone. Another possibility is to vary the distance between the PV-active areas and so enhance the transparency of the BIPV module and simultaneously reduce the area-specific efficiency. With windows, one can find different approaches, eg, angle-selective systems such as PVShade® (Fig. 8.5) (https://www.ise.fraunhofer.de/en/press-and-media/press-releases/presseinformationen-2013/fraunhofer-ise-inaugurates-new-energy-efficient-laboratory-building) or nearly invisible ones like Pythagoras Solar (http://www.glasswebsite.com/divisions/energy/profiles/pythagoras.asp), where the PV-active layer is inserted horizontally into the space between two panes attached to a special prism.

Due to the limited area on roof and façade areas, the BIPV system efficiency should be as high as possible. To optimize the energy yield of BIPV systems in mostly heterogeneous irradiation situations, a detailed electrical design is needed (see next subsection).

Fig. 8.5 The transparent PV module (PVShade®) contains PV strips in two planes, which allows it to provide angle-selective solar control, glare protection and visual contact in addition to generating electricity.
© Fraunhofer ISE.

From this broad range of technologies, it is possible to define a wide variety of BIPV systems. In future, BIPV will be a commodity building product, including new technologies with advanced design possibilities and clearly defined multi-functionality (colouring, patterns, advertising, etc.). Most of the new technologies allow inherently different colours and semitransparency, so they have a large design potential. The use of this large design potential will open and boost new market segments for BIPV products, with its main focus on design requirements.

8.2.3 Electrical design of BIPV systems

The installation of BIPV systems needs a lot of electrical planning. Until now, the electrical design of BIPV systems has to be defined by (expensive) specialists. In a BIM implementation in the future, these calculations may be automated to a large extent.

There are several differences between a free-standing PV plant or a rooftop PV application and a BIPV system. First, the irradiation on the PV cells is inhomogeneous in almost all cases. There may be parts of the building that cause partial shading, there may be different orientations of the involved BIPV modules, and shading and even reflections from surrounding objects (buildings) can cause inhomogeneous irradiation on the PV cells. If this is the case, the problem of electrical mismatch arises. Any electrical connection of PV cells in series will lead to the situation that the PV cell with the lowest irradiation defines the current of all other PV cells connected in series to it. If the BIPV elements involved have different orientations and are connected in series with each other, the BIPV elements receiving lower irradiation values will define the current of the whole PV module string, leading to serious power losses (Fig. 8.6).

Most BIPV systems are connected to the grid. This makes it necessary to convert the DC power to AC, which is done with an inverter. These inverters again have limited ranges of acceptable electrical DC specifications. If the power of a PV system drops to a value significantly below the nominal MPP power, the inverter may change

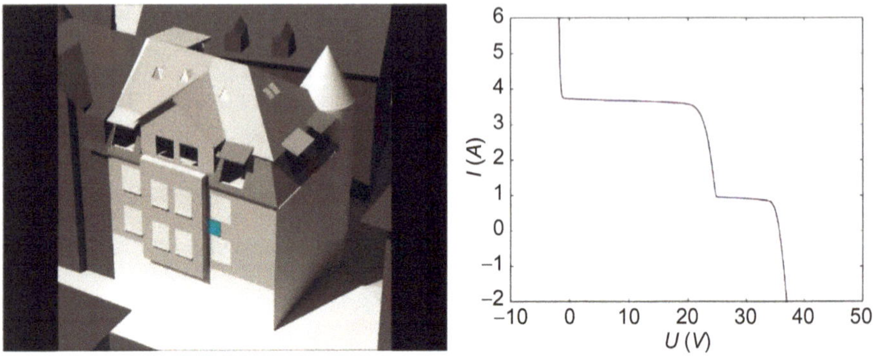

Fig. 8.6 Calculation of IV curves for each module and time step based on ray-tracing simulation incl. irradiation, temperature, cell technology, ambient, etc.

its operation mode to constant-voltage, leading to an additional power loss. However, if the DC quantities exceed the inverter specifications, the inverter may be damaged. For this reason, it is very important to understand the electrical details of the BIPV electricity production. The electric circuit of the BIPV system (especially combining the PV modules in series and parallel circuits; the choice of the appropriate inverter specifications) needs to be calculated in advance. Automating this procedure within the BIM context will not only simplify the planning process dramatically, but it will also simplify the procedures that are necessary if any BIPV-relevant changes in the construction process occur. Last, but not least, the measured power generated by the PV system can be compared with simulations to check whether the BIPV system is working properly.

8.3 BIPV requirements and constraints

8.3.1 BIPV regulations and standardization efforts

When applying PV as a construction product, two worlds have to be combined. On the one hand, the cost-effective production of a PV module requires a highly automated, technical process. For this reason, PV manufacturing companies need to be involved. Their main market, however, is the production of PV modules that are used for free-standing PV plants or rooftop PV applications. On the other hand, every construction product in the EU needs to comply with the Construction Products Regulation (CPR), leading to many additional requirements for the PV modules, such as higher demands on fire safety or post-breakage behaviour. Furthermore, some basic properties of PV modules, especially the dimensions, need to be adapted to the situation of the building.

Not many manufacturers of PV modules are able to produce PV modules with customized dimensions. Due to their main market, the automated processes that are needed for the production of PV modules do not allow the size of the PV modules to be changed arbitrarily. Furthermore, their automation process does not always allow PV modules to be manufactured with the appropriate mechanical specifications that construction products need to have.

There are several companies worldwide that have specialized their manufacturing process for BIPV applications. They are based on a number of different technological approaches. Similar to the main PV market of free-standing and rooftop PV applications, the wafer-based PV technologies (monocrystalline and polycrystalline silicon, heterojunction technology, etc.) are the leading PV technologies for building products, although the aesthetic options that these PV technologies provide are limited. Besides the high efficiency of the wafer-based PV technologies and their status of development, it is the strong similarity of glass–glass modules to laminated glass panes that simplifies their application as a construction product. Large parts of the already existing standardization procedures for glass-based construction products can be easily adapted for glass-glass PV modules.

For BIPV manufacturers, it is of major importance that standardized testing procedures are developed to judge the suitability of their products to defined applications. In Europe, each country still has its own legal procedures for building products. For BIPV, this is a serious constraint due to the limited demand for BIPV products within the limited market of a country. In 2015, the EU member states agreed on the European standard EN 50583 with the title ‘Photovoltaics in buildings’. Due to legal procedures, it will take several additional years until the European standard will be valid within the EU member states.

The standard EN 50583 is the furthest developed standardization effort worldwide on the topic of PV in buildings to date. Besides listing the standardized procedures for electrical testing of the PV modules, it summarizes a long list of relevant mechanical testing procedures that have to be taken into account when using PV modules as construction products.

The EN standard defines five different categories of BIPV (Fig. 8.7).

The aforementioned BIPV standard lists all the building standards that need to be taken into account for PV modules that are used as a construction product, depending on the intended category of application.

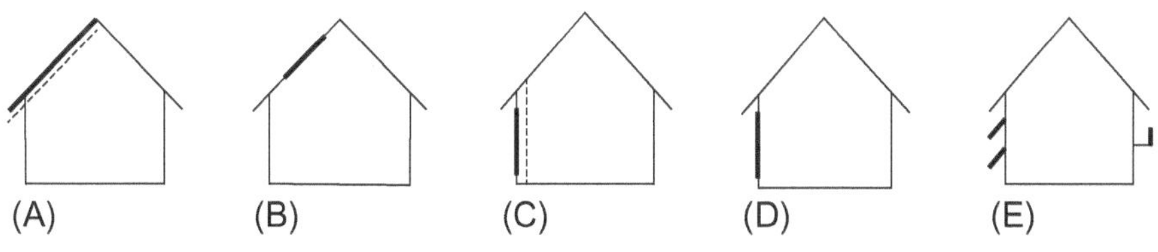

Fig. 8.7 The five BIPV categories defined within EN 50583. (A) Sloped, roof-integrated, not accessible from within the building. (B) Sloped, roof-integrated, accessible from within the building. (C) Non-sloped (vertically) mounted, not accessible from within the building. (D) Non-sloped (vertically) mounted, accessible from within the building. (E) Externally integrated, accessible or not accessible from within the building.
Source: EN 50583.

On an international level, there are standardization efforts from ISO (eg, BS ISO 18178) as well as from IEC (IEC 62980). Both standards, however, are still under development and do not reach the degree of detail of EN 50583 at the moment.

8.3.2 Photovoltaics in the construction sector

Until now, building projects with BIPV elements installed are within a niche market and do not necessarily comply with common procedures of the building sector. Especially for large building projects, the involved companies are used to working with a 3D object-based representation of the whole building project. Companies already offer commercially available software programs with databases of construction products, each with all the information that is necessary for the construction process of the building. On the basis of these objects, the software programs offer a wide range of automated calculations from cost calculations to earthquake stability. Building elements that are not part of the available databases only have a very limited chance to be considered in sophisticated construction projects.

As soon as BIPV modules are available as database objects, including all the information that is needed in the construction process, BIPV systems will be able to take the step from niche market to standardized commodity product. With the development of BIPV standards and the availability of corresponding BIPV products, the next step to be taken is to create IT representations of these BIPV products for the building sector. Unfortunately, the number of PV researchers that are simultaneously aware of the needs and common procedures of the building industry is very limited.

When using PV modules as building elements, the effort for planning is very large. The electrical design of the BIPV system often needs time-consuming simulations. However, BIPV is not the only challenging topic in the building sector. The larger a building project is, the larger is the number of different companies involved, and different simulation programs are applied. During the last decade, there has been widespread discussion on how to simplify the IT procedures when topological representation formats of different simulation programs have to be combined.

The solution, called BIM (building information modelling), arises from agreement on a software-independent data structure that includes all the relevant information for the construction process and can be extended by all kinds of proprietary software programs. This independent data structure is mainly developed by the buildingSMART organization (http://www.buildingsmart.org). The IFC format that has been codified in ISO 16739 has already been incorporated into many commercial CAD programs.

The general development of BIM is also a huge opportunity for the BIPV topic. The development of a standard procedure to define the relevant mechanical, electrical or economic properties of BIPV products can decide whether they will be implemented in the IT representation of the construction process or not. It will be decided by the efforts of the PV community whether building integration will become a relevant PV application or if it will be limited to a niche market in the future as well.

For smaller building projects, the communication between people involved in the planning and installation of BIPV systems can be optimized without the generation of a general data structure. All planning and exchange of information should always include the period of operation and maintenance of the building and the installed BIPV system.

8.4 Measurement and assessment

8.4.1 Monitoring of electricity yield

As for conventional PV systems, continuous monitoring of BIPV systems is valuable as a prerequisite for determining whether the installation is generating the amount of electricity anticipated during the planning process. In the building context, not only is the return on investment resulting from the generated electricity of interest, but also the detection of performance deviation that may indicate safety or service lifetime-relevant issues such as hot-spot formation.

Recommended electrical and meteorological quantities for monitoring

The primary electrical quantity to be monitored is the output AC power from each inverter. Measurement of the DC current and voltage of each module string allows the system performance to be analysed in greater detail and enables better localization of faults, should they occur.

The solar irradiance in the module plane and the ambient temperature are the most important meteorological quantities to be monitored. If the BIPV system combines modules with many different orientations, an alternative concept is to measure the global irradiance in a horizontal plane and either the diffuse irradiance on a horizontal plane or the direct irradiance, and use these measured quantities as input data to sky luminance models, from which the irradiance on differently oriented planes can be calculated in turn. Direct measurement of the module temperature, using sensors attached to the back surface of the modules, is recommended if accurate performance analysis is intended. Because cooling of the BIPV modules by wind has only a second-order effect on the generated power, the wind speed and direction would generally be monitored only in systems that are to be analysed at a very detailed level.

A 10 s sampling interval and recording of 1 or 5 min averages of the monitored quantities with data loggers have proved to provide a good data basis for further analysis in most cases.

Presentation and analysis of monitored data

For a quick visual check that the BIPV is performing within plausible limits, a numerical or graphical display of the instantaneous irradiance on the module plane and the generated AC power may be adequate. Time series of the same two quantities or their hourly, daily, monthly and annual averages or time-integrated values also allow quick plausibility checks.

To assess the electrical performance in more detail, time series of the performance ratio (PR) for periods of a day, a month or a year are widely used, as the PR is defined to compensate the first-order effect of the varying solar irradiance in the module plane on the electricity yield. As the irradiance in the module plane is a fundamental parameter for calculating the PR, it should be calculated separately for each significantly different module orientation in the BIPV system, using the relevant irradiance value. A PR of 100% indicates that the analysed system is generating the same amount of electricity as the installed modules would under STC (STC: irradiance of 1000 Wm^{-2}; solar spectral distribution corresponding to AM 1.5, global; ambient temperature of 25°C). The frequent occurrence in BIPV installations of significantly lower irradiance values, resulting from the orientation defined by the building envelope, and higher module temperatures in better thermally insulated positions than for STC, both having a negative effect on the module efficiency, mean that the PR is usually below 80%. In the BIPV context, time series of the PR are most useful to identify sudden changes, which may indicate significant component failures, or gradual changes, which may be caused by soiling or degradation.

At the most sophisticated level, time series of the monitored electrical yield can be compared with the results of simulations based on the monitored meteorological quantities, as described in Sprenger [12]. Due to the effect of shading, which is likely to be more frequent and more complex in BIPV installations than in ground-based PV systems, this type of detailed comparison is needed both to determine whether the system is performing according to expectations and to detect component failures. Otherwise, false alarms concerning the system output may be frequently triggered by partial shading of module arrays.

8.4.2 Practical aspects

It is worth explicitly noting that providing an accessible location for the inverters, being essential components of any PV system, should be part of the standard electrical planning for the building—just as for a fuse box.

In contrast to conventional PV systems, the positions for locating most of the meteorological sensors are no longer easily accessible once the BIPV modules have been installed (and the scaffolding removed). Thus, it is strongly recommended to take the monitoring system into account during the early phases of the planning process and to ensure that the required sensors and cabling are available for installation together with the BIPV system installation. As the sensors for the electrical quantities are usually located near the inverters, the problem of accessibility is usually not as acute as for the meteorological sensors, but space must still be planned for installation of additional junction boxes and electrical cabinets housing the data acquisition and transmission systems.

If a significant fault in a BIPV system is identified, deciding on the appropriate course of action is less straightforward than in ground-based systems. Not only the effect on the electrical yield but also aesthetic aspects (colour matching), accessibility and human safety will affect the decision. Thermographic images recorded with an IR camera may be useful in this context to localize hot spots and determine whether they present a fire risk.

8.4.3 Requirements for BIPV modules as building components

As construction products, BIPV modules in Europe must fulfil the so-called 'essential requirements' specified in the European Construction Product Regulation CPR 305/2011 [4].

The essential requirements defined in the CPR 305/2011 are:

1. mechanical resistance and stability,
2. safety in case of fire,
3. hygiene, health and the environment,
4. safety and accessibility in use,
5. protection against noise,
6. energy economy and heat retention and
7. sustainable use of natural resources.

The characteristics required of a specific BIPV module will depend on the function it assumes within the building envelope, replacing transparent or opaque building components for the walls or the roof, and the location of the building. Following the approach of the recently approved EN 50583, titled 'Photovoltaics in Buildings', BIPV modules can be classified as those including glass panes or those based on polymer or metal substrates. The characteristics specified for similar building products, eg, glass laminates, polymer-based waterproofing sheets or extruded metal roofing sheets, must then also be determined for BIPV modules that provide the same function within the building.

For example, if a semitransparent glass-glass PV module is to be used as the outer pane of a triple glazing unit, measurements of the transmittance and reflectance spectra and the thermal emissivity of that module will be needed, so that the light transmittance, the total solar energy transmittance (g value), the colour rendering and the thermal transmittance (U value) of the glazing unit can be calculated. If the same semitransparent glass-glass PV module is a component of overhead glazing for a bus shelter, its mechanical strength must comply with local authorization requirements based on specified snow, wind and hail loads.

The reader is referred to EN 50583 for a comprehensive (but not exhaustive) set of standards detailing the construction-related characteristics which must be determined for BIPV modules acting as different types of building components. These standards typically define characteristics which can be determined under stationary conditions in the laboratory. Although the transmittance characteristics listed here as examples are actually dynamic quantities which depend on the prevailing environmental conditions, long-term monitoring of their variation in the installed component requires rather more elaborate measurement technology than to determine the amount of electricity generated by a BIPV installation.

8.4.4 Economic assessment of BIPV systems

The life-cycle costing model developed at Fraunhofer ISE for BIPV systems considers three main phases of the system lifetime from the point of view of the system owner: the investment/installation phase, the operation phase and the demolition/disposal phase [13]. Throughout the system lifetime, which is typically assumed to be 20 or 25 years, the annual cash flows (income and expenditure) are determined, taking financing costs and inflation rates into account, to determine the net present value of the whole system at any given time.

The cash flows during the first phase are mainly outward, including the investment for system components (BIPV modules, cabling, inverters, mounting structure, monitoring equipment) and planning, authorization and installation costs. However, a bonus representing the cost of substituted building components may also be taken into account. In doing so, it is useful to define the system boundary to exclude components such as substructure that could equally well be used for mounting BIPV or the substituted, conventional building components.

During operation, ongoing expenses for cleaning, maintenance, monitoring and administration are accompanied by income resulting from the electricity generated by the system. This may be in the form of real income from feeding electricity into the public grid, or saved expenditure for externally generated electricity due to in-house usage of the electricity generated on site. Depending on their thermal and optical properties and their function within the building envelope, BIPV modules may also affect the building's thermal and lighting energy balance; the financial effect can also be included in the cash flows for the operation phase.

Finally, demolition and disposal will again incur expenses, but income may also be generated from selling the component materials (eg, copper cables) for recycling.

Depending on the relative value of the inward and outward cash flows, BIPV systems can amortise themselves within their system lifetime, a property not otherwise expected of or found in conventional, 'passive' building components.

8.5 Product development

8.5.1 Complementary technical approaches

Presently, two technical approaches are being applied to reduce costs and increase application of BIPV in renovation of existing buildings and new construction.

One is to produce customized BIPV modules in a highly automized production line that allows differences in size, colour and power output. For the electrical design of BIPV systems combining modules with such diverse parameters, including orientation and tilt angle, experts and expert systems are needed to simulate an optimized electrical configuration for a given geometrical configuration, taking the surroundings and neighbouring buildings into account. Currently, automated simulation systems that are directly linked to CAD and BIM systems are being developed.

Another approach is to identify building classes with standard constructions and dimensions by detailed analysis of the building stock. The resulting detailed data basis allows the definition of standardized modules that can be prefabricated indoors and are equipped with plug-and-play technology. These BIPV modules therefore feature a high-quality standard on the one hand and can be installed more quickly on the other.

8.5.2 BIPV applications

For existing buildings and new construction, there are several possibilities to integrate PV technologies into buildings, eg, as roofing material, in façade systems, in window systems (eg, the angle-selective PVShade® product), as horizontal and vertical shading systems, integrated into venetian blinds or window shutters (see Fig. 8.7). Many design possibilities are given by the range of possible colours, forms, patterns and flexible dimensions.

Some examples that are entering the market include:

- SBskin, a highly insulating glass block with integrated DSC, for roofs and façades.
- TIFAIN, adaptable shading system with DSC.
- SwissInso, new colour system for whole modules with minimal reduction of PV efficiency.
- Multiwire (SCHMID Group)/SmartWire Connection Technology (SWCT, Meyer Burger) concepts, making intercell connectors less visible.

A detailed product overview of market available BIPV systems for roofs and façades has been published by Frontini et al. [7].

For each application, it is important to understand, design and handle BIPV products as construction products, not only as PV modules. Combined roof and façade elements with additional layers and functions will increase the functionality and application of BIPV modules. Further business is stimulated by the return of investment resulting from the electricity generated by BIPV components, a unique property for building products.

Extended application and interpretation will be generated by the use of BIPV in the urban context as urban furniture such as bus shelters and roofs for parking lots.

Today, prices for BIPV are often calculated and published as €/Wp, or €/kWh; in future, €/m^2 will be more important to concur with construction industry practice and the thinking of architects and planners. The objective prices for standard PV modules are helpful as a guide; they are expected to decrease from 0.53€/Wp today to 0.20–0.30€/Wp in 2030 [14]. Assuming the development of standardized BIPV modules for selected building categories, one can extrapolate additional costs for BIPV systems of approx. 100€/m^2 to the near future.

References

[1] BMWi, http://www.bmwi.de/DE/Themen/Energie/Erneuerbare-Energien/sonne-wind-und-co.html (accessed 07.22.15).

[2] H.-M. Henning, A. Palzer, What will the energy transformation cost? https://www.ise.fraunhofer.de/en/publications/veroeffentlichungen-pdf-dateien-en/studien-und-konzeptpapiere/study-what-will-the-energy-transformation-cost.pdf.

[3] Statistische Bundesamt, FS 3 Land- und Forstwirtschaft, Fischerei, R. 5.1 Bodenflächen nach der tatsächlichen Nutzung 2011. Statistisches Bundesamt (DESTATIS), Wiesbaden, 2012.

[4] EN 50583 Photovoltaics in Buildings. Part 1: BIPV Modules. Part 2: BIPV Systems, 2016.

[5] Construction Product Regulation CPR 305/2011, Regulation (EU) No. 305/2011 of the European Parliament and of the Council, 9 March 2011.

[6] C. Renken, Strategies to Increase the Deployment of PV in Façades, in: Proceedings of 10th Conference on Advanced Building Skins, Berne, Switzerland, 3–4 November 2015.
[7] F. Frontini, et al., Building Integrated Photovoltaics—Report 2015, SUPSI-SEAC BIPV, Lugano, 2015.
[8] T. Weber, et al., Impact and consequences of soiling and cleaning of PV modules, in: NREL Photovoltaic Reliability Workshop, 2015, Denver, USA, 2015.
[9] M. Mani, R. Pillai, Impact of dust on solar photovoltaic (PV) performance: research status, challenges and recommendations, Renew. Sust. Energ. Rev. 14 (9) (2010) 3124–3131.
[10] National Renewable Energy Laboratory (NREL), Best Research-Cell Efficiencies. NREL, Golden, CO, 2016. http://www.nrel.gov/ncpv/images/efficiency_chart.jpg.
[11] B. Burger, et al., Fraunhofer ISE, PSE AG: Photovoltaics Report, Fraunhofer Institute for Solar Energy Systems, Freiburg, 2015.
[12] W. Sprenger, Electricity Yield Simulation of Complex BIPV Systems, Fraunhofer Verlag, 2013, ISBN 978-3-8396-0606-3.
[13] K. Fath, A. Hartmann, H.R. Wilson, C. Hemmerle, T.E. Kuhn, J. Stengel, F. Schultmann, B. Weller, Life-cycle cost assessment of photovoltaic façade panels, in: Conference Proceedings of ENERGY FORUM on Solar Building Skins, 06–07 December 2011, Bressanone, Italy, 2011.
[14] PV LCOE in Europe, 2014–2030.
[15] http://us.sunpower.com/utility-scale-solar-power-plants/solar-energy-projects/solar-star-projects/, accessed 2016–03-31.

Stand-alone photovoltaic systems

9

V. Salas
Charles III University of Madrid, Madrid, Spain

9.1 Introduction

Stand-alone (off-grid) systems were the origin of photovoltaic (PV) systems. The world's first PV companies were launched in the early 1970s to develop products for remote power applications like navigation aids and telecommunications, and in developing countries. The first oil shock in 1973 led to an increase in interest in PV for electricity generation, new R&D and new companies. In 1978 the first PV water pumps in Africa, working with Mali Aqua Viva, were installed. In 1978 the World Bank/UNDP also launched the Global Solar Pumping Programme.

Since then, there has been a niche market in some small markets. However, due to reasons such as the end of FiTs (feed-in tariffs) in some countries, the high cost of extending a utility line or the wish to obtain independence of homemade energy production, as well as the advantage of having a silent, emission-free energy source with a 25-year warranty and, above all, with the considerably decreased cost of PV systems, the market in PV off-grid applications has become interesting on the continents of Africa, Asia and Latin America, which have by far the largest off-grid potential [1].

Nowadays, different markets are becoming viable, mainly in developing countries: small applications, PV pumps and rural electrification. But also PV hybrid systems offer huge potential for the off-grid mining sector. Diesel-based power generation is widely used around the world. Diesel fuel-based plants are used mainly to compensate for inadequate power supply from centralized generation, such as in weak grids, or in remote areas not connected at all to a large grid (eg, mining sites, islands). However, the rapidly declining cost of PV-based power generation in recent times provides a clear motivation to offset their diesel consumption. This can be done by means of the PV–battery–diesel hybrid systems. For achieving high shares of solar energy, battery systems are required to store the intermittent solar energy and to assure the reliability of the hybrid system.

For instance, due to a lower operation and maintenance cost, the market demand for global PV pumps is rapidly escalating. PV pumps are neither dependent upon diesel nor grid-supplied electricity, thus ensuring that people in remote areas can receive the benefit which in turn will result in increased efficiency and productivity. PV pumps are free from difficulties such as supply cuts, feeder damage and rate hikes.

On the basis of application, the global solar pump market can be segmented into drinking water, agriculture purposes and other (chemical, oil and gas). The global solar pump market is geographically segmented into seven key regions: North

The Performance of Photovoltaic (PV) Systems. http://dx.doi.org/10.1016/B978-1-78242-336-2.00009-4

America, South America, Eastern Europe, Western Europe, Asia Pacific, Japan and Middle East and Africa.

Therefore, in this chapter, different solutions related to PV stand-alone systems will be considered as well as the power converters (PCs) to be implemented in them. In addition, important issues such as the design and performance will be discussed.

9.2 Solutions for stand-alone PV systems

Stand-alone or off-grid PV systems can be defined as those systems that are not connected to the public grid. They can be distinguished between systems with batteries and those without. The design depends on the application.

PV systems without batteries are called *directly coupled* PV systems. These are usually simple systems where the PV generator is connected directly to a motor or pump which matches the voltage and amperage output of the panel. When the sun shines and the PV panel produces electricity, the device runs. When the sun is not available, the device stops. This system is often used for livestock where a well-pump lifts water out of the ground to a watering trough in remote locations. Other applications include solar-powered attic fans, irrigation systems and small day-time garden waterfalls or fountains.

In any case, a stand-alone power system is always designed so it covers the electricity needs of the application, and when it needs to supply energy all of the time, some amount of stored electricity has to be allowed for high drain uses or periods of low generation.

According to the load profile, off-grid PV systems can be divided as follows: with direct current (DC) load, with alternating current (AC) load or both.

In general, the elements of an off-grid PV system are the PV generator (responsible for the photoelectric conversion), the BoS (balance of system) that consists of the wires, circuit breakers, mounting structure, protection devices, PCs (inverters, charge controllers) and the energy storage if necessary (batteries). Essentially, it is through the BoS components that the cost is controlled. Thus, an appropriate integration of the BoS is vital for the proper functionality and the reliability of the off-grid PV system. However, it is often overlooked and poorly integrated.

In addition, stand-alone PV power systems are the best suited for remote village and rural electrification projects, and can provide a more affordable and reliable source of electricity than conventional solutions, particularly in more remote rural areas utilizing only diesel generators. This topic will be discussed in the last section of this chapter.

Then, the following categories can be distinguished in the off-grid PV systems: pico photovoltaic (pico PV) systems, solar home systems (SHS) and PV hybrid systems (where micro-/mini-grids are also included). Each category will be considered separately in this chapter.

9.2.1 Pico PV systems

Pico PV systems [2–4] are very small-scale PV systems usually large enough for the basic electricity needs of one household, ranging from 1 to 10 W of PV capacity. These systems focus on lighting, but can often be designed to also run other small electrical appliances such as mobile phone charging, radios and MP3 players. They offer an alternative for providing energy access to basic lighting and communications. They can be permanently installed at home or can be used as mobile systems. They can charge a radio or cell phone and lead to savings in everyday spending for low-income families.

They are powered by a PV module and use a battery for electricity storage. In general, the size of the module and the storage capacity of the battery determine the range of electric appliances that can be connected to the system.

Pico PV systems offer a wide range of advantages: easy installation (plug and play), user-friendly application, low investment costs, little maintenance required, and high degree of expandability and flexible use.

Pico PV systems can be categorized according to their appliances and the respective energy services that they provide: classic solar lantern, multifunctional system and classic Pico PV system. Consequently, the classic solar lantern constitutes one category that provides mainly a lighting service. It is to be distinguished from multifunctional systems that provide lighting as well as additional services. Bigger pico PV systems allow for operation of external devices as well.

Alternatively, pico PV systems can be categorized according to their main lighting use (task light, outdoor light and room light) or according to whether the PV module is integrated into the lamp, linked by a cable or centralized in a nearby charging station for many lamps at a time.

The primary purpose of pico PV systems is to provide lighting services. Then, a pico PV system can be equipped with various lighting technologies: compact fluorescent lamps (CFLs) or light-emitting diodes (LEDs). On the one hand, CFLs are based on the same functional principle as standard fluorescent lamps, but are more energy efficient than the latter as they work at a higher internal pressure. They are smaller and have a higher illuminating power. However, at present the LED technology seems to be the most forward-looking option [5]. Fortunately, LED lighting technology has experienced dramatic gains in efficacy and affordability over the past 5 years. Data [6] indicate that LED package prices have dropped by an order of magnitude since mid-2009 while the efficacy of light output has doubled from about 70 lm/W to over 135 lm/W for warm white LEDs. On the other hand, it should be taken into account that, to reach a satisfying level of light in one household, a minimum of 300 lm (comparable to a 30 W incandescent bulb) is recommended, which may require more than one lantern. Currently, more than 50 PV lamp models are available on the market. The majority of them are manufactured in China, followed by India, the United States and Germany. A majority of the lights are equipped with low power LEDs ($<$100 mW/LED) instead of power LEDs ($\geq$1 W/LED).

The PV panels that supply energy for the lights are mostly made of polycrystalline or monocrystalline silicon. The majority of the systems are equipped with panels from 1 to 3 W nominal power. And with respect to the batteries connected to the pico PV system, nowadays there are different types of batteries used in those systems, lead-acid and NiMH batteries being the most common types, while Li-ion batteries are presently used only in few cases. Battery capacity ranges from 20 mAh in a crank-charged system to 7200 mAh for a combined solar lantern/solar home system. Many of the systems are not equipped with a charge controller.

9.2.2 Project examples for pico PV systems

Solar Pico PV Market Potential in Nepal, current trend and future perspective.
Photovoltaic water pumping technology, SNV Netherlands Development Organisation, Nepal [7].

9.2.3 Off-grid directly coupled PV

A direct-coupled stand-alone PV system is one where the DC output of a PV array is directly connected to a DC load, as in Fig. 9.1. Since there is no electrical energy storage in these direct-coupled systems, the load only operates during sunlight hours. Its application is suitable for the supply of ventilation fans, water pumps and small circulation pumps for solar thermal water heating systems. Matching the impedance of the electrical load to the maximum power output of the PV array is a critical part of designing a well-performing direct-coupled system. For certain loads, such as positive-displacement water pumps, a type of electronic DC–DC converter, called a maximum power point tracker (MPPT), is used between the array and load to help better utilize the available array maximum power output.

The most useful directly coupled PV system is the water-pumping system. This consists of a PV generator, a DC or AC surface-mounted/submersible/floating motor pump set and power electronics converter (AC or DC converter) [7]. The PV generator is mounted on a suitable structure (fixed or with tracking). Then, water is pumped during the day and stored in tanks for being used during daytime, night or under cloudy

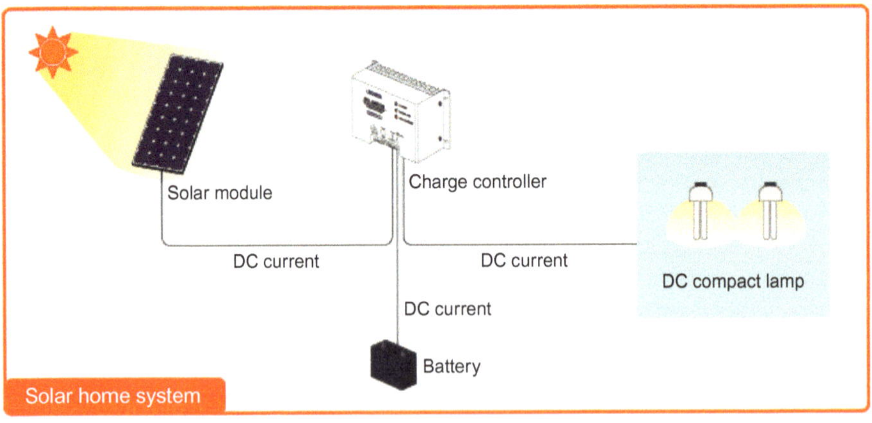

Fig. 9.1 Photovoltaic solar home systems for DC loads.
Courtesy: Leonics.

conditions. The water tank acts as storage. PV water pumping systems have shown significant improvements in the last decade, above all from the efficiency and cost points of view. The capacity of a solar pumping system to pump water is a function of three main variables: pressure, flow and power to the pump.

Solar pumps are classified into three types according to their application: submersible, surface and floating water pumps. The water pumps may be driven by DC or AC motors. The earlier PV pumping systems were principally based on DC motors. DC motors present the drawback of maintenance expenses due to frequent brush replacements. Brushless permanent magnet DC motors have been introduced in some systems. Recent developments in induction motor technology made this option attractive among the AC motor-based pumping set-ups. The induction motor is more robust, requires much less maintenance and is available at lower cost than a DC motor.

Then, there are three types of PV water pumps according to their electrical configuration: direct-coupled PV DC water pumping system, PV AC water pumping system and PV water pumping system with battery storage. The capacity of a solar pumping system to pump water is a function of three main variables: pressure, flow and power to the pump.

The DC motor-based PV water pumping systems consist of a PV array, with or without an intermediate PC, and a DC motor coupled with a pump. If a PC is used, then often some MPPT algorithm will be implemented [8].

However, for an AC motor water pumping system, a DC–AC inverter, which can be either multistage or single stage, is required [9]. A multistage water pumping system comprises a PV array, a DC/DC converter, a DC/AC inverter and an induction motor coupled to a pump. In the first stage, a DC/DC converter can be used to search the peak power point using any one of the algorithms, as has been described before, and then the DC/AC inverter is used to change the frequency, with a variable frequency power source.

Conversely, a single-stage converter is a simple, low-cost and efficient architecture which minimizes losses and reduces maintenance. In this case, a DC/DC converter is not required because some MPP (maximum power point) tracking algorithm is implemented in the DC/AC inverter [10] in addition to generating the variable voltage/variable frequency output required in order to feed the induction motor. Some discussion about the performance of different converters can be found in the literature [11].

The performance of solar photovoltaic (SPV) water pumping systems in the field is not always as estimated because of poor design of the array size or BoS and improper pumps.

9.2.4 Solar home system

The term *solar home system*, and its acronym SHS, refers to a stand-alone system, suitable for residential applications, such as home appliances, lighting, computers and water pumps. Normally, the SHS is low power, less than 100 W [12]. The SHS is generally designed and sized to supply DC and/or AC electrical appliances. It consists of PV modules connected to a PV charge controller, stand-alone inverter

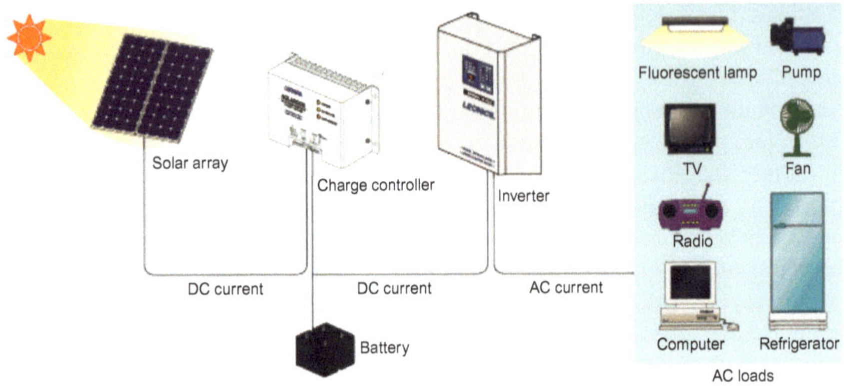

Fig. 9.2 Photovoltaic solar home systems for AC loads.
Courtesy: Leonics.

and battery system. The generated DC power is stored in the battery and converted to AC power for supplying to AC loads; see Fig. 9.2.

9.2.5 Off-grid PV hybrid solutions for residential, industrial and rural electrification

In the following part of this chapter, the current solutions applied to residential, industrial and rural electrification are discussed.

Residential solutions

Nowadays, there are different options for implementing any PV system applied to residential solar power systems: grid-tied solar, off-grid solar, self-consumption, grid tied with backup, backup power and rural electrification. Many homeowners prefer buried power lines. Other homeowners face large expenditures to extend grid power to their remote location. In both situations, the expense associated with connecting to the traditional electric grid might be better spent building your own, independent power plant. Off-grid PV systems are a highly reliable, effective means of providing power without the regular costs associated with electric utilities and generator power.

Off-grid projects can be designed in many ways. One of these can be to use a traditional diesel (or gas) generator running continuously. Petrol/diesel generators (gensets) are widely available and can be purchased in almost any country in the world. They are easy to install and set up. However, this option is highly polluting, noisy and smoky, and presents a high variability of the fuel price, among other disadvantages.

Another option can be to install a PV system consisting of a PV generator and a battery system, but if consumption needs to be provided in the night the battery system will be very expensive. However, it can be demonstrated that if a PV generator is combined properly with a battery system and a backup generator, the cost of the total system, LCC cost, is lower than for other options. This solution is called a PV hybrid system [13] or an AC micro-/mini-grid system (Fig. 9.3).

According to Fig. 9.3, there are two buses: DC and AC (where any AC load can be connected). Then, as this is an off-grid system, the AC bus must be generated (grid forming), as discussed in the next section.

Depending on the connection of the elements, there will be two different configurations: DC coupling (charge controller, off-grid inverter and battery system—see Fig. 9.4), and AC coupling (grid-connected inverter, bidirectional inverter (BDI) and battery system—see Fig. 9.5).

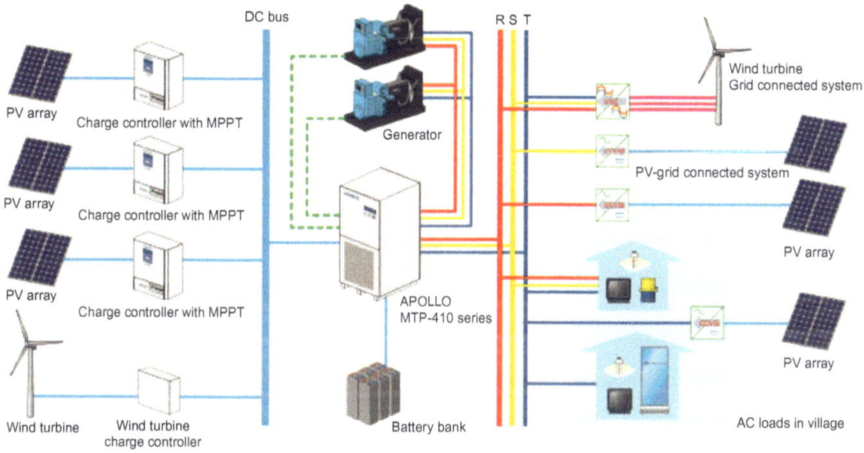

Fig. 9.3 General scheme of a PV hybrid system.
Courtesy: Leonics

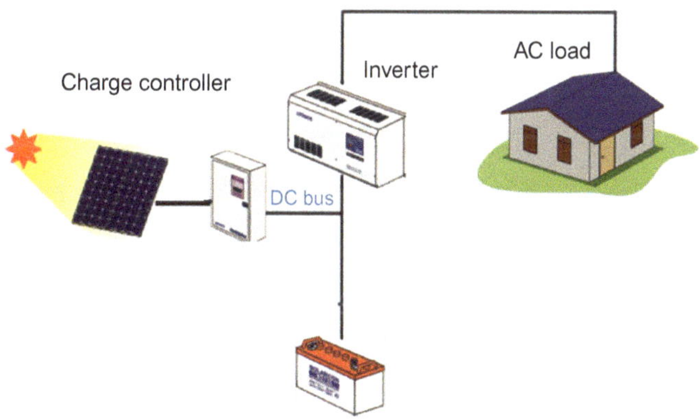

Fig. 9.4 DC-coupled configuration implemented in some PV hybrid systems.
Courtesy: Leonics.

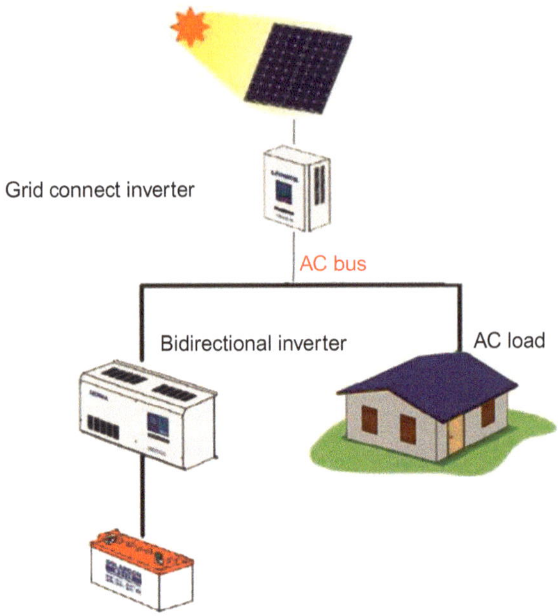

Fig. 9.5 AC-coupled configuration implemented in some PV hybrid systems. *Courtesy:* Leonics.

9.2.6 Grid forming

The possible grid-forming architectures that can be implemented into PV hybrid systems are the following: genset, PC, single switch (genset and PC) or multimaster inverter-dominated system.

Any of the methods reviewed previously are capable of regulating mini-grid voltage and frequency to meet current user needs. The choice of strategy employed will depend on the architecture of the mini-grid (centralized or decentralized), the generation mix (diesel dominated or high penetration of PV generator) and the economics of incorporating auxiliary stabilization mechanisms, such as energy storage systems or dispatchable loads.

The *genset-dominated system* (single master) is a typical configuration for a PV–diesel hybrid system (mini-grid) that has multiple AC sources (gensets and PV inverters) connected to the mini-grid and simultaneously supplying power. The gensets (one of various) are the only components responsible for grid forming by regulating the voltage/frequency, while the other sources are synchronized to the mini-grid. This is the typical architecture for low PV penetration, but requires the genset to be operating at all times.

The *single switched master* PV hybrid system architecture can have multiple AC sources connected to the system (typically inverters and a genset), but the grid-forming control is switched between the genset, Fig. 9.6, and the BDI(s), Fig. 9.7.

This is the typical AC-coupled configuration. This allows the genset to be turned off and the system to operate only from the PV generator and storage. This architecture

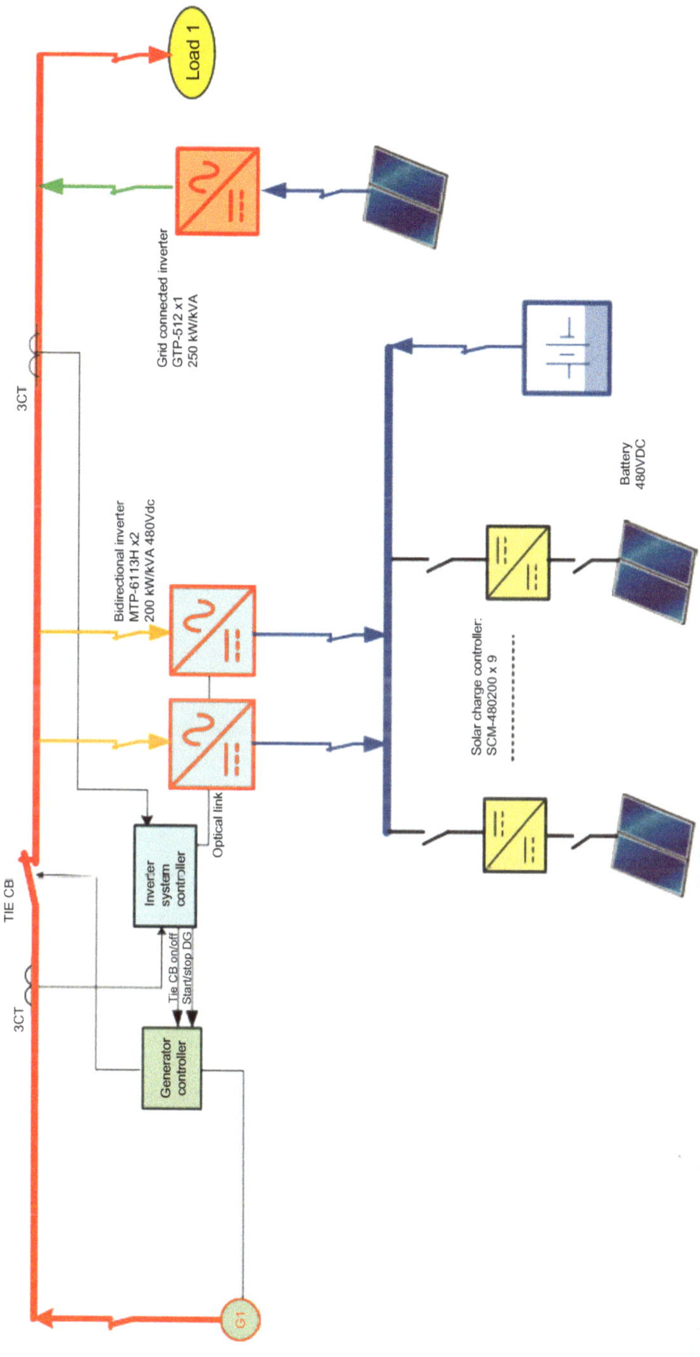

Fig. 9.6 Diesel generator is the master of system.

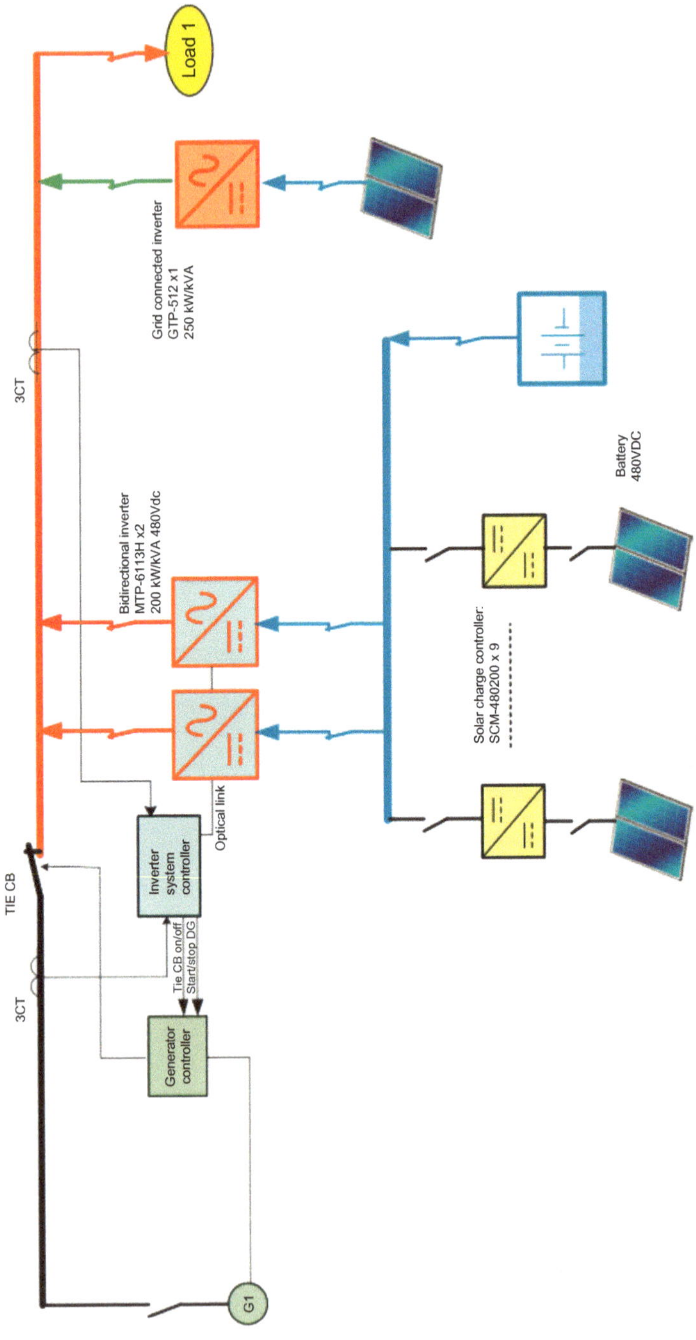

Fig. 9.7 Bidirectional inverter is the master of system.

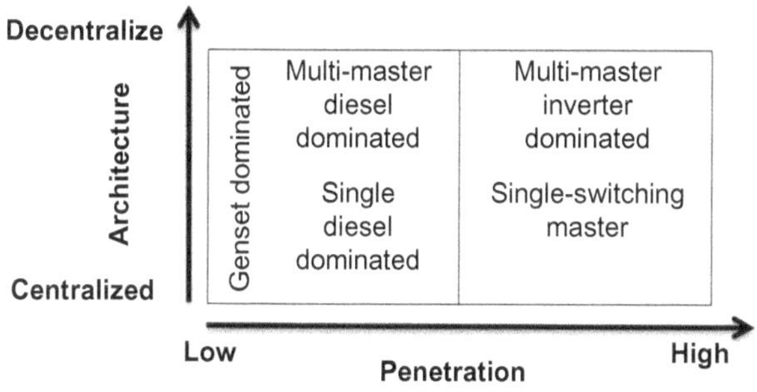

Fig. 9.8 Architecture versus PV penetration (%).

is typically used where the application does not have sufficient base load to support continuous genset operation or multiple genset plants. With sufficient energy storage and PV capacity, very high PV penetration can be achieved. In this system, the genset is used only as a backup.

The *multimaster inverter-dominated* architecture has a decentralized control architecture in which several generating sources distributed within the PV hybrid system cooperate to act as grid formers. Generally, most of these sources are interfaced to the PV hybrid system by power electronic interfaces (inverters), but a conventional genset using droop control may be integrated as well.

In the *multimaster genset-dominated* architecture, one or more gensets maintain an AC grid to regulate voltage and stabilize frequency. Renewable energy sources (RES) such as SPV, small wind turbines and/or run-of-river hydropower sources also supply part of the load when the energy source is available.

At least one of the gensets operates continuously, while the other(s) can be dispatched based on variations in load demand and changes in the RES availability. In most cases, a diesel dominated mini-grid does not have any sort of energy storage or only utilizes 'short-term' energy storage to assist with transient dynamics of the system. The RES usually operates as passive generation units, injecting as much power as possible, with little or no participation in the control strategy. The renewable sources can be seen as negative loads that reduce the amount of power demanded from the gensets.

In Fig. 9.8 is shown the different applicability of the different forming strategies with respect to the architecture (centralized or decentralized) and the PV penetration.

9.2.7 Control in the single-switching master

As has been seen before, the single-switching master is the most useful architecture. As two sources actively participate in grid stabilization and voltage regulation, the PV generator and the genset, then frequency-droop and voltage-droop control strategies are used to share real and reactive power components. In this case, the voltage and frequency of the system may deviate from the rated values, within acceptable limits, depending on the load level and the droop characteristics.

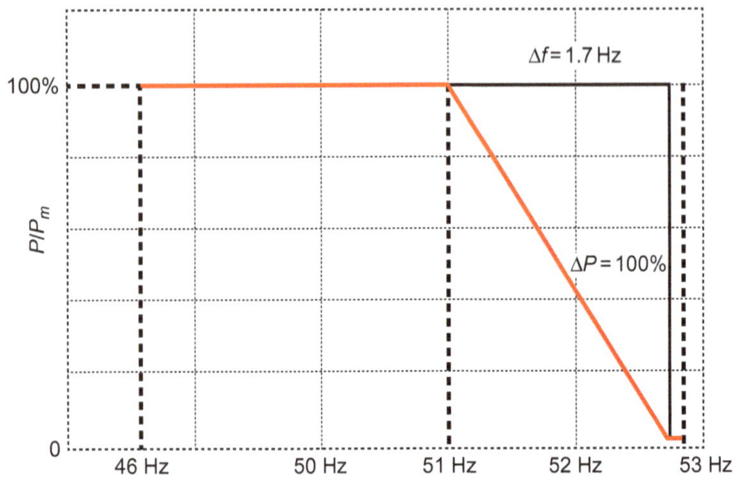

Fig. 9.9 Frequency-droop function implemented in the Fronius grid-connected inverters. *Courtesy:* Fronius

Droop control is a well-known technique used for operation and power sharing of power generators connected in parallel. The relationship between frequency and power can be shown as in Fig. 9.9, implemented in the control of the inverters.

The PV unit (PV generator and the grid-connected inverter) is commonly controlled as in grid-connected configurations, where the interfacing voltage-sourced converter is controlled as a current source to inject the available PV power into the PV hybrid system bus (the power quality (PQ) control strategy). This technique was developed originally for grid-connected configurations. Therefore, the operation of the PV unit, the battery storage, and other units in the PV hybrid system, such as droop-controlled units, must be coordinated to balance the power in the off-grid system while respecting the battery storage limits [14].

9.2.8 Power management in PV hybrid systems

When the batteries are implemented in the system, four elements must be controllable: PV generator, genset(s), energy storage and load (see Table 9.1).

Depending on the element, there will be different controllable characteristics and control possibilities.

A good operation of any PV hybrid system can be achieved only by suitable power management between the elements. This is fundamental for the optimization of the use of all elements: optimize the PV energy, minimize the wear on the batteries and consume the smallest possible quantity of fossil fuel. This will manage power flows among the different energy sources and the storage unit in the system.

A control strategy consists of certain predetermined control settings implemented when the system is installed. Such settings concern the timing of certain values representing the system state, such as the battery state of charge (SOC) and demand placed on the system. Normally, the PV generation is the primary source, the battery is the supplement and the diesel generator is the backup source of energy.

Table 9.1 Controllable characteristics and means of control for components typically found on mini-grids

Elements	Controllable characteristic	Control possibilities
PV generator	Power	Communication line Frequency
Batteries	Depth of discharge Number of cycles per day Charge/discharge power	Communication line Frequency Energy management
Genset	Number of start/stop cycles Generation power	Communication line Start/stop signal Frequency
Consumption	Power consumption (on/off)	Communication line Frequency User

9.2.9 PV penetration in PV hybrid systems

The fraction of PV power installed with respect to the load profile, PV penetration, is a key parameter in the PV hybrid system. So the integration of the PV into diesel systems can be divided into: low penetration systems (PV penetration <20%); medium penetration systems (where the control is needed), active PV penetration 20–65%; and high penetration systems with PV penetration between 65% and 100% and where energy storage systems must be installed.

For residential applications, the medium penetration (20–65%) is the most useful and interesting solution to be implemented.

9.2.10 Medium penetration: Intermittent genset operation

Under medium penetration operation, two possible solutions can be implemented: continuous operation genset (single or multiple genset) or intermittent operation genset (multiple generators under master-slave control). In this type of operation, the on/off control of the genset is imposed. Also, two different systems can be distinguished: without energy storage and with energy storage. For PV residential options, intermittent genset operation with energy storage is the best solution.

9.2.11 Integrated photovoltaic air-source heat pump

Another possible residential solution is the use of an aerothermal system powered by PV. Then, due to the decreasing prices of PV systems, the generation for households of cool/hot air and hot water can be powered by PV and often it will be the most economical method.

Compared to a PV system, a solar thermal system has several disadvantages: unlike a PV system, most solar thermal systems have moving parts (pumps and solenoid valves); in freezing climates, solar thermal systems are sometimes subject to freeze damage; solar thermal systems require regular maintenance, including antifreeze replacement; unlike owners of a grid-connected PV system, who can be credited for their excess electricity production during the summer, or the excess can be handled in some way, owners of a solar thermal system cannot sell the excess summer production of their hot water systems or manage it. And while a pole-mounted PV array can include a tracking mechanism to follow the sun's path across the sky, it is virtually impossible to install solar thermal collectors on a tracker.

Nowadays, the most energy-efficient alternative for generation of the cool/heat is by heat pump, which can be by means of the air (air-source heat pump, ASHP) or the ground (ground-source heat pump) [15,16]. During the heating season, heat pumps move heat from the cool outdoors into your warm house and during the cooling season, heat pumps move heat from your cool house into the warm outdoors.

For homes without ducts, ASHPs are also available in a ductless version called a minisplit heat pump. In addition, a special type of ASHP generates hot and cold water rather than air, allowing it to be used with radiant floor heating systems in heating mode.

An integrated PV-ASHP system is a system where the heat pump is powered by PVs [17]. The system can be made up of five parts as follows: PV generator, a PV controller, batteries, a DC–AC inverter and an ASHP system.

ASHPs extract heat from the outside air using a reverse refrigeration cycle (imagine if you took a window air conditioner and flipped it around). By extracting heat from the outside air and moving it inside, rather than directly heating the indoor air, the heat pump runs two to three times more efficiently than an electric baseboard heater. This is commonly referred to as a coefficient of performance of 2–3. Modern heat pumps are able to heat and cool two to three times more efficiently than older electric technologies, making them much less expensive to operate than nearly every other form of heat, and the best air conditioner you can buy.

There are a number of different system designs, but the most common installation is a 1:1 ductless 'mini split' unit. This system design consists of an outdoor unit with heat exchanger and compressor, which performs the actual heat extraction, and an indoor blower unit which blows warm (or cool) air directly into the home. Mini split means there is a 1:1 ratio: it is one indoor unit for every outdoor unit. Also it is possible to install multi splits, one outdoor unit to two or more interior units, and ducted units when appropriate to site specifics.

Until this point, there have been very few examples of that integration [18,19] and a lot fewer for geothermal heat pumps [16].

9.2.12 Industrial PV solutions

A relevant PV application is when large units like hospitals, schools, tourism facilities, telecom stations and small manufacturers operate their machinery (eg, water pumps, desalination systems or water purification systems) using power from a PV system, but off-grid. The more typical solutions use a low PV penetration for PV–diesel hybridization (when originally the energy is powered by diesel genset, operating continuously) or implement a PV hybrid system under medium penetration, which will be shown in Section 9.6.1.

9.2.13 PV diesel hybridization in low penetration (<20%)

The integration of the PV system in preexisting or new diesel installations is another type of stand-alone system with renewable energy. PV–diesel hybrid solutions may hold the answer to reducing fuel costs and increasing energy security for different sectors, including mining, tourism, real estate, agriculture, retail business and chemical, food and textile industry. Bloomberg New Energy Finance estimated a 150-GW installed base of diesel gensets in 2011 in developing countries alone [20].

A low penetration off-grid solar connection refers to a PV system with an output capacity of approximately equal to 20% compared to the capacity of the load. Then, all PV energy can be used in the system and the diesel generator is always operating at its nominal conditions, that is, the load is equal to the nominal load of the genset(s) and the contribution of the PV to the load coverage is negligible.

In this solution the genset is the main power supply and is designed to operate daily, full time, along with the PV generator, to supply power to the load. The function of the PV generator is to reduce fuel usage. There is no special management or equipment required as the PV system acts as a simple negative load on the diesel mini-grid. The genset is used always for grid generation and the grid frequency is within operational limits. The genset is the master of the control, acting as the main governor of the system running under normal operation mode. The elements for this solution are PV generator, PV grid-connected inverter, genset(s) and controller (Fig. 9.10). The PV inverter, coupled by the AC bus, monitors the frequency and adjusts its power injection in order to both maximize the fuel saving and ensure minimum genset load and spinning reserve (it can be defined as the unused capacity which can be activated on the decision of the system operator and which is provided by devices synchronized to the network and able to affect the active power). The PV inverters always operate in the MPPT mode of operation.

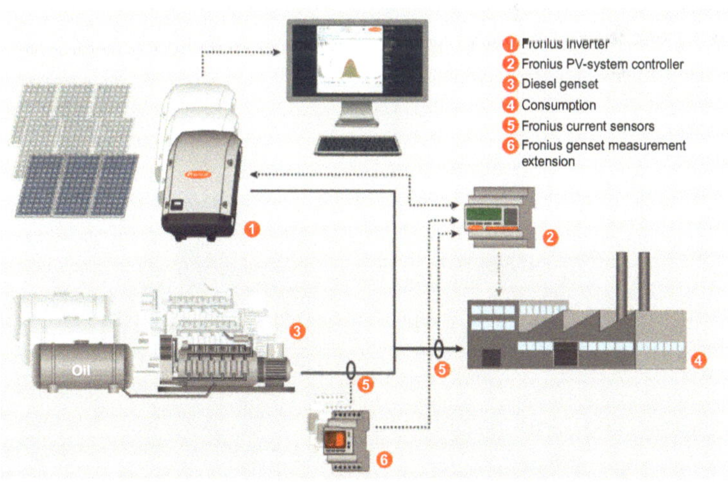

Fig. 9.10 Example of one solution for PV low penetration (PV–diesel hybrid). *Courtesy:* Fronius.

When the electricity production of the PV system reduces, for instance due to clouds, it is similar to turning on a piece of equipment whereby the spinning reserve of the generator will now supply more electricity to meet the demand. The power plant control in diesel genset operated systems can accommodate all natural fluctuations.

Benefits of low penetration include: no energy storage required (low cost); no specialized control system required (easily installed), many utility scale reference sites (well proven) and lower investment/commitment.

Some improvements in the performance of the system can be made with respect to the traditional genset, which is slow to start up, cannot be run in low-load mode, and is not very flexible for balancing the unstable solar component [21].

9.2.14 PV hybrid systems for industrial or service activities (10–100 kVA)

Industrial or service activities (10–100 kVA) can be powered by means of a PV hybrid system, PV–battery-backup genset. The solution that can be implemented can be decentralized (by using single-phase PCs—Fig. 9.11), or centralized (by using three-phase medium PCs—see Fig. 9.12). For instance, in Fig. 9.11 is shown a three-phase solution where single-phase BDIs are implemented. At least one BDI must be installed on each phase.

Nowadays, there are two possibilities for implementation: closed solution (where the same manufacturer supplies the grid-connected inverters and the BDIs) and open solution (where one manufacturer supplies the grid-connected inverter and a different manufacturer supplies the BDI). In the first group would-be manufacturers such as SMA, Schneider, Aros and Leonics, and in the second group, Fronius, Victron (eg, see Figs 9.13 and 9.14).

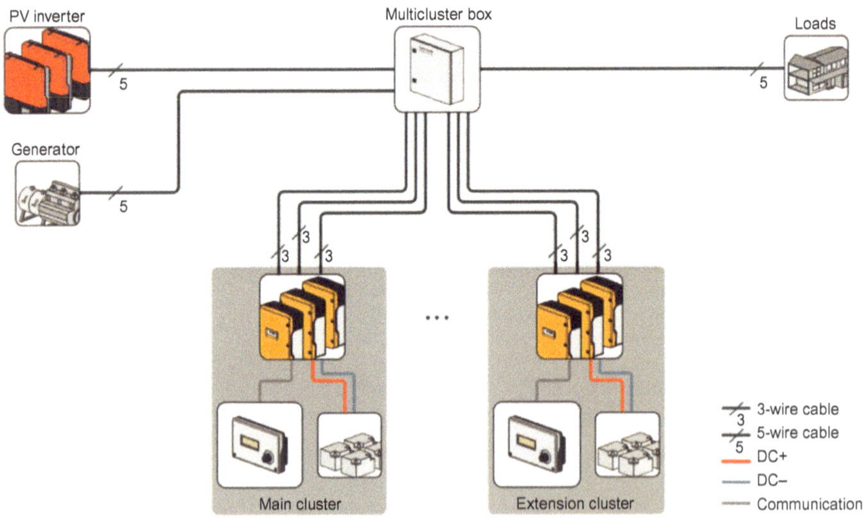

Fig. 9.11 Block circuit diagram for a three-phase multicluster off-grid system. *Courtesy:* SMA.

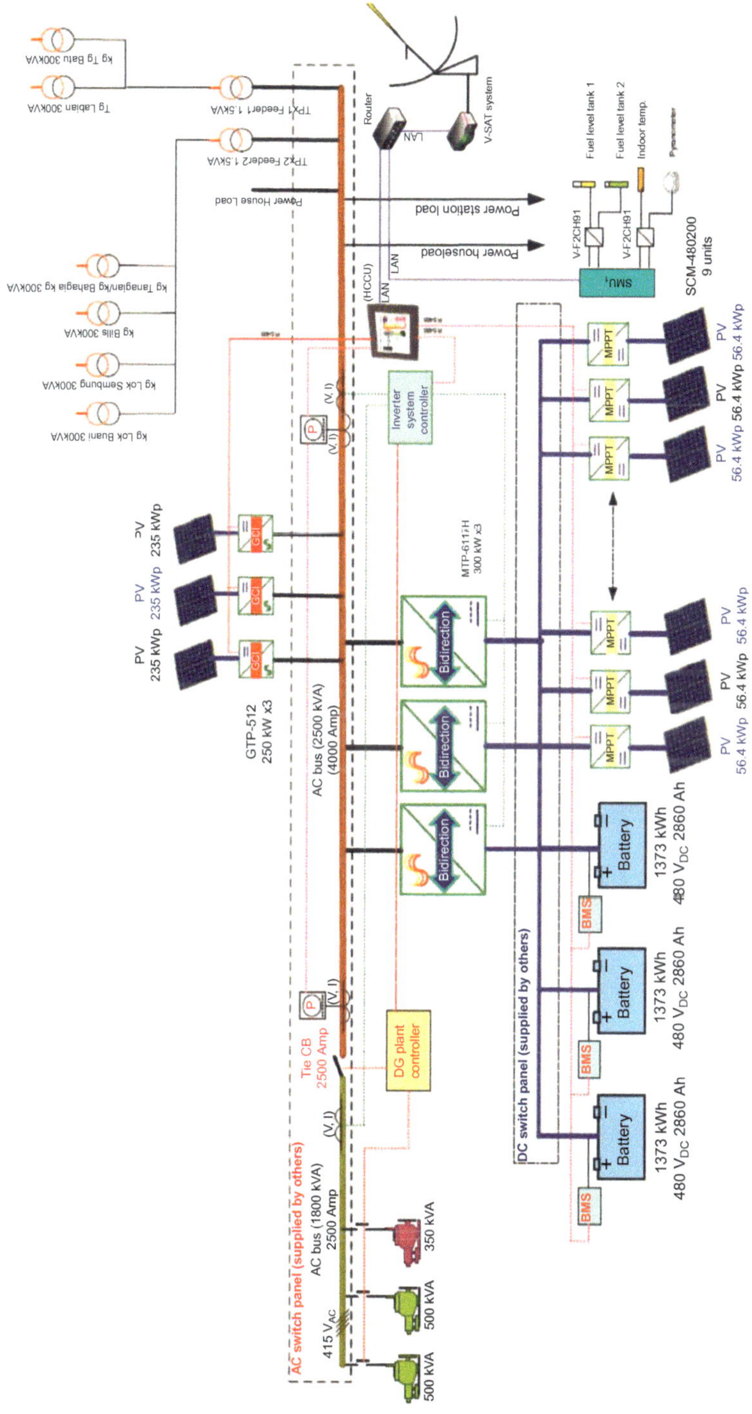

Fig. 9.12 Mini-grid centralized implementation with central inverters.
Courtesy: Leonics.

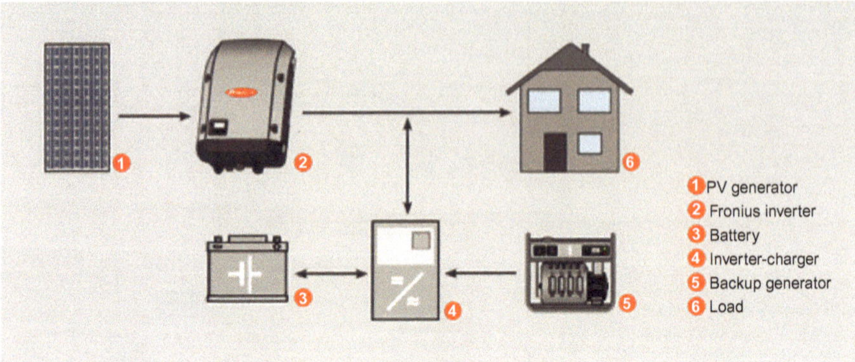

Fig. 9.13 Solution for industrial or service activities for PV hybrid system with range power between 10 and 100 kVA.
Courtesy: Fronius.

Fig. 9.14 Example of implementation for industrial activity implemented by using a PV hybrid system with single-phase grid-connected inverters and bidirectional inverters.
Courtesy: Fronius and Victron.

9.3 Power converters implemented in PV stand-alone systems

In this part of the chapter, the different PCs that are often installed in PV stand-alone systems will be discussed.

9.3.1 Battery charging systems

These systems are based around the charge controller. A battery charge controller (BCC) regulates the flow of electricity from the PV generator to the battery. Its function is to regulate the voltage and current from the PV array in order to prevent overcharging and also overdischarging of the battery.

There are four general types of charge controllers, categorized by the method used to regulate the charge from the solar modules to the batteries: shunt type charge controllers; series type charge controllers; pulse-width modulation charge controllers and MPPT charge controllers.

The shunt charge controller was the first type of charger developed and is the simplest one still on the market today. The solar panel is short-circuited to prevent further energy flow when the battery reaches the full charge. In general, shunt type charge controllers are low cost and reliable, with a simple design suitable for small off-grid PV systems.

Series type charge controllers are very similar to shunt type charge controllers, but instead of short-circuiting the solar panel output, they open the circuit, interrupting the path to the batteries. For disconnecting the circuit, this type of controller uses a relay or a solid-state switch. After the batteries reach a set voltage value, the solar module is disconnected from the batteries. When the battery SOC decreases, the switch is reset and the panel is once again connected to the batteries.

Pulse-width modulation charge controllers provide a variable charging current, given by variable power output of the solar panel or by the battery SOC, switching a series element at a high frequency for variable periods.

The MPPT controllers are DC–DC converters, providing a matching interface between the PV generator and the battery. The main function of the MPPT controller is to adjust the PV generator output in order to transfer maximum energy to the batteries. This offers a high efficiency over a wide range of operating points.

Regarding their electrical specifications, four important parameters must be taken into consideration: the nominal battery voltage, the number of MPPT inputs, the input voltage and the shade-tolerant MPPT. A review of current MPPT charge controllers on the market considering those parameters is shown in Table 9.2.

Table 9.2 Some important parameters related to MPPT charge controllers

Manufacturer	XW MPPT 80 600 model	MPPT input number	Nominal battery voltage (V)	Maximum battery current (A)	Maximum open-circuit voltage (V)
Leonics	SCM-480200	1	480	200	<550
Morningstar	TS-MPPT-60	1	48	60	<150
Outback	Flexmax 60	1	12–60	60	<150
Schneider	XW MPPT 80 600	1	24–48	80	<500
SMA	Sunny Island Charge Controller 50	1	12–24–48	50	<140
Steca	Tarom MPPT 6000-M	2	10–60	60	<200
Studer	VT-80	1	12–48	80	<150
Victron	BlueSolar MPPT 150/85	1	12–48	85	<150

Regarding the nominal battery voltage, two ranges are found: the first one with 12, 24 or 48 V_{DC} (low voltage) and another one with 120, 240 and 480 V_{DC} (high voltage). The input voltage is related to the maximum open-circuit voltage of the PV array admissible to be connected to the equipment. So, two ranges can be distinguished: low voltage and high voltage (600 V_{DC}), also named as HVI-MPPT. The benefit of using a high-voltage charge controller is the ability to have one long series string of solar panels wired together, or fewer strings in a system. This reduces the number of cables, reduces wire size and voltage drop issues and reduces the number of breakers in the system for easier wiring and faster installation.

Another important parameter is the ability to obtain the MPPT under shaded conditions of the PV array. However, this topic is focused in practice on domestic applications.

In addition, these products often have up to four stages of charging: bulk, absorption, float and optional equalization; they often include temperature compensation at the controller or at the battery when using the optional remote temperature sensor, together with protection against reverse polarity, short circuit, overcurrent, lightning and transient surges, high temperature and reverse current at night.

Nowadays, charge controllers provide several communication options. They can use a proprietary protocol and/or nonproprietary open standard such as Modbus and Modbus TCP/IP protocols for RS-232, EIA-485 and ethernet networks. Additionally, HTTP, SMTP and SNMP are often supported for web page, e-mail and network message support.

9.3.2 PV inverters implemented in PV hybrid systems

The PV inverters can be divided into two categories: stand-alone and grid-connected inverters. Stand-alone inverters are further subdivided into stand-alone inverter, grid-interactive inverter, BDI and multiport inverter (also called hybrid inverter).

Stand-alone inverter

This converter is designed for remote stand-alone applications, or off-grid power systems with battery backup where the inverter draws its DC power from batteries charged by PV array and converts this to AC power. On the market today, there are three different power inverters: square wave, modified sine wave and pure sine wave inverters. These inverters differ in their output, providing varying levels of efficiency and distortion that can affect electronic devices; see Table 9.3.

These inverters are used only in the DC-coupled configuration, Fig. 9.15, as will be shown later.

Grid-interactive inverter

This inverter can operate on both grid-tied and stand-alone off-grid operations, according to the configuration shown in Fig. 9.16.

Table 9.3 Characteristics of the different types of off-grid inverters

Square wave inverters	Modified sine wave inverter	Sine wave inverter
Simple technology	Better efficiency	Technique similar to inverters used for connecting to grid but with simpler circuits and no protection from or synchronization to grid
Risk of generating odd harmonics	Fewer harmonics than square wave inverters. Precise regulation of voltage	Higher efficiency than other inverters
No regulation of outgoing voltage (varies with charge and incoming voltage)	Appropriate for powering many devices (televisions, engines, table saws)	Suitable for practically every type of use

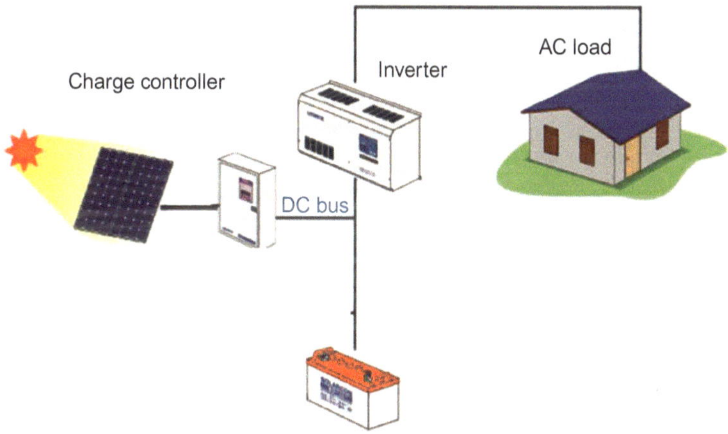

Fig. 9.15 DC-coupled configuration implemented in some PV hybrid Systems.
Courtesy: Leonics.

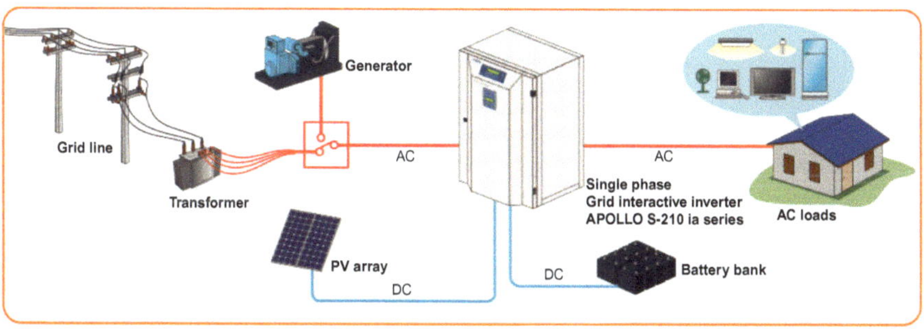

Fig. 9.16 PV grid-interactive inverter.
Courtesy: Leonics.

When the utility power is available, the inverter can operate as a grid-tied inverter that converts DC power generated by a PV generator into AC power for supplying to load and feeds the excess energy back to the utility grid line. When the utility power is not available, the inverter can operate as a backup power source to supply power from the PV panels and battery; see Fig. 9.17.

In a grid-interactive system, that inverter is a much smarter, more agile device and capable of doing three things as opposed to the one-trick grid-tied inverter: like a grid-tied inverter, a grid-interactive type can convert solar-generated DC power to AC power; it can also function as a battery charger and store energy in a battery system designed for residential and commercial applications, and it can convert battery-produced DC during an outage into usable AC power and charge those batteries during the day from the PV generator or a genset. It can even turn a generator on as needed. Having battery backup taking care of power needs in the evening means the generator's run time and fuel consumption can be greatly reduced during an outage.

Bidirectional inverter

The BDI has two ports: AC and DC ports. The DC port is connected to a battery bank, so this type of inverter is also called a battery inverter. Some models have extra inputs to which the output of the charge controller can be connected. The AC side is connected to the AC line, where the utility grid or a genset(s) (AC configuration) can be coupled, together with AC loads. Sometimes there are two independent AC outputs, one of these acting as an auxiliary output (noncritical load).

The BDI has dual control and operates in two modes: VCVSI (voltage control voltage source inverter) mode, as a rectifier, and CCVSI (current control voltage source inverter) mode, operating as an off-grid inverter [22]. In addition, it has intelligent battery management for charging and discharging the batteries.

The BDI can be integrated into the AC-coupled topology; see Fig. 9.18.

Currently, the BDI is marketed as a single compact enclosure for ranges between 0.8 kW up to 1800 kW. Lithium batteries are approved for almost all models.

According to Table 9.4, it can be seen that there are two types of BDI with regard to battery voltage: low voltage (12, 24 and 48 V) and high voltage (330 and 820 V). In addition, we can distinguish between single-phase (up to 20 kW) and three-phase (up to 1800 kW), although there are many more models for single-phase (with brands such as Leonics, Outback, SMA, Studer, Schneider, Victron and Zigor) than for three-phase (with only three brands: Ingeteam, Leonics and Zigor).

The BDI permits a modular design of any PV off-grid plant. So, it is possible to design a three-phase system with single-phase units; see Table 9.5.

For example, SMA uses their multicluster technology where the off-grid systems can be extended at any time as the energy demand increases, from 3 to 300 kW. Other manufacturers have similar products; see Fig. 9.19.

A BDI can operate as a simple converter and also as controller of grid-connected inverters in an AC-coupled configuration. Typically the active power of the grid-connected inverter can be controlled without any communication. In this case, a frequency-droop method can be implemented in the BDI and the grid-connected

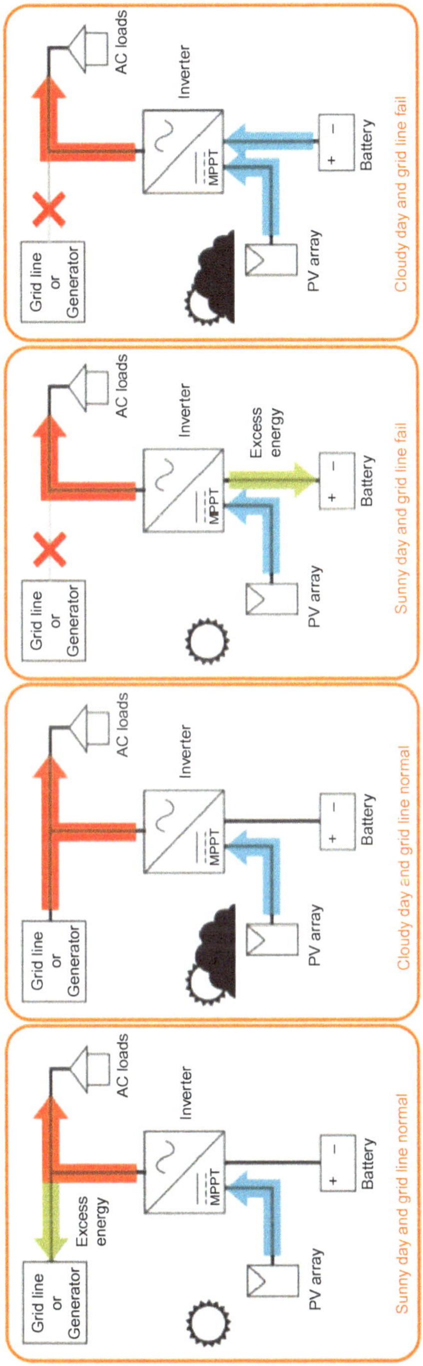

Fig. 9.17 Operation of a PV grid-interactive inverter.
Courtesy: Leonics.

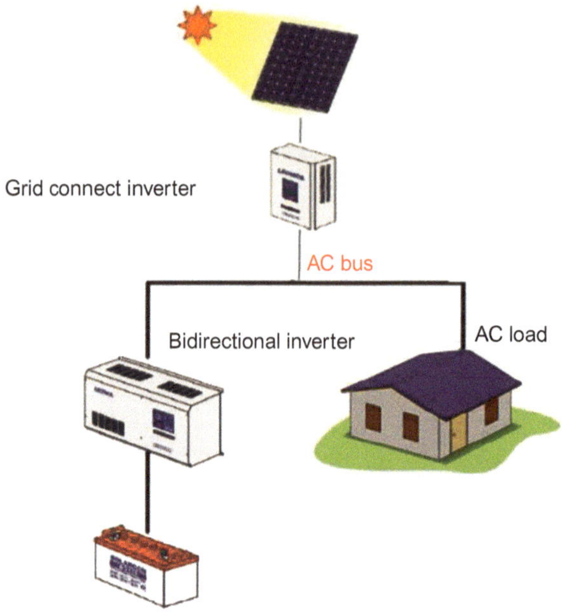

Fig. 9.18 AC-coupled configuration implemented in some PV hybrid systems. *Courtesy:* Leonics.

Table 9.4 Off-grid bidirectional inverters for different brands, single-phase and three-phase

Manufacturer	Model	Phase	Range of power (kW)	Battery voltage (V)
Ingeteam	Ingecon Hybrid AC Link	3	60–250	330–820
Leonics	APOLLO S	1	3–30	48
	APOLLO STP	3	15–100	48
	APOLLO M	1	5–20	120–240
	APOLLO MTP	3	15–2400	120–480
Outback	FX and VFX	1	2–3	12–48
SMA	Sunny Island	1	5–8	48
Studer	Compact and Xtender	1	0.5–7	12–48
Schneider	Conext XW	7	7	48
Victron	Multiplus	1	0.8–5	12–48
Zigor	HIS1	1	1.2–8	24–48
	HTDI	3	150–1800	672

Table 9.5 Three-phase off-grid PV hybrid system options for AC-coupled configuration

Three-phase system	String inverter	Central inverter	Bidirectional single-phase	Bidirectional three-phase
Option 1	X		X	
Option 2	X			X
Option 3		X	X	
Option 4		X		X

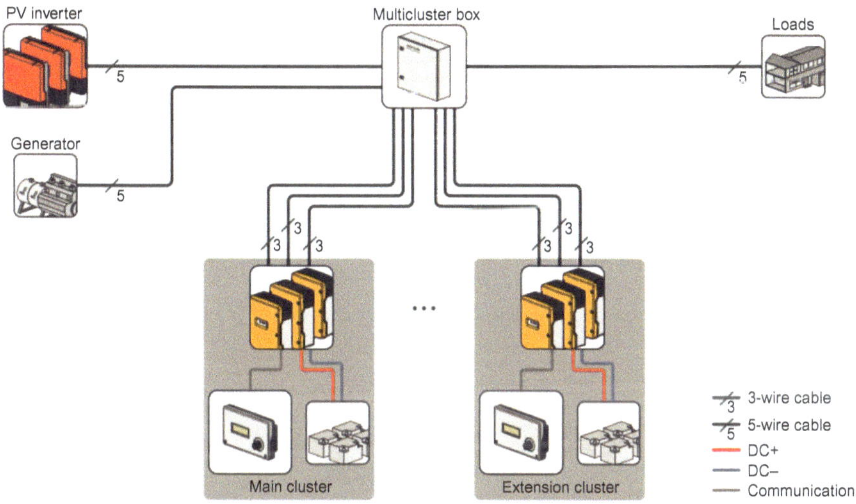

Fig. 9.19 Block circuit diagram for a three-phase multicluster off-grid system. *Courtesy:* SMA.

inverter. In addition, a function for voltage dependent power reduction can be implemented to prevent tripping caused by overvoltage.

There is also a possibility to implement communication between the bidirectional and the grid-connected inverter or to implement load measurement to limit the maximum power that is to be consumed by the BDI. This can be done by using an external controller.

Finally, it is also possible to control BDI with a proprietary protocol and/or non-proprietary open standard such as Modbus and Modbus TCP/IP protocols for RS-232, EIA-485 and ethernet networks. Additionally, HTTP, SMTP and SNMP are often supported for web page, e-mail and network message support.

Multiport input/multioutput inverter

In a PV hybrid system with batteries, the PV and the batteries need a power electronic interface to be connected to the load. One solution is to use a system formed by a hybrid inverter that consists of two inverters operating in parallel and whose outputs are tied to the AC output through a multiwinding step-up transformer.

Another possible configuration can be a multiinput DC/DC converter with an additional AC/DC inverter stage for feeding the AC loads or with a power stage.

However, in both cases, there are multiple power processing stages that contribute to lowering the efficiency of the overall system.

Finally, an alternative to the inverter topologies previously presented, that simplifies and reduces the cost of those inverters while increasing reliability, can be a single-stage multiinput high-frequency topology. It is capable of both stepping up or down the voltage; see Fig. 9.20.

It is based on the single input/single-output partial resonant AC link converter.

In all the topologies shown, the final inverter implemented is a multiport inverter with three ports: two for input and one for output, implemented in a single compact enclosure. Then, those inverters incorporate a PV input where the PV field can be connected directly to the inverter without using another extra output inverter. The PV input also has MPPT management. Also, as in the BDIs, the multiport inverter implements a battery management technology that ensures the maximum life of the storage system, by constantly controlling the battery temperature (often using a three-wire PT100 sensor).

From the energy point of view, the inverter will have four possible power flow scenarios, depending on the power generated by PV modules, the battery SOC, and the load requirements: (1) from the PV modules to the load; (2) from the battery to the load; (3) from the PV modules to the load and the battery; and (4) from the PV modules and the battery to the load.

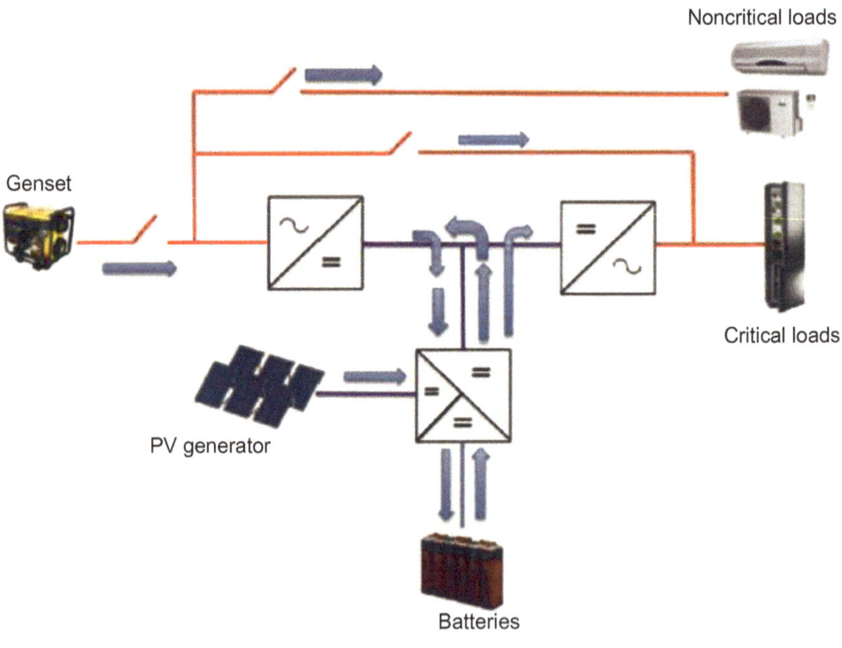

Fig. 9.20 The schematic block diagram of the multiport input/multioutput inverter.

If the grid is considered, there may be another scenario in which power flows from the grid to the battery.

Regarding these hybrid inverters, nowadays two different products are marketed (Fig. 9.21): single-phase (2–5 kW) and three-phase (30–300 kW) hybrid inverters.

The single-phase products are dedicated to home applications whereas the three-phase products are implemented for industrial applications. In turn, the small size hybrid inverters can be distinguished without integrated batteries (Ingeteam) and with integrated batteries (ABB, Bosch, Fronius, GoodWe, Kostal, PowerRouter and SMA, for instance). The batteries most often used in these inverters are Li-ion, except the Kostal product where only the OPzV (Ortsfest PanZerplatte Verschlossen), sealed tubular plate battery type can be connected. At any rate, by using single-phase equipment, a three-phase off-grid system can also be created, increasing the system power by connecting inverters in parallel.

Fig. 9.22 shows an example of topology for those inverters.

In general, they are applicable for both on-grid and off-grid PV systems and can control the flow of energy in a hybrid system, with its working mode able to be switched automatically or manually.

During the daytime, the PV plant generates electricity that can be provided to the loads, fed into the grid or used to charge the battery. The power stored can be released when the loads require it during the night. In Table 9.6 is shown a list of some battery hybrid integrated inverters. According to Table 9.6, the battery nominal voltage varies between 40 and 288 V.

Medium-size hybrid inverters are always three-phase inverters where the power rating is from 120 to 300 kW for each inverter and 50 to 200 kW for the MPPT charge controller. In addition, they can be used with parallel operation of inverters for total inverter power up to 1200 kW and up to 2000 kW of PV. They are designed to operate in micro-grid operation with the utility line or to operate as a stand-alone PV–DG hybrid (Fig. 9.23). They can provide backup power to the load when the utility grid line is not available and supply the load and charge the battery from the PV panels at the same time [23].

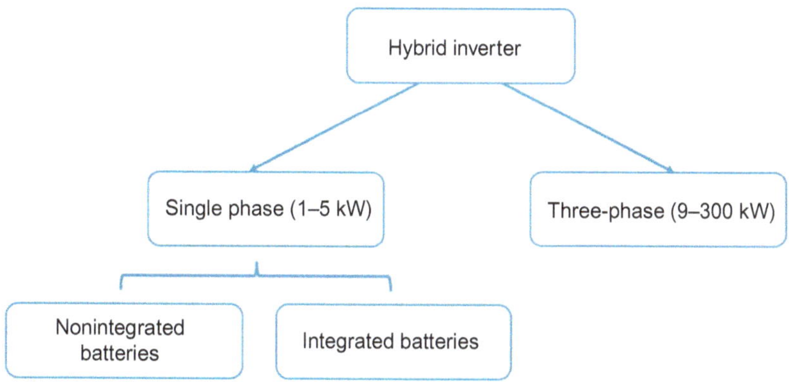

Fig. 9.21 Classification of the PV hybrid inverters.

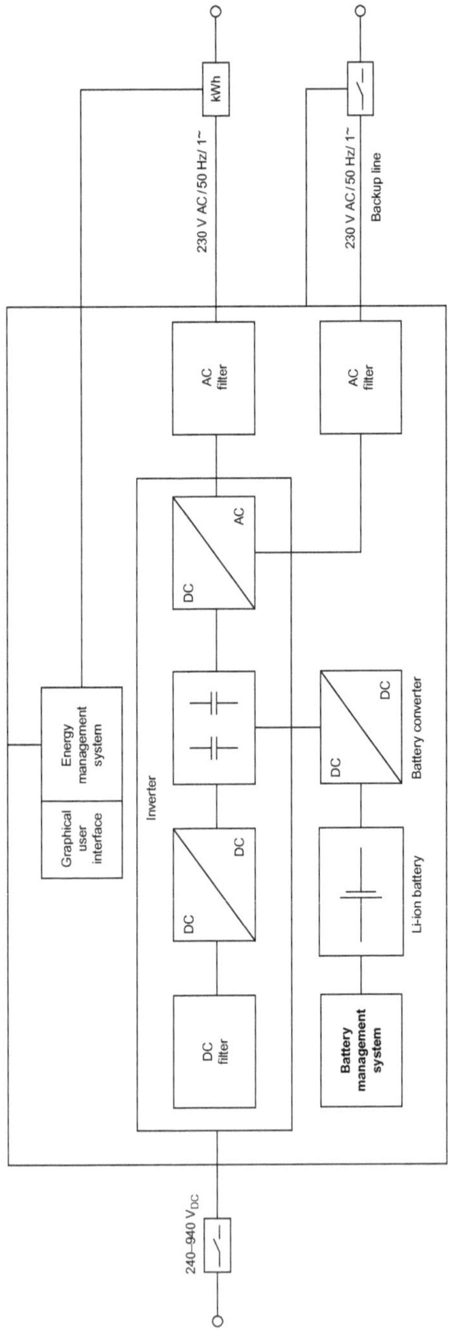

Fig. 9.22 BPT-S 5 hybrid battery-storage solar inverter system.
Courtesy: Bosch.

Table 9.6 **Battery hybrid integrated inverters**

Brand	Model	System power (kWh)	Battery nominal voltage (V)
ABB	REACT-3.6-TL	2	–
	REACT-4.6-TL	2	–
Bosch	BPT-S 5 Hybrid	4.4	96
		6.6	144
		8.8	192
		11	240
		13.2	288
GoodWe	GW5048D-ES	0.5	40–60
	GW3648D-ES	0.5	40–60
Kostal	PIKO BA System	2.8	228
PowerRouter	Power Router Solar Battery	3.6	48
SMA	Sunny Boy 3600 Smart Energy	2	150
	Sunny Boy 5000 Smart Energy	2	150

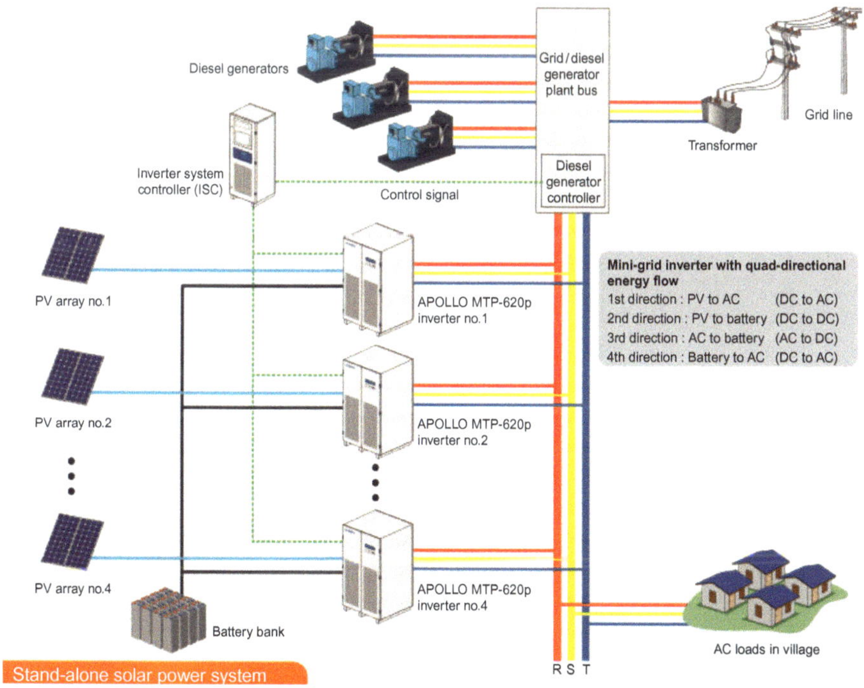

Fig. 9.23 Multiport inverter (quad-directional inverter) implemented by Leonics.

Grid-connected inverters

These inverters have a well-known technology [24–26] and they can only be implemented in the PV hybrid systems along with a BDI in an AC-coupling configuration. Then, they are always running as a current source. Stand-alone operation is not possible for these inverters.

The AC bus is provided by the BDI or by the genset [2,3,27].

This inverter is cheaper than other converters used in PV hybrid systems, and is very mature and tested equipment. In accordance with the guidelines followed, the BDI or external controller has the ability to modify the energy generation (active power) from the PV generator by means of frequency [22,28].

In addition, nowadays, there are a lot of other capabilities that some inverters can provide in PV hybrid systems: passive anti-islanding functions (V,f), active anti-islanding functions (frequency shift), reactive power supply (Q), low voltage ride-through (LVR), short circuit current supply (Isc), tertiary control (active power depending on frequency) and remote control functions to set power output and power factor.

Solutions for commercial and industrial applications

For this type of applications two implementations are the most useful: low PV penetration (PV–diesel power plant) and medium penetration with intermittent genset operation and energy storage.

9.4 Design: Influence of application and user behaviour

A stand-alone power system is always designed so it covers the required electricity needs and has a calculated amount of stored electricity for high drain uses, or periods of low generation. The sizing of stand-alone PV systems is an important task of the PV system. Sizing of the PV panel is the selection of the accurate size of components in a system so that cost can be reduced and efficiency can be increased. If the system is larger in size than required, it can cause an increase in cost. If it is smaller in size, shortage of demand is the result.

Sizing decides the number of panels, types of panel required to capture solar energy, the capacity of the battery that will store energy for the days of low solar energy and the required characteristics of the rest of the elements that integrate the system. Optimal sizing must be done because the size of the elements of the installation must be balanced. However, optimal sizing of stand-alone PV systems is a very difficult task that needs the development of mathematical models for the components and the use of optimization techniques. In addition, careful consideration of the appliances or loads in a home is crucial in off-grid system design.

The objective of system optimization is to size the system components sufficiently to meet load requirement with possible minimum life-cycle cost. There have been many studies of the sizing and optimization of renewable energy systems since their recent popular utilization. Optimal sizing methods developed in the literature [20] can be classified as in Fig. 9.24.

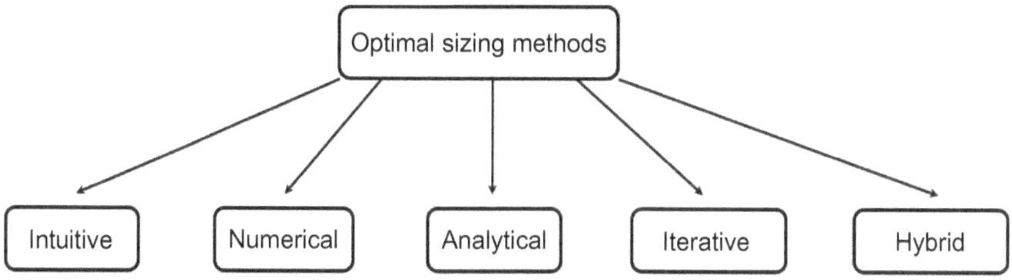

Fig. 9.24 Classification of optimal sizing methods.

9.4.1 Intuitive method

The intuitive method is defined as a simplified calculation of the size of the system carried out without establishing any relationship between the different subsystems or taking into account the random nature of solar radiation. These methods can be based on the month that gives the worst solar radiation. The major demerit of this technique is that difference in sizing of the PV panel can reduce the reliability of the system.

The simplest and most widely used methods are intuitive and calculate dimensions according to 'worst month' or yearly average values; therefore the system determined with these methods can be oversized or undersized.

9.4.2 Numerical method

Those methods use a system simulation for each time period and calculate the energy balance of the system and SOC of the storage unit. Numerical methods are applied due to their accuracy. Nevertheless, these methods require long computation time and time-series data for long periods. In the case of a lack of past time-series data, the solar irradiation values have to be calculated synthetically.

9.4.3 Analytical method

In analytical methods, the size of the PV system is expressed by mathematical equations taking into consideration the reliability. The analytical sizing methods could be categorized into three types: loads and irradiation, available areas and LOLP (loss of load probability). There are different approaches to analytical optimization methods: linear or nonlinear techniques.

The inconvenience of this type of method is the requirement to predict the coefficients and location-dependent factors. The determination of the optimal configuration can be realized according to many indicators. Sometimes a graphical analysis of different configurations is necessary. So, some authors utilized linear programming and solved using the Simplex method (a well-known algorithm used for linear programming in mathematical optimization) in order to minimize loss of power supply (LPS), whereas others have used mixed linear integer programming to minimize costs in stand-alone systems. Also, this method has been applied to dynamic programming to optimally size a PV pumping system in a systematic way, considering all relevant elements of water intake, local climate, soil moisture, evapotranspiration, crops and method of irrigation [20].

9.4.4 Iterative methods

Numerical methods require long time-series data and analytical methods involve complex equations and assumptions. Therefore, iterative methods can be an alternative to those methods by utilizing artificial intelligence techniques. In such iterative methods, a procedure is repeated until an optimal configuration is found according to design specifications, including indicators such as capital cost, annualized cost, total net present cost, life-cycle cost or loss of power supply probability (LPSP) [29].

9.4.5 Stochastic method

Stochastic methods use the random nature of solar radiation as the data input for PV system sizing. Usually this method is carried out by calculating LPSP, that is, the total time when the power supply is not able to supply the load, as the design parameter representing the reliability of the system.

However, others authors use different terms such as loss of power probability (LPP), loss of load probability (LOLP/LLP) or loss of energy probability (LOLE).

The stochastic method is frequently used and studied by many researchers [30].

9.4.6 Hybrid methods

Hybrid methods are a combination of different heuristics. They provide a flexible and dynamic sizing process and solve multiobjective optimal sizing problems.

9.4.7 Sizing and optimization in PV hybrid systems

The objective of any design where a PV hybrid system must be implemented is to find the best combination between the elements with a low LLC. In the literature [31], mostly those are optimized with respect to two (technical and economic) or three (technical, economic and environmental) objectives using simulations or multi-objective evolutionary algorithms. Different algorithms are used for the optimization of these systems, such as the differential evolution algorithms, genetic algorithms, particle swarm optimization, simulated annealing and ant colony algorithms [32].

The solutions of the sizing problem with respect to multiobjectives are generally obtained by simulations (software tools) [33]. Many software programs exist for hybrid systems design, such as HOMER, HYBRID2 (Hybrid Power System Simulation Model), HOGA (Hybrid Optimization by Genetic Algorithms) and TRNSYS (Transient Energy System Simulation Program).

9.5 Measuring and assessing performance

Testing (meters and data monitoring) for an off-grid PV system is in general even more critical than for a grid-connected system, although it depends on the complexity of the system. The target is to test the functionality of the overall system. For instance, if the system has batteries, it is interesting and important to track the trends of loads versus charging over time and also monitor the battery SOC (state of charge). Measuring and assessing performance also helps to gain insight into future system needs [34].

Data monitoring can show trends in battery charging and load profiles and help to identify potential problems. Currently, most converters implemented in the systems, inverters and charge controllers can log the daily energy consumption and production data, and the minimum and maximum battery SOC.

Measuring and assessing performance should be adapted to different parameters such as PV system size, user requirements and user objectives [22,35]. So, performance can be classified depending on the levels of precision: basic, medium or high precision.

Larger PV systems should have more monitoring points and higher precision sensors than smaller PV systems.

The requirements for performance monitoring of a PV system can be found in Ref. [36]. Nowadays a new edition of IEC 61724 is in progress, including new measured parameters, additional irradiance values, soiling, and PQ, addressing of curtailment and clipping (to be applied to PV grid-connected systems) and new metrics including temperature-corrected performance ratios (PRs).

In measuring and assessing performance, different components are involved. In general, the measured parameters can be the following:

- For irradiance: In-plane irradiance (POA—plane of array).
- Environmental factors: PV module temperature, ambient air temperature, wind speed, wind direction and humidity.
- DC electrical output: PV array voltage, PV array current, PV array power, output voltage, output current, output power, output energy, output power factor and voltages and current from the battery bank.
- External system requests (if applicable): reduced load demand and system output power factor request.

Similar to analysis done in Ref. [37], Fig. 9.25 shows the measurements gathered for a small-scale PV hybrid system with total power of 75 kW, installed in a school in Sabah, in Malaysia. The PV–DG hybrid system consists of 20 kW of PV modules (divided into three arrays), a BDI 25 kW, a lead-acid battery 1500 Ah 240 V_{DC}, and a diesel generator of 30 kW.

The data that have been gathered are the following:

- Meteorological data: irradiance (W/m^2), irradiation (W/m^2 day), PV temperature (°C) and room temperature (°C).
- For every PV array: output voltage (V), current (A) and power (kW).
- For every charge controller: temperature (°C), output current (A) and power (kW).
- The total output coming from the three charge controllers: output current (A) and power (kW).

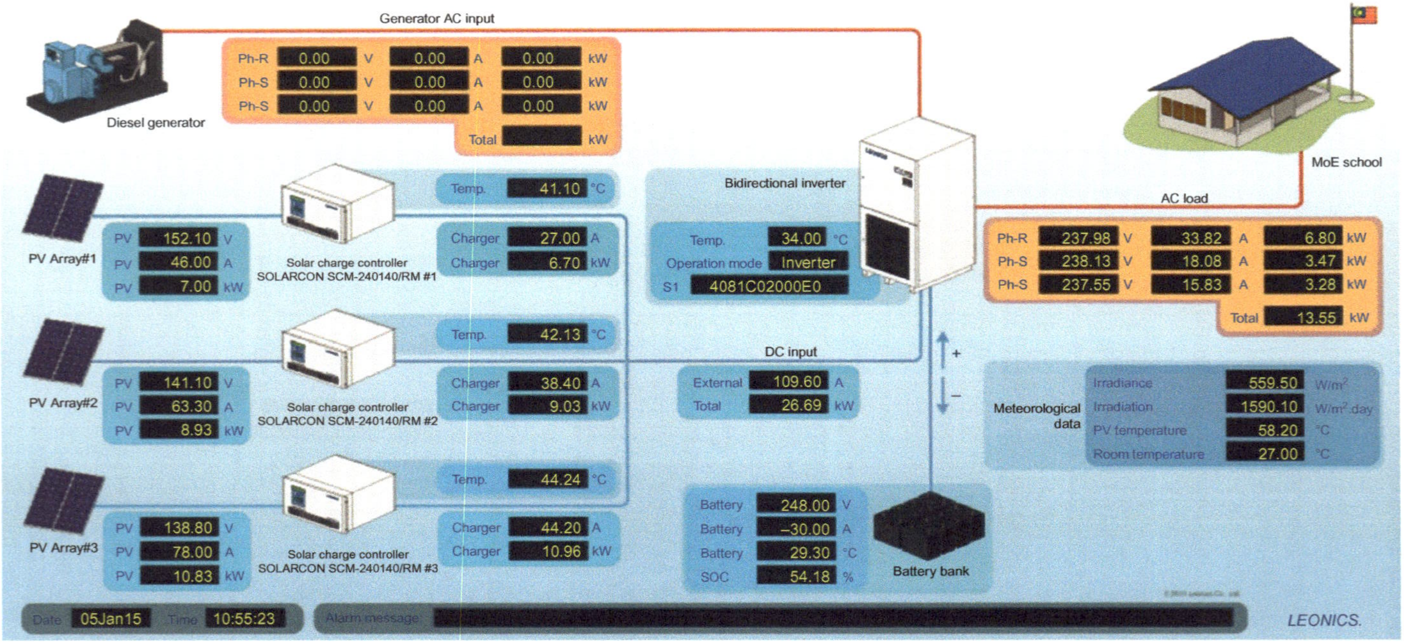

Fig. 9.25 Case study of small-scale three-phase PV hybrid system total power 75 kW is the system installed in a school in Sabah, in Malaysia. *Courtesy:* Leonics.

- For the bank of batteries: bus voltage (V), output current (A), temperature (°C) and SOC (%).
- For all diesel generators, for every phase: output current, voltage and power.
- For every phase of the load: output current, voltage and power.

9.5.1 Performance analysis

The performance analysis of stand-alone PV systems (SAS) is not as straightforward as it is for grid-connected PV systems [34,38].

As has been seen in Section 9.1, there are different types of systems and the performance depends on and differs for every type. Reference [35] gives an example of an evaluation for decentralized rural electrification, whereas the performance of water pumping systems is discussed in Ref. [7].

It is not the same approach to check the performance of small systems without batteries as for a MW system including energy storage and genset, for instance.

Also, there are some examples applied to PV hybrid systems [23].

9.5.2 Performance metrics

A number of metrics can be defined for expressing the system performance. In grid-connected PV plants [24] the most useful has been the PR, defined in [36], although it can be used for any PV system. It measures how effectively the system converts sunlight collected by the PV panels into AC energy delivered to the off-taker relative to what would be expected from the panel nameplate rating. This metric quantifies the overall effect of losses due to inverter inefficiency, wiring, cell mismatch, elevated PV module temperature, reflection from the module front surface, soiling, system downtime, shading and component failures.

The annual PR, PR_{annual}, is the PR evaluated for a reporting period of 1 year.

However, some of these factors are also weather dependent [25].

So, the traditional PR neglects array temperature, resulting in seasonal variation when calculated for time periods less than 1 year [26]. This aspect will be included in the new edition of the IEC 61724 standard that is expected to be published in 2016. Then, new temperature-corrected PRs can be defined. Seasonal variation of the traditional PR is removed by calculating a temperature-corrected PR, PR′. If 25°C is used as the reference temperature, then it gives PR'_{STC}.

Corrections are made for the difference between actual temperature and the STC temperature used for power rating. Values of PR'_{STC} are closer to 1 than for traditional PR.

Using irradiance-weighted annual average module temperature for T_{ref} gives $PR'_{annual\text{-}eq}$. This approximates the value that would be obtained for traditional PR evaluated over one full year, by compensating for seasonal variation.

It is worth mentioning the fact that PR is often not useful for PV stand-alone systems because they are not grid-interactive and therefore have to have excess nameplate capacity of the PV generator to provide low LOLP [2]. Therefore, for those PV systems low PR does not necessarily mean a poorly performing system (perhaps just one with a high degree of safety margin in the sizing).

9.5.3 Performance for a PV pump

The performance of a PV water pump mainly depends on the water flow rate, which is influenced by weather conditions at the location, especially solar irradiance and air temperature variations. The performance of the PV system depends on the water requirement, size of water storage tank, head (m) through which water has to be lifted, volume of water to be pumped (m^3), PV array virtual energy (kWh), energy at pump (kWh), unused PV energy (kWh), pump efficiency (%), and system efficiency (%) and diurnal variation in pump pressure due to change in irradiance and pressure compensation. There are different parameters that influence the performance of PV water pumping. One of these is the efficiency of PV technology used in the PV generator. Another will be the degradation of the PV panels [7].

Then, to summarize, the performance of a PV water pumping system depends on the following parameters: solar radiation availability at the location, the total dynamic head, the sum of suction head (height from suction point till pump), discharge head (height from pump to storage inlet) and frictional losses, the flow rate of water, the total quantity of water requirement and the potential energy required in raising the water to discharge level.

9.6 Micro-/mini-grids: PV for rural electrification

The best practice implementation for rural electrification is the use of mini-grids. For that concept there are no meaningful definitions. From a technical point of view, a mini-grid is a stand-alone PV hybrid system. Meanwhile, the term micro-grid is mostly applied to mini-grids connected to the grid, although sometimes they can operate off-grid. There are mini-grids for different sizes: from kilowatts to megawatts. For installations smaller than a few tens of kilowatts, the installations are named nanogrids.

9.6.1 Integration of the photovoltaic into genset systems

The integration of PV arrays into genset (diesel) systems without energy storage, PV–diesel hybrid solutions, may hold the answer to reducing fuel costs and increasing energy security for different applications related to low electrification regions, for instance.

As has been presented in previous sections, according to the PV penetration (the fraction of PV power installed with respect to the load profile), the integration of the PV into preexisting or new diesel systems can be divided into the following (see Fig. 9.26): low penetration systems (PV penetration <20%); medium penetration systems (where the control is needed), action PV penetration 20–65%; and high penetration systems with PV penetration between 65% and 100% and where energy storage systems must be installed.

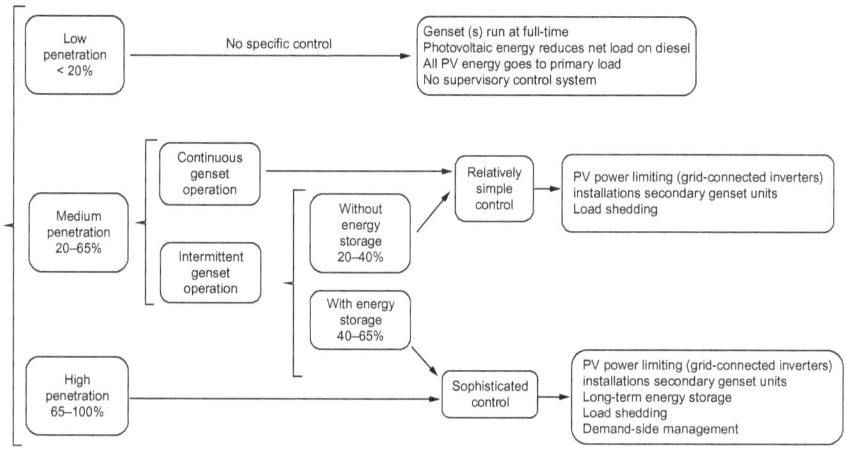

Fig. 9.26 Classification of the PV hybrid systems according to PV penetration.

The penetration level of PV energy in a hybrid power system mainly depends upon load profile, local weather conditions and the customer's preference. With a given PV penetration level, system components can be sized using various approaches.

9.6.2 Low penetration

Many diesel gensets have been upgraded with solar installations in order to reduce diesel consumption and improve the cost of electricity generation. There are numerous examples of newly built PV–diesel hybrid power plants. Bloomberg New Energy Finance estimated a 150-GW installed base of diesel gensets in developing countries. Typical cost savings from adding a PV power plant are in the range of 25–30%. In very remote locations with elevated diesel prices, the savings can amount to more than 70%. These projects use traditional gensets that are slow to start up, cannot be run in low-load mode, and are not very flexible for balancing the unstable solar component.

9.6.3 Medium penetration (20–65%)

Two possible solutions can be implemented for this PV penetration: continuous operation genset (single or multiple genset) or intermittent operation genset (multiple gensets under master-slave control).

In continuous operation, the gensets are not switched off. Under intermittent operation some genset(s) are switched off with low load. In both cases, additional control equipment must be included.

In continuous operation the genset(s) is always used for grid generation and establishes the grid voltage and frequency. It operates as a voltage source. The PV generator acts to reduce load but may not work all the time.

Two different ranges can be distinguished: PV penetration between 20% and 40% and PV penetration between 40% and 65%. In both cases, an additional control must be included.

For a PV to diesel ratio between 20% and 40%, high irradiance and low load could lead to an unacceptable operating point for the genset. In that case, fast transients due to changes in irradiance could result in unacceptable voltage or frequency fluctuations. Therefore, active control is required to keep the grid operating within specified limits. This is technically possible but has implications for the economic model, since the PV generator will sometimes operate below the theoretical power limits and thus reduce the kWh/kW ratio. An active control should allow maximizing of the PV to diesel ratio while also maximizing the PV kWh/kW ratio.

For a PV penetration between 40% and 65% additional components are required, such as short- and long-term energy storage.

9.6.4 Continuous genset operation

Any PV hybrid system under continuous operation in medium penetration has the following elements: PV generator, genset(s), PV grid-connected inverter, wattmeter and control system. The control is required for protecting the diesel generator(s) from low loading and limiting generator cycling. It must be taken into account that it is not recommended for a genset to be operated at a load below 15% of its nominal power. This value may also be recommended to secure the electrical stability of the grid formed by the gensets. As a consequence, for systems without storage, PV power has to be dumped in the case that the effective load (load–PV power) is pushed to this boundary. This again affects the fuel savings. However, a reduction of these losses can be achieved by splitting the total DGS (diesel generators) capacity into several units with, at best, unequal sizes, together with the appropriate scheduling. By this, the required minimal load can be reduced to that for the smallest of the diesel units.

Then, the control system establishes communication with the power meter, PV inverters and load control managing demand-based PV feed-in to the diesel-powered grid. This control system helps the system in the following ways: minimum diesel load (it guarantees that the diesel generator does not operate under a minimum load level, in order to ensure a longer engine life); reverse power protection (it avoids a power flow from the inverter to the genset, for example, in the event of a total load disconnection); stability against irradiance variations (this is guaranteed by the spinning reserve of the genset, previously sized for the maximum possible load); and load control (it allows switching loads on when the genset load level is low and there is a PV generation surplus).

Advantages for this solution are reduction of diesel consumption, lower cost of energy, unlimited resource and resource availability.

Some of the biggest PV–diesel hybrid projects that have been finished lately or are under construction use storage solutions, for example:

- 5.2 MWp PV with 2.2 MW storage in Cobija, Region Pando, Bolivia for rural electrification with an additional genset capacity of 15.2 MVA, Fig. 9.27.
- 10.6 MWp PV with 6 MW storage in Australia for the Sandfire Degrussa mining project with an additional genset capacity of 20 MVA.

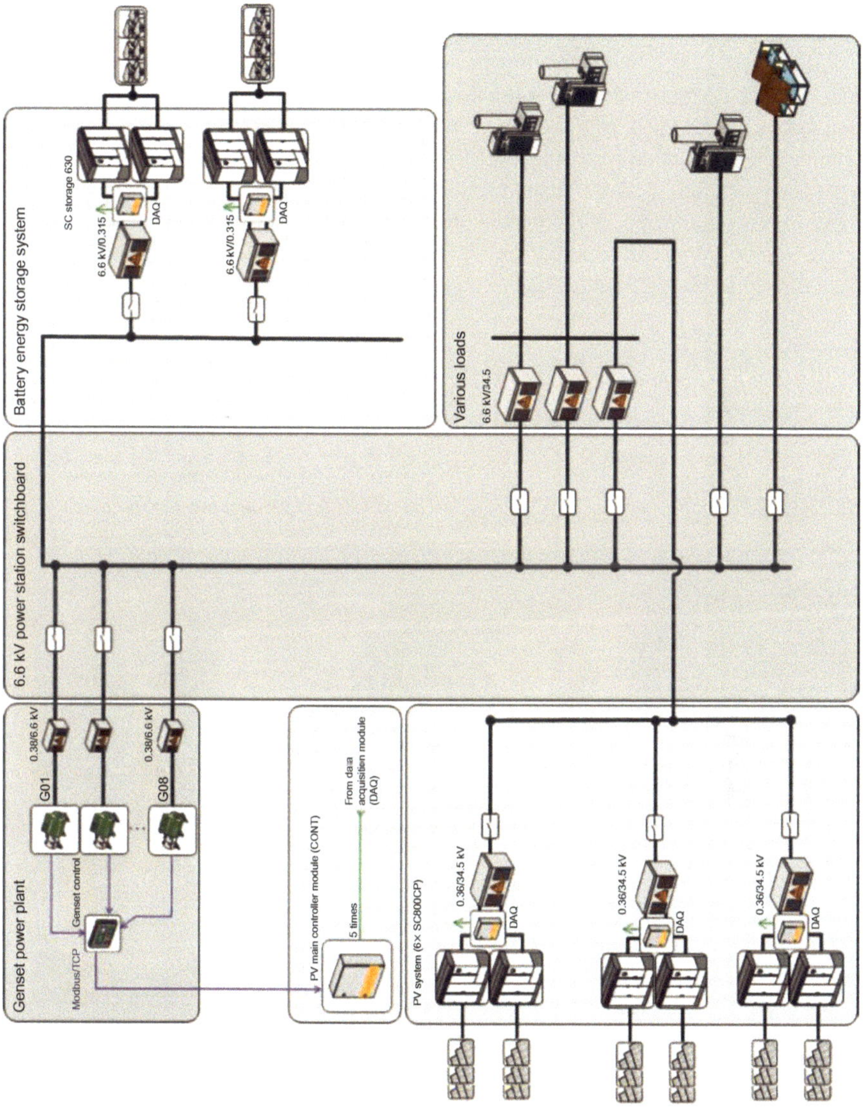

Fig. 9.27 Scheme of the PV–diesel with energy storage project implemented in Cobija (Bolivia) under continuous genset operation [3]. *Courtesy*: SMA

9.6.5 Mini-grids with intermittent operation genset operation

From the technical point of view, this type of system is one of the most interesting solutions to be designed and implemented. As has been seen in previous sections, it can be achieved by either DC coupling or AC coupling. In recent years it seems that AC coupling is the option more in demand and implemented in the projects independently of the load profile. However, it is not always the best option from the energy/cost point of view [27]. The best option should be chosen depending on the load profile. So, it can be demonstrated that if the load profile is only during the night, the best solution to be implemented is DC coupling. However, if the load profile is only during the daytime, the best option would be the AC-coupling implementation. What happens when the load profile is all day? Then, possibly the best option is to divide the total PV generator into two parts: one of them to be connected to the DC coupling and the other one to the AC coupling. This can be found in real megawatt off-grid PV hybrid projects, according to Figs 9.28 and 9.29.

Fig. 9.28 MW off-grid PV hybrid systems (mini-grids) with energy storage implemented in Malaysia.
Courtesy: Leonics.

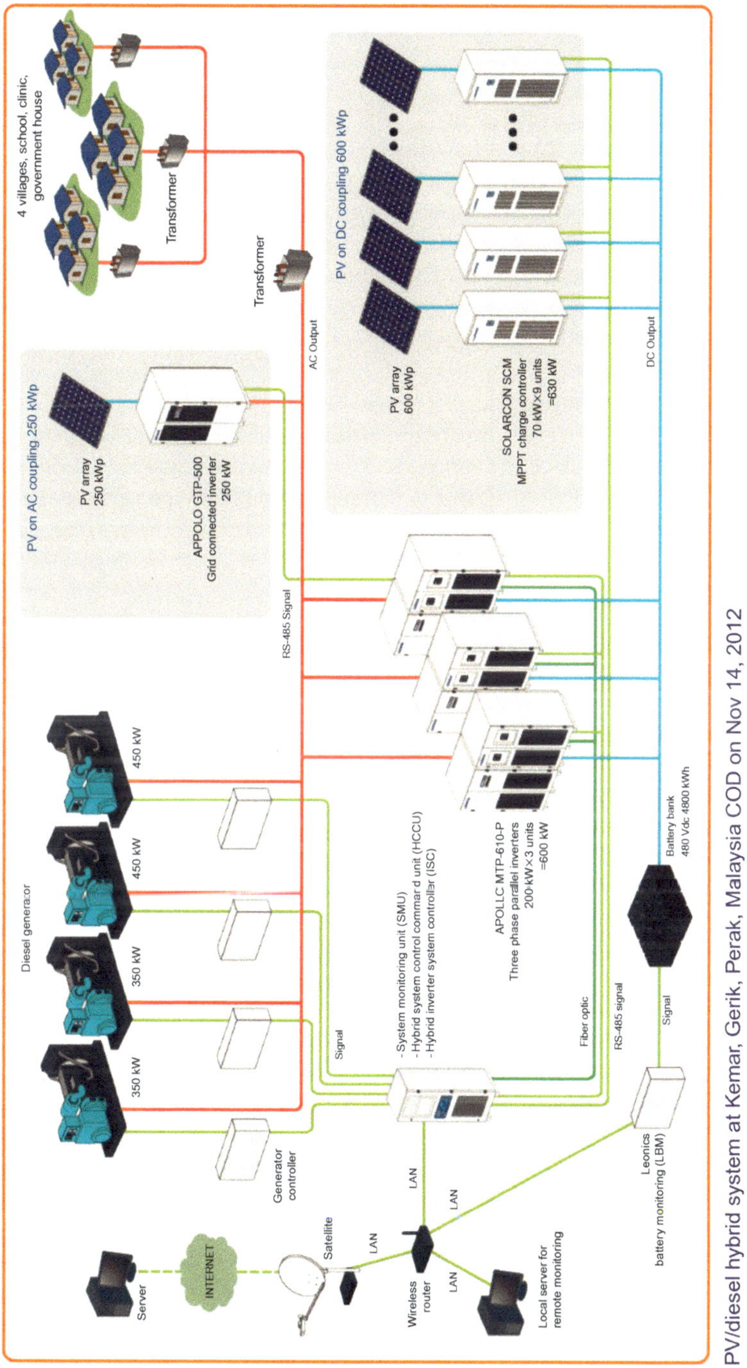

Fig. 9.29 Schematic of the 3.3 MW off-grid PV hybrid system (mini-grid).
Courtesy: Leonics.

9.7 International standardization

The IEC standards related to PV hybrid systems are mainly linked to the JWG 1 (JWG TC 82/TC 88/TC 21/SC 21A) of the Technical Committee (TC) 82 of IEC (International Electrotechnical Committee). That group has the scope to prepare guidelines for Decentralized Rural Electrification (DRE) projects which are now being implemented in developing countries. However, nowadays in IEC there are only specific publications about the small renewable energy and hybrid systems for rural electrification, in the IEC/TS 62257 series, and not for large systems. The IEC/TS 62257 series are technical specifications (TS) and not standards. The purpose of this series is to assist renewable energy project managers, engineers and system designers as well as operators to choose the right system for the right place and to design, operate and maintain the system.

Also, the IEC 62124 (Photovoltaic (PV) stand-alone systems—Design verification) verifies system design and performance of stand-alone PV systems. The performance test consists of a check of the functionality, the autonomy and the ability to recover after periods of low state-of-charge of the battery, and hence gives reasonable assurance that the system will not fail prematurely. The testing conditions are intended to represent the majority of climatic zones for which these systems are designed.

In addition, IEEE provides standards for this topic, for instance the *IEEE Guide for Design, Operation, and Integration of Distributed Resource Island Systems with Electric Power Systems*. It provides alternative approaches and good practices for the design, operation and integration of distributed resource (DR) island systems with electric power systems (EPS). This includes the ability to separate from and reconnect to part of the area EPS while providing power to the islanded EPS. This guide includes the DRs, interconnection systems and participating EPSs.

9.7.1 Standards related to PV modules

Table 9.7 shows the standards related to PV modules, including topics such as quality (IEC 61215 and IEC 61646) and safety (IEC 61730). IEC 60904-1 provides procedures for the measurement of current–voltage characteristics of PV devices in natural or simulated sunlight.

9.7.2 Standards for BCCs

For safety, the manufacturers often comply with IEC 60335-1 (Household and similar electrical appliances—Safety—Part 1: General requirements) and IEC 60335-2-29 (Particular requirements for battery chargers). The rated voltage of the appliances should not be more than 250 V for single-phase appliances and 480 V for other appliances.

The specific IEC standard for BCCs is IEC 62509 (Battery charge controllers for photovoltaic systems—Performance and functioning), which establishes minimum requirements for the functioning and performance of BCCs used with lead-acid

Table 9.7 Standards related to PV modules

Standard	Title
IEC 61215	Crystalline silicon terrestrial photovoltaic modules—Design qualification and type approval
IEC 61646	Thin-film terrestrial photovoltaic modules—Design qualification and type approval
IEC 61730	PV module safety qualification Part 1: Requirements for construction Part 2: Requirements for testing
IEC 60904-1	Photovoltaic devices. Part 1: Measurement of photovoltaic current–voltage characteristics PV modules field test

batteries in terrestrial PV systems. The main aims are to ensure BCC reliability and to maximize the life of the battery. This standard should be managed in conjunction with IEC 62093 (Balance-of-system components for photovoltaic systems—Design qualification natural environments).

With respect to safety, the systems should comply with IEC 62109-1 (Safety of PCs for use in photovoltaic power systems—Part 1: General requirements).

9.7.3 Standards for inverters (stand-alone, bidirectional, interactive and grid-connected)

Inverters installed in these systems are covered by the following standards. With respect to safety, they should comply with IEC 62109-1 (Safety of PCs for use in photovoltaic power systems—Part 1: General requirements) and IEC 62109-2, which covers the particular safety requirements relevant to inverter products as well as products that perform inverter functions in addition to other functions, where the inverter is intended for use in PV power systems. Inverters covered by this standard may be grid-interactive, stand-alone, or multiple mode, may be supplied by single or multiple PV modules grouped in various array configurations, and may be intended for use in conjunction with batteries or other forms of energy storage.

Efficiency is another important parameter to be taken into consideration. This is related to two standards: IEC 61683 and IEC 62891, Table 9.8.

With respect to EMC (electromagnetic compatibility) there are now a lot of possible standards related to this topic. There is a current standard development project for a

Table 9.8 Standards related to efficiency in PV power converters

Standard	Title
IEC 61683	Photovoltaic systems—Power conditioners—Procedure for measuring efficiency
IEC 62891	Maximum point power point efficiency in photovoltaic converters

specific standard, IEC 62920, where the EMC requirements and test methods for power conversion equipment applying to PV power generating systems will be defined.

9.7.4 Standards for batteries

There are many standards for batteries but Table 9.9 summarizes those related to PV hybrid systems.

9.7.5 Standards related to consumption

In Table 9.10 are depicted the standards related to lamps powered by PV systems. Those are a very good alternative to kerosene for lighting, for instance.

Table 9.9 **Standards for batteries**

Standard	Title
IEC 60896-11	Stationary lead-acid batteries: Part 11: Vented types—General requirements and method tests Part 21: Valve regulated types—Methods of tests Part 22: Valve regulated types—Requirements
IEC 61427	Secondary cells and batteries for photovoltaic energy systems (PVES)—Part 1: Photovoltaic off-grid application
IEC 62485-2	Safety requirements for secondary batteries and battery installations—Part 2: Stationary batteries
IEEE 1361	Practice for determining performance characteristics and suitability of batteries in photovoltaic systems-field test

Table 9.10 **Standards related to consumptions**

Standard	Title
IEC 60969	Self-ballasted lamps for general lighting purposes—Performance requirements
IEC 61347-1-4	Lamp control gear: Part 1: General and safety requirements Part 3: Particular requirements for AC-supplied electronic ballasts for fluorescent lamps Part 4: Particular requirements for DC-supplied electronic ballasts for general lighting

9.8 Summary

In this chapter, different aspects of PV stand-alone systems have been analysed. Solutions have been discussed for pico PV, PV pump, residential, industrial and services. Also, the PV stand-alone solution based on the hybrid solar system has been described.

This is an off-grid power system that combines a PV system with diesel generators and/or other renewable energy systems (eg, wind turbines, biogas units, small-scale hydropower, etc.) to supply continuous electric power. Hybrid systems are designed to operate in remote areas or islands where the cost of grid expansion is not economically feasible. These systems are also called micro- or mini-grids. Hybrid energy systems (renewable energy coupled with conventional energy sources) have been shown to significantly reduce the total life-cycle cost of stand-alone power supplies in many off-grid locations while simultaneously providing a reliable supply of electricity from a combination of energy sources.

References

[1] Solar pumps market: global industry analysis and opportunity assessment 2015–2025, Market Research Report, Future Market Insights (FMI), 2016.

[2] IEA PVPS, Pico solar PV systems for remote homes. A new generation of small PV systems for lighting and communication, Report IEA-PVPS T9-12:2012, Jan 2013.

[3] IEC recommendations for small renewable energy and hybrid systems for rural electrification—part 9–5: integrated system—selection of stand-alone lighting kits for rural electrification, IEC 62257-9-5, 2013.

[4] K. Reiche, et al., What Difference Can a PicoPV System Make, Deutsche Gesellschaft, Eschborn, 2010.

[5] A. Phadke, Powering a home with just 25 watts of solar PV: super-efficient appliances can enable expanded off-grid energy service using small solar power systems, Ernest Orlando Lawrence Berkeley National Laboratory, Berkeley, CA, 2015. LBNL-175726.

[6] U.S. Department of Energy, Energy Efficiency & Renewable Energy, Solid-State Lighting Research and Development: Multi-Year Program Plan, United States Department of Energy, Washington, DC, 2014.

[7] S.S. Chandel, M. Nagaraju Naik, R. Chandel, Review of solar photovoltaic water pumping system technology for irrigation and community drinking water supplies, Renew. Sust. Energy Rev. 49 (2015) 1084–1099.

[8] V. Salas, et al., Review of the maximum power point tracking algorithms for stand-alone photovoltaic systems, Sol. Energy Mater. Sol. Cells 90 (11) (2006) 1555–1578.

[9] P. Periasamy, N.K. Jain, I.P. Singh, A review on development of photovoltaic water pumping system, Renew. Sust. Energy Rev. 43 (2015) 918–925.

[10] A. Hmidet, N. Rebei, O. Hasnaoui, Experimental studies and performance evaluation of MPPT control strategies for solar-powered water pumps, in: 2015 Tenth International Conference on Ecological Vehicles and Renewable Energies (EVER), IEEE, 2015.

[11] K. Yadav, et al., Performance comparison of controllers for solar PV water pumping applications, Sol. Energy 119 (2015) 195–202.

[12] J. Urpelainen, S. Yoon, Solar home systems for rural India: survey evidence on awareness and willingness to pay from Uttar Pradesh, Energy Sustain. Dev. 24 (2015) 70–78.

[13] V. Salas, W. Suponthana, R.A. Salas, Overview of the off-grid photovoltaic diesel batteries systems with AC loads, Appl. Energy 157 (2015) 195–216.

[14] J. Lopes, C.L. Moreira, A.G. Madureira, Defining control strategies for microgrids islanded operation, IEEE Trans. Power Syst. 21 (2) (2006) 916–924.

[15] F. Reda, et al., Energy assessment of solar technologies coupled with a ground source heat pump system for residential energy supply in Southern European climates, Energy 91 (2015) 294–305.

[16] A. Franco, F. Fantozzi, Experimental analysis of a self consumption strategy for residential building: the integration of PV system and geothermal heat pump, Renew. Energy 86 (2016) 1075–1085.
[17] J. Ji, et al., Experimental study of photovoltaic solar assisted heat pump system, Sol. Energy 82 (1) (2008) 43–52.
[18] C. Wang, et al., Dimensionless and thermodynamic modelling of integrated photovoltaics-air source heat pump systems, Sol. Energy 118 (2015) 175–185.
[19] C. Wang, et al., Efficacy of integrated photovoltaics-air source heat pump systems for application in Central-south China, Renew. Sust. Energy Rev. 49 (2015) 1190–1197.
[20] T. Khatib, A. Mohamed, K. Sopian, A review of photovoltaic systems size optimization techniques, Renew. Sust. Energy Rev. 22 (2013) 454–465.
[21] Low-Load Gensets for Solar–Diesel Hybrid Plants in the Mining Industry, Danvest Hybrid Power, Hellerup, 2015.
[22] IEA PVPS T3-13:2003, Guidelines for monitoring stand-alone photovoltaic power systems.
[23] K.Y. Lau, M.F.M. Yousof, et al., Performance analysis of hybrid photovoltaic/diesel energy system under Malaysian conditions, Energy 35 (8) (2010) 3245–3255.
[24] Y. Ueda, et al., Performance analysis of various system configurations on grid-connected residential PV systems, Sol. Energy Mater. Sol. Cells 93 (6) (2009) 945–949.
[25] T. Dierauf, et al., Weather-Corrected Performance Ratio, NREL, Golden, CO, 2013.
[26] Project IEC 61724-1 Ed. 1.0. Photovoltaic system performance—part 1: monitoring.
[27] V. Salas, W. Suponthana, Hybrid DC and AC coupling design: verification by using HOMER PRO, in: Microgrid 2015, October 14–15, Canberra, Australia, 2015.
[28] K.Y. Lau, M.F. Yousof, S.N. Arshad, M. Anwari, A.H. Yatim, Performance analysis of hybrid photovoltaic/diesel energy system under Malaysian conditions, Energy 35 (8) (2010) 3245–3255.
[29] C. Gopal, et al., Renewable energy source water pumping systems—a literature review, Renew. Sust. Energy Rev. 25 (2013) 351–370.
[30] W.X. Shen, Optimally sizing of solar array and battery in a standalone photovoltaic system in Malaysia, Renew. Energy 34 (1) (2009) 348–352.
[31] A. Hina Fathima, K. Palanisamy, Optimization in microgrids with hybrid energy systems—a review, Renew. Sust. Energy Rev. 45 (2015) 431–446.
[32] O. Erdinc, M. Uzunoglu, Optimum design of hybrid renewable energy systems: overview of different approaches, Renew. Sust. Energy Rev. 16 (3) (2012) 1412–1425.
[33] V. Khare, S. Nema, P. Baredar, Optimisation of the hybrid renewable energy system by HOMER, PSO and CPSO for the study area, Int. J. Sustain. Energy (2015) 1–8.
[34] P. McNutt, et al., Procedures for determining the performance of stand-alone photovoltaic systems, in: Conference Record of the Twenty-Eighth IEEE Photovoltaic Specialists Conference, 2000, IEEE, 2000.
[35] A. Chaurey, T.C. Kandpal, Assessment and evaluation of PV based decentralized rural electrification: an overview, Renew. Sust. Energy Rev. 14 (8) (2010) 2266–2278.
[36] IEC, Photovoltaic system performance monitoring—guidelines for measurement, data exchange and analysis, Data acquisition timing and reporting periods, IEC 61724, 1998.
[37] A. El Fathi, et al., Performance parameters of a standalone PV plant, Energy Convers. Manag. 86 (2014) 490–495.
[38] S. Rehman, I. El-Amin, Performance evaluation of an off-grid photovoltaic system in Saudi Arabia, Energy 46 (1) (2012) 451–458.

Concentrating photovoltaic systems

10

M. Steiner*, T. Gerstmaier[†,a], A.W. Bett*
*Fraunhofer Institute for Solar Energy Systems ISE, Freiburg, Germany, [†]Soitec Solar GmbH, Freiburg, Germany

10.1 General overview and description of CPV systems and technologies

Concentrating photovoltaic (CPV) technology is a recognized path to lower the cost of solar-generated electricity. The basic idea behind this technology has been pursued for many years, ie, reduce the comparatively expensive semiconductor material in a module. While in a standard flat-plate photovoltaic (PV) panel, the area of the solar cell collects and converts the sunlight in one unit, in CPV these two functions are separated into two units. The sunlight is first collected on an area which is an optical element. This optical element guides and therefore focuses the sunlight onto a smaller area, which is a tiny solar cell acting as a conversion unit. In this way, and as shown in Fig. 10.1, the light is concentrated and thus the intensity of the sunlight on the solar cell increases. In CPV the concentration factor is of importance. It indicates roughly how much semiconductor area is substituted for by an optical element. Two definitions are used for the concentration factor. One is based on the geometrical concentration (see Eq. 10.1) the other on the power concentration (see Eq. 10.2). Both are connected by the optical efficiency.

$$C_{\text{geo}} = \frac{A_{\text{lens}}}{A_{\text{cell}}} \tag{10.1}$$

$$C_{\text{elec}} = \frac{P_{\text{in, lens}}}{P_{\text{in, cell}}} = \eta_{\text{opt}} \times C_{\text{geo}} \tag{10.2}$$

The optical efficiency accounts for losses in the optical element. These losses could be due to reflection, but also due to refraction, chromatic aberration or other optical losses. Typical practical optical efficiencies are in the range from 78% to 90%. It must be noted that the optical efficiency depends also on the concentration level. Higher concentration factors reduce the optical efficiency for a given optical element. However, there is a wide variety of different optical elements. In commercially available CPV systems today, the preferred primary optical elements are Fresnel lenses or mirrors. Frequently, a secondary optical element is implemented and this is typically mounted directly on the PV receiver area; see Fig. 10.2.

[a] Formerly Soitec Solar GmbH

The Performance of Photovoltaic (PV) Systems. http://dx.doi.org/10.1016/B978-1-78242-336-2.00010-0

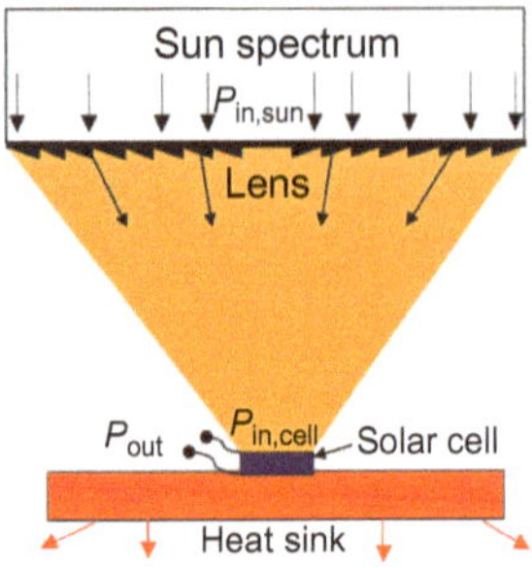

Fig. 10.1 Principle of concentrating photovoltaics: an optical element collects the direct sunlight and focuses it onto a small area solar cell. The geometrical concentration factor is defined by the ratio of the area of the optical collector to the solar cell area. The electrical concentration is the ratio of sun power to power received at the solar cell area.

Fig. 10.2 Left: Photo of a FLATCON module which uses a Fresnel lens as primary optics and a reflective secondary optic on the solar cells. The reflective secondary mounted on the solar cell is shown in the inset of the graph. Right: Photo of CPV set-up which uses a parabolic mirror as the primary optical element.

This secondary element helps to either increase the concentration factor further or to increase the acceptance angle of the CPV system. The acceptance angle of a CPV system defines the sector of the sky from which the sun's rays are collected and converted into electricity. For low concentration factors $<30\times$ it is up to 5 degrees, whereas for high-concentration ratios $>500\times$ it is in the range of 0.3–1.0 degrees. Thus, CPV uses preferably the light coming out of the direct solar disc, ie, the direct sunlight. Therefore, for rating CPV cells, modules and systems the AM1.5d(irect) reference spectrum, as defined in International Electrotechnical Commission (IEC) 60904-3, is used. Since the module can only use direct sunlight, the module area must be always vertically aligned to the sun's disc. Therefore, the CPV technology needs a tracking system to follow the sun precisely; see also Figs 10.4 and 10.5. The need for tracking increases the system complexity but results also in the benefit of a favourable, more rectangular daily energy production profile, which is visualized in Fig. 10.3.

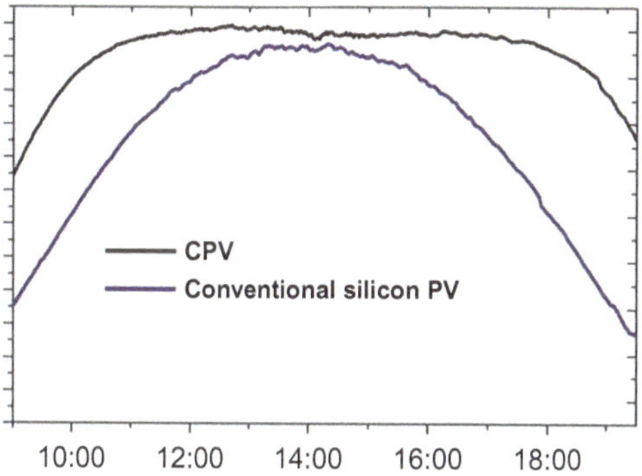

Fig. 10.3 Typical daily energy production profile of a tracked CPV system compared to a fixed mounted flat-plate photovoltaic system. Due to the tracking the production profile becomes more rectangular.

The high variability in the choice of optical elements, tracking strategies and solar cells leads to a wide diversity in proprietary system designs. At any rate, the concentration factor is a possibility for categorizing the CPV systems adequately. Currently, two categories must be mentioned:

1. Low concentration system with concentration factor above 1 up to $30\times$ (LCPV).
2. High-concentration systems with concentration factors $>300\times$ (HCPV).

It might be surprising that no systems with concentration factors between $30\times$ and $300\times$ are on the market, but as it turns out, these systems are not yet cost-competitive.

The LCPV systems typically use modified Si solar cells, one-axis tracking and often a parabolic trough mirror. However, copper indium gallium selenide solar cell-based cells and linear Frensel lens collectors are also used. A typical LCPV system

Fig. 10.4 Photo of an LCPV system provided by the company Sunpower Inc. The geometrical concentration factor is $7\times$. A parabolic trough as optical concentrator focuses the sunlight onto Si solar cells. Cell and concentrator are mounted on a one-axis tracking system. In the inset of the figure the layout of the system is schematically shown.

developed by the company Sunpower is shown in Fig. 10.4. For HCPV, two-axis tracking is mandatory. High-efficiency multijunction solar cells are used in these systems. The vast majority of the HCPV systems use Fresnel lenses as primary optics. This leads to the so-called point-focus design, ie, each Fresnel lens is placed directly over a small solar cell. The small solar cell is mounted on a heat spreader which distributes the heat very effectively. This cooling technique is called passive cooling. Fig. 10.5 shows a photo of a Fresnel lens type point-focus system and the schematics of the working principles.

Another approach for HCPV systems relies on active cooling. Active cooling increases the system complexity but a good benefit results from using the excess heat. So-called HCPVT (T means thermal) systems cogenerate heat and electricity. In this manner, sunlight conversion efficiencies beyond 70% are achieved [1]. Such systems use several tenths of a square-metre sized parabolic-shaped mirror to concentrate the sunlight onto PV receiver areas of several tenths of a centimetre squared. As receivers are commonly water cooled, densely packed PV arrays are used. A photo of such a system is shown in Fig. 10.6. Even larger PV receivers are used in solar towers, where many mirrors focus the light on a tower. The first solar tower demonstration systems have been set up [2] and prove the principles of this technology.

Today, the CPV market is clearly dominated by HCPV technology. As reported [3], the total installed capacity of HCPV is 314 MW whereas the capacity of LCPV is 26 MW. Obviously, the total installations of CPV systems are three orders of magnitude below standard flat-plate PV technology. Thus, CPV technology is just starting to enter the market. A very essential milestone to support the market entrance was achieved when the first IEC standard 62108, Design Qualification, was published in 2007. Further IEC standards have been developed in Technical Committee 82, Working Group 7 (TC82, WG7) and are now introduced. Some very important standards are the definition of the standard testing conditions for CPV technology and the procedure for rating of the systems, ie, the IEC62670 series. The background for these standards is more thoroughly explained in the following sections.

Fig. 10.5 Photo of a Fresnel lens type point-focus HCPV system. This type of system uses Fresnel lenses to concentrate the sunlight onto small area high-efficiency multijunction solar cells. Each solar cell is mounted on a heat spreader which allows for a passive cooling concept (see right photo). The system uses a precise two-axis sun-tracking system.

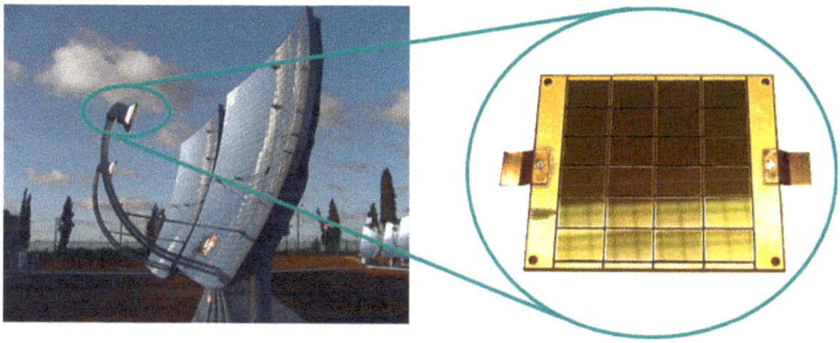

Fig. 10.6 Left, photo of a parabolic mirror system installed in the Kibbutz Javne, Israel. It concentrates the light on a densely packed photovoltaic array, see photo right. The dense array must be actively cooled, typically with water. If correspondingly designed, such a system allows cogeneration of heat and electricity.

One specific feature in the HCPV system is the use of multijunction solar cells, which provide the highest solar conversion efficiency. Solar cells with efficiency of 46% have been elaborated in the laboratory [4] and industrially fabricated solar cells with efficiencies beyond 40% are commercially available in large quantities [5]. The industrial standard cell is a triple-junction device based on the material composition of GaInP/GaInAs/Ge. The 46% record efficiency value was achieved with a four-junction cell based on GaInP/GaAs//GaInAsP/GaInAs using wafer bonding technology. There are many triple, quadruple, and up to sixtuple multijunction solar cell architectures under research applying a variety of different production technologies. An overview can be found in Ref. [6].

However, the internal series connection of these solar cells makes precise measurement and characterization more challenging and requires new sun simulators [7]. For application in concentrator modules, not only are the laboratory measured solar cell characteristics of importance, but also the optical transmission function of the concentrating optics must be respected. In total, it has been shown for CPV modules that the sensitivity for changes in the spectral distribution of the sunlight is increased compared to flat-plate standard PVs. But eventually the yearly energy production is the only real measure that counts. Extremes can be found, but due to many daily and seasonal changes of the sun spectrum, the statistics lead to a smoothing effect. Several studies have been conducted investigating the influence on system performance of different sun spectra associated with different sites and modified (ie, site optimized) cell architectures [8]. The maximum difference in the yearly energy yield was found to be 3%. Thus, site-specific adapted solar cell designs can lead to a small improvement. At any rate, the existing excellent performance prediction models consider the influence of the spectrum; see also Section 10.3.

Aside from many specific details, CPV cells, modules and systems achieve the highest power output per area. Assuming the same aperture area, HCPV technology produces nearly twice the power compared to flat-plate PV. This proves that CPV technology is a high-efficiency technology. Moreover, further progress and higher output power can be expected in the future; see Fig. 10.7.

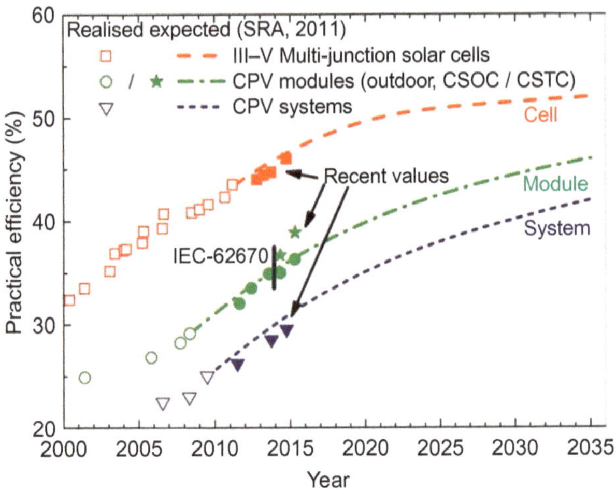

Fig. 10.7 Development of record efficiencies of III–V multijunction solar cells and CPV modules. Progress in top-of-the-line CPV system efficiencies is also indicated. The *dots* represent cell, module and system efficiencies as measured. The cell values are AM1.5d lab records according to Green et al., Solar Cell Efficiency Tables (from 1993 [9] to 2015 [4]). The CPV module and system efficiencies are collected from various publications. The *trend lines* show expected efficiencies from the Strategic Research Agenda developed by the European Photovoltaic Technology Platform in 2011 [10]. Recent efficiency values (*full symbols*) follow the trend very well.

10.2 Impact factors on CPV performance

In silicon flat-plate PV the two highest impact factors on module performance are irradiance and solar cell temperature. For CPV the irradiance is the highest impact factor as well. But the influence of the cell temperature on CPV module performance is much lower compared to silicon flat-plate PV [11]. The second highest impact factor in CPV is not the cell temperature; it is the spectral distribution of the irradiance. The impact of the spectral distribution on module performance results from the use of multijunction solar cells. Multijunction solar cells consist of several subcells with different bandgap energies which are internally series connected. Each subcell absorbs light in a different wavelength range. In this manner, the sunlight can be more effectively used, ie, the solar cell efficiency is drastically increased. However, a change in spectral distribution of the irradiance, which occurs during every day and across the year, immediately alters the current generation in the individual subcells. For instance, assume two spectral distributions A and B with the same integral value of irradiance. Under spectral distribution A, all subcells in the multijunction solar cell produce the same amount of current, whereas under spectral distribution B one subcell produces much less current while the others produce more. Since the subcells in the multijunction solar cells are connected in series, the short-circuit current of the multijunction solar cell is limited by the subcell with the minimum current generation. As a consequence, the current of a solar cell measured under spectral distribution A can have a much higher value than for the same solar cell measured under spectral distribution B, even though spectra A and B have the same irradiance value. This is the reason why in a CPV application the impact of the spectral irradiance distribution has to be considered.

The integral value of the spectral irradiance distribution is measured in a similar manner as that for flat-plate PV by using a pyrheliometer. A pyrheliometer has the same operating principle as a pyranometer, but measures the irradiance only in an opening angle of ± 2.5 degrees. Thus, a pyrheliometer has to be mounted on a sun-tracking unit. The output of the pyrheliometer is called direct normal irradiance (DNI), which is the absolute integral value of the spectral irradiance distribution. Fig. 10.8 shows the dependence of module power output on DNI, using as an example a FLATCON-type CPV module [12] with lattice-matched triple-junction [5] and silicone-on-glass lenses [13]. The data were measured in one full year of operation at Fraunhofer ISE in Freiburg, Germany. Fig. 10.8 shows a linear fit to the data to underline the linear dependence of power output and DNI. The variation of the measured data points from the straight line of the linear fit is around $\pm 10\%$. This $\pm 10\%$ variation is due to other impact factors such as the spectral irradiance distribution, temperature of cell and lens, module and tracker alignment.

The relative variations of the spectral irradiance distribution have to be measured with additional equipment as well as the pyrheliometer. The ideal method to measure this distribution would be to use a spectral radiometer with high resolution. The disadvantage of spectral radiometers is their high costs in combination with high measurement uncertainty when performing long-term measurements in a wide range of

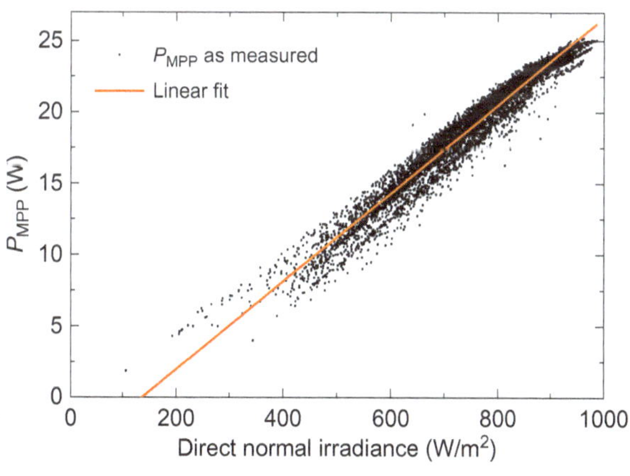

Fig. 10.8 Maximum power output (P_{MPP}) of a FLATCON-type CPV module as a function of direct normal irradiance (DNI). The module was measured for one full year on a sun-tracking unit at Fraunhofer ISE in Freiburg, Germany. The *straight line* shows a linear fit to this data to underline the linear correlation of P_{MPP} and DNI. The variation of the measured data points from the straight line of the linear fit is around $\pm 10\%$. This $\pm 10\%$ variation is due to other impact factors such as the spectral irradiance distribution, temperature of cell and lens, module and tracker alignment.

wavelengths. For CPV, typically the wavelength range of interest is between 250 and 1800 nm. Therefore, in CPV, component/isotype cell sensors [14–16] are used to quantify the impact of the spectral irradiance distribution on module performance.

The component sensor consists of multijunction subcells, which have the same optical characteristics as the full solar cell but the electrical characteristics of the single subcell. Typically the three subcells of a lattice-matched triple-junction solar cell [5] are used. These subcells have bandgap energies of 1.9 eV (GaInP, top subcell), 1.4 eV (GaInAs, middle subcell) and 0.7 eV (Ge, bottom subcell). They absorb light in the wavelength range 300–600 nm (top), 600–900 nm (mid) and 900–1800 nm (bot). The component sensor is not used to quantify the absolute value of the spectral irradiance distribution but its relative shape. For this reason, spectral matching ratios (SMRs) are calculated as defined by the following equation:

$$\mathrm{SMR}(i,j) = \frac{\left(J_{SC,i}/J_{SC,i}(\mathrm{AM1.5d})\right)}{\left(J_{SC,j}/J_{SC,j}(\mathrm{AM1.5d})\right)} \tag{10.3}$$

where i and j are assigned to distinct subcells in the multijunction cell and J_{SC} is the short-circuit current density as measured under prevailing spectral conditions or at AM1.5d standard spectral irradiance. Typically, for the lattice-matched triple-junction cell, three SMR values are calculated: SMR(top,mid) = SMR(1,2), SMR(top,bot) = SMR(1,3) and SMR(mid,bot) = SMR(2,3).

If all three SMR values are close enough to unity, the prevailing spectral condition is considered as equivalent to the standard spectral condition. The current IEC draft standard 62670 considers SMR within $1.0 \pm 3.0\%$ as close enough.

Fig. 10.9 shows the short-circuit current density and the fill factor (FF) of a FLATCON-type CPV module as a function of SMR(1,2). The J_{SC} is normalized to

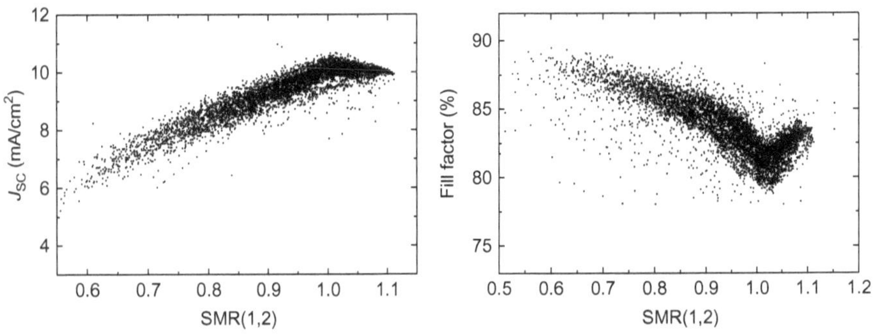

Fig. 10.9 Short-circuit current density is normalized to a DNI of 1000 W/m^2 (J_{SC}) and fill factor (FF) as a function of SMR(1,2). The J_{SC} peaks at around unity, which means that the impact of these spectral conditions is similar to the standard spectral irradiance. The fill factor has its minimum value where the J_{SC} peaks. The reason for this is the current matching of the subcells. The more similar the current output of the individual subcells is, the higher is the J_{SC} and the lower is the FF. The data were measured over a 1-year period at Fraunhofer ISE in Freiburg, Germany.

a DNI of 1000 W/m^2. The dependence of J_{SC} on SMR(1,2) is a linear increase until a maximum is reached and the J_{SC} is then linearly decreasing. At this maximum the current generation of the subcells shows the lowest difference. For SMR values below 1 the spectral irradiances are reduced in the short wavelength region (red rich) compared to the reference spectrum and for SMR values above 1 the long wavelength region is reduced (blue rich). The dependence of the module power output on SMR and thus on spectral irradiance distribution is less pronounced than that of the J_{SC}. The reason for this is the FF, which increases for SMR values higher or lower than the SMR value at the maximum J_{SC}. But the increase of FF is less pronounced than the decrease of J_{SC}.

In CPV technology the collection area and cell area are separated, as described in Section 10.1. Therefore, if the influence of temperature on performance is analyzed, both the cell and lens temperature must be considered. Both temperatures have an influence in the same order of magnitude, but it has to be mentioned that it depends on the applied CPV technology. The cell temperature influences mainly the V_{OC} of the module whereas the lens temperature influences mainly the I_{SC} of the module. Fig. 10.10, left, shows the V_{OC} measured on a module as a function of ambient temperature. Between 0 and 40°C the V_{OC} drops by about 0.15%/K in this case. This value compares very well with the typical temperature coefficient for multijunction solar cells [17].

The influence of lens temperature on module power output is rather different for different CPV module technologies. It depends on the primary and secondary optics used and also on the concentration factor. Today many CPV modules use silicone-on-glass lenses which show a high lens temperature dependence. Fig. 10.10, right, displays a typical paraboloidical dependence of a CPV module I_{SC} on lens temperature for such a module.

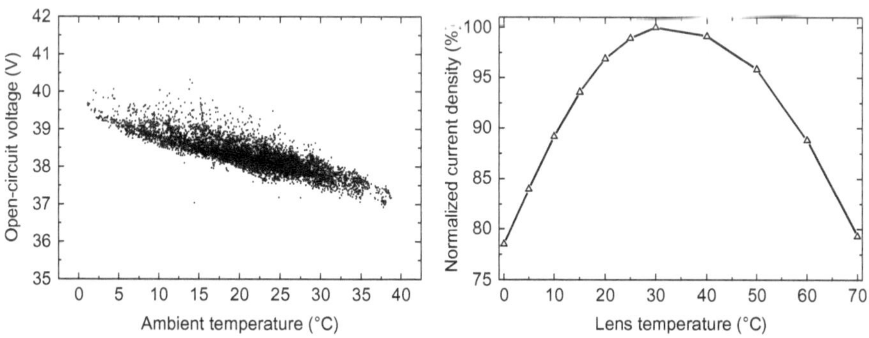

Fig. 10.10 Left: Open-circuit voltage (V_{OC}) of a FLATCON-type CPV module as a function of ambient temperature. The V_{OC} changes with ambient temperature by 0.15%/K in this case. This corresponds to typical temperature coefficients of multijunction solar cell [17]. The data is measured over a 1-year period in Freiburg, Germany. Right: I_{SC} as a function of lens temperature for a silicone-on-glass lens. The I_{SC} is divided by the maximum I_{SC} at 30°C. The data shown in the right graph is calculated using the model described in Ref. [18].

Table 10.1 Energy yield losses for FLATCON-type CPV module using silicone-on-glass lenses and lattice-matched triple-junction solar cells

Impact factor	Energy yield losses (%)
Spectral	3.0–5.9
Cell temperature	2.7–3.5
Lens temperature	1.7–2.7

The previously shown figures describe the impact of spectral irradiance distribution, cell and lens temperature on the instantaneous power output of CPV modules. The impact of these factors on the energy yield of CPV is much less pronounced. The impact on energy yield depends on the variation of these factors within one full year of operation. This variation is strongly related to a distinct location. Philipps et al. [19] and Hornung et al. [20] already investigated this impact on energy yield at four locations relevant to the CPV market. Table 10.1 summarizes their findings as amount of losses on energy yield separated into spectral irradiance distribution, cell temperature and lens temperature. The spectral losses are calculated from the ratio of energy harvesting efficiency to the efficiency at standard spectral irradiance. The spectral losses of energy yield are in the range of 3–6%. Philipps et al. [8] investigated the development of solar cells with bandgap combinations specifically designed for a distinct location to decrease the spectral losses. The outcome of this investigation is that specific designs of cells do not vary significantly and do not lead to a significant increase in energy yield.

The energy yield losses due to cell temperature are calculated from the ratio of energy yield with varying and increased cell temperatures to the energy yield with a constant cell temperature of 25°C. These losses are in the order of 3%. The energy yield losses due to changes in the lens temperature are calculated from the ratio of the energy yield with varying lens temperature to the energy yield with a constant lens temperature of 20°C. This is the temperature for which the lens has the highest optical efficiency. The lens temperature losses are in the order of 2%.

10.3 Standardization of CPV system and module performance measurement

In CPV several international standards have been recently developed or just released to define how to determine the power output and energy yield of CPV cells, modules and systems. The outcome of this standardization is reliable data for the comparison of different CPV power plant technologies. For performance and energy yield measurement of CPV systems and modules, the series of IEC 62670 draft standards is the most important. In IEC 62670 the conditions and methods for rating the performance of CPV are defined. This draft standard consists of four parts:

(I) IEC 62670-1 defines the ambient conditions for the rating: concentrator standard test conditions (CSTC) and concentrator standard operating conditions (CSOC).
(II) IEC 62670-2 defines how to measure the energy of a CPV system.
(III) IEC 62670-3 defines the methods and procedure to determine the power output of CPV modules at CSTC and CSOC.
(IV) IEC 62670-4 defines how to rate the energy produced by a CPV system.

The draft standards IEC 62670-2 and IEC 62670-4 will be discussed in Section 10.4 in more detail.

This section focuses on IEC 62670-1 and IEC 62670-3.

As described in Section 10.2 the highest impact factors on the performance of CPV modules are DNI, spectral irradiance distribution, and cell temperature. This is the reason why these impact factors are defined in the standard IEC 62670-1 as the set of standard conditions for rating of CPV module performance. The two sets of standard conditions are listed in Table 10.2. The principal idea of having two sets of standard conditions is that power output rated at CSOC gives realistic values for the CPV performance in the field, whereas CSTC are conditions found in the laboratory which are more suitable to directly compare different CPV technologies and to ease the performance comparison to standard flat-plate PV. That is why CSOC defines an ambient temperature, whereas CSTC defines the cell temperature. Thus, the cell temperature at CSOC is much higher than 25°C and depends also on the specific CPV technology. In the CSOC dataset the cell and lens temperature are defined indirectly by the DNI, the ambient temperature and the wind speed. The reason for this indirect definition is that the cell temperature cannot be directly measured in CPV, because the cells are located inside the housing, which is common for most CPV technologies.

IEC 62670-3 defines how to do performance testing at the standard conditions listed in IEC 62670-1.

The aim of IEC 62670-3 is to provide a measurement procedure which gives repeatable and comparable results when measuring CPV modules at different test periods at different test labs. The procedure is a consequence of the trade-off between effort and accuracy.

Fig. 10.11 gives a schematic overview of the complete rating procedure. This rating procedure is also described in Ref. [21]. The rating procedure consists of three measurement methods: outdoor *I–V* measurements, outdoor thermal transient measurement (TTM) and indoor *I–V* measurements.

Table 10.2 Standard conditions for CPV as defined in IEC 62670-1

	CSOC	CSTC
DNI (W/m^2)	900 W/m^2	1000 W/m^2
Temperature (°C)	20°C (ambient)	25°C (cell)
Wind speed (m/s)	2 m/s	n.a.
Spectrum	Direct normal AM1.5 spectral irradiance distribution consistent with conditions described in IEC 60904-3	

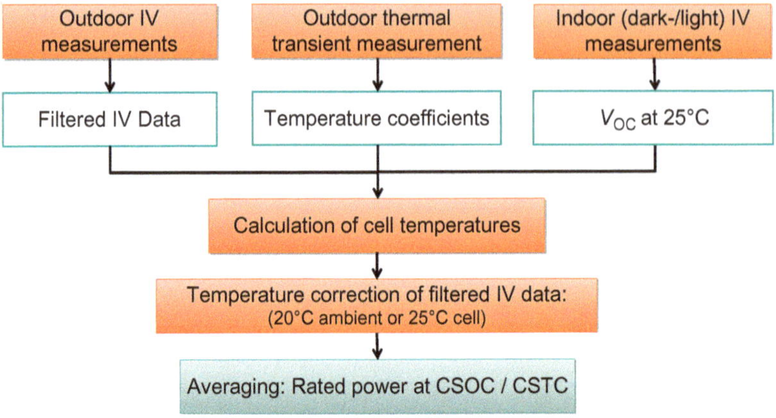

Fig. 10.11 Schematic overview of the rating procedure.

The outdoor *I–V* measurement of CPV modules provides a dataset of *I–V* curves measured at various prevailing ambient conditions. This *I–V* dataset is translated to CSOC and CSTC in two steps: filtering and temperature correction.

The filtering step removes all *I–V* curves measured at prevailing ambient conditions which are not stable and differ too much from the standard conditions. Therefore, a dataset of *I–V* curves is required which include datasets fulfilling the filtering criteria. For the step of temperature correction, the representative temperature for the cells inside the module is required. This representative cell temperature is calculated using module temperature coefficients and the module open-circuit voltage (V_{OC}) at 25°C. The temperature coefficients are determined from TTM. The V_{OC} at 25°C is calculated from indoor *I–V* measurements. In the following the rating procedure is described in more detail.

The basis of the CPV module power rating is the measurement of *I–V* curves outdoors in a test period comprising varying prevailing ambient conditions. The module is mounted and aligned on a sun-tracking unit while its *I–V* curves and the ambient conditions are recorded automatically. The ambient conditions have to be recorded in an interval of less than 1 min and the duration of the *I–V* sweep should be less than 15 s. The *I–V* dataset shall include at least three separate SMR crossing points. The SMR values change during one day due to atmospheric composition and due to air mass (AM). SMR values of unity indicate standard spectral conditions. An SMR crossing point is reached when the SMR values get close enough to unity to fulfil the filtering criteria. The *I–V* dataset is filtered for stable conditions and for ambient conditions similar to CSOC. The most important criteria are as follows:

(i) DNI shall not vary more than 10% in 10 min before the *I–V* sweep and not by more than 1% directly before and after the *I–V* sweep.
(ii) The portion of DNI from global normal irradiance shall be above 80%.
(iii) All SMR values shall be within unity by ±3.0%.

The first criteria shall assure that the cell and lens temperatures are stable. The second criteria assures a clear sky and the third one permits only prevailing spectral conditions close to the AM1.5d reference conditions. Especially the SMR filtering criteria is a trade-off between effort and accuracy. A tighter filtering around unity will decrease the measurement uncertainty, whereas a wider filtering increases the amount of days throughout the year where test labs can fulfill the filtering criteria and thus reduces the length of the test period.

Fig.10.12 shows an example of the module efficiency histogram of *I–V* curves measured at a FLATCON-type CPV module at Fraunhofer ISE within one full year of operation. Additionally, Fig. 10.12 shows the distribution of module efficiency after applying the filter criteria for stable conditions and for CSOC similar conditions. The remaining dataset of module efficiencies is used to calculate the CSOC and CSTC power outputs after applying a temperature correction.

The temperature correction of the module efficiency is done using module temperature coefficients and a representative cell temperature. The temperature coefficients are obtained by TTMs. TTM means mounting and aligning a CPV module on a sun-tracking unit. Cover the module and wait until it is close to ambient temperature. Measure the module back-plate temperature. Then uncover the module and measure *I–V* curves while the cells in the module are heating up. The heating process commonly takes around 10–20 min and the back-plate temperature is increased by 10–20 K in this time. Fig. 10.13 shows the V_{OC}, I_{SC}/DNI, FF and efficiency as a function of back-plate temperature derived from TTM. For the outcome of the TTM it is important that the back-plate temperature is changing in a one-to-one rate with the cell temperature. This one-to-one rate is given if deviations from linear correlation of V_{OC} versus back-plate temperature are small. In the first seconds after uncovering the module such a one-to-one rate is typically not found and this data has to be discarded. The temperature coefficients for V_{OC}, I_{SC}, FF and efficiency are then obtained by linear fits to the TTM data as shown in Fig. 10.13. The temperature of the cell in the module is

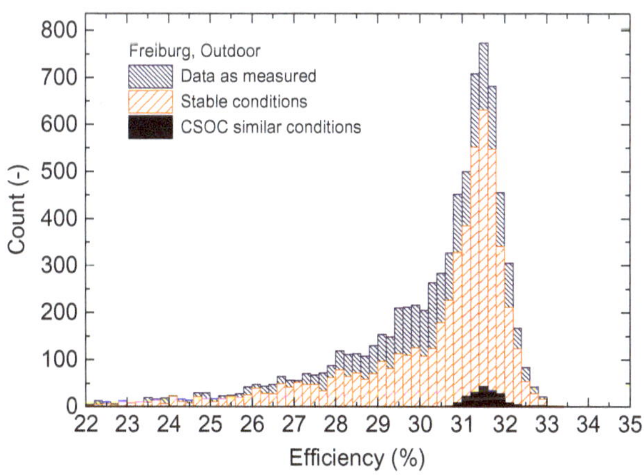

Fig. 10.12 Histogram of module efficiencies measured in 1 year in Freiburg, Germany. The histograms show efficiency values as measured, filtered for stable ambient conditions and additionally filtered for CSOC similar ambient conditions. The efficiency values filtered for stable and CSOC conditions are used to calculate the power rated at CSOC and CSTC.

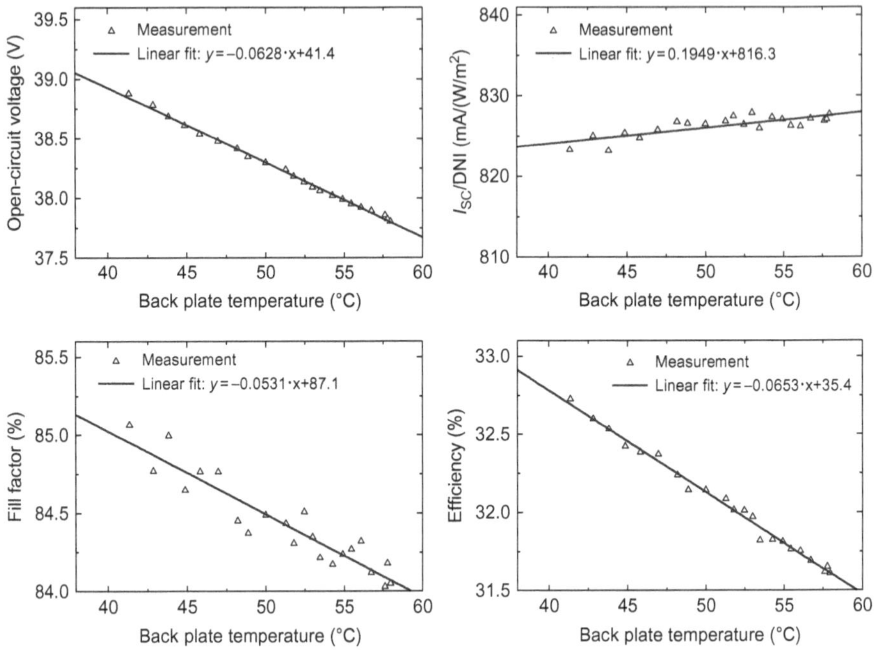

Fig. 10.13 Open-circuit voltage, normalized short-circuit current (I_{SC}/DNI), fill factor and module efficiency as a function of module back-plate temperature. The short-circuit current is divided by the DNI. The data is obtained by thermal transient measurement (TTM). TTM means measuring *I–V* curves while the module is heated when being exposed by sunlight after a covering/uncovering event. The relative temperature coefficients of the module are derived from linear fits as shown in the graph. The resulting fit equations are given in the inset of the graphs.

typically higher than the back-plate temperature of the module. Thus, TTM can only provide relative temperature coefficients. The absolute temperature of the cells is calculated using the module V_{OC} at 25°C. The difference of the V_{OC} at increased cell temperature to the V_{OC} at 25°C in combination with the relative temperature coefficient of the V_{OC} gives the absolute cell temperature. This calculation of cell temperature is based on the one-diode model and described in Ref. [21].

The module V_{OC} at 25°C is derived from *I–V* curves measured indoors. There are two options for measuring the required *I–V* curves. First, measuring the module dark *I–V* curve and second using a sun simulator for CPV modules to measure the light *I–V* curve. In both cases the module *I–V* curve is measured at 25°C room temperature.

The advantage of the dark *I–V* option is that a CPV module sun simulator is not required. In this case, the dark *I–V* of the CPV module is measured at 25°C room temperature. The dark *I–V* curve needs to be corrected for series resistance as explained in Ref. [21]. The module V_{OC} at 25°C corresponds to the voltage of the corrected dark *I–V* curve where the current equals the module I_{SC}.

Extracting the V_{OC} from the sun simulator light *I–V* curve requires setting up the intensity of the sun simulator. For this reason, the module I_{SC} is derived from outdoor

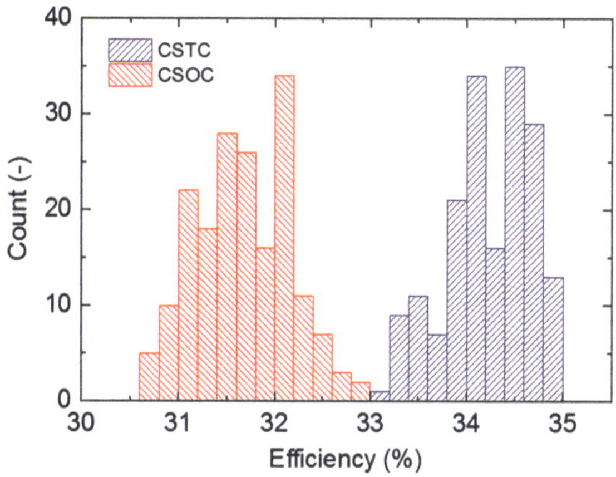

Fig. 10.14 The CSOC efficiencies are obtained from *I–V* curves measured outdoors and corrected for ambient temperature. The CSTC efficiencies are calculated from CSOC efficiency by translation to 25°C cell temperature. The rated CSOC and CSTC efficiency are obtained from averaging the efficiencies shown in the graph.

I–V measurements and translated to 25°C cell temperature using the module temperature coefficients. The intensity of the sun simulator is varied until the measured module I_{SC} indoor matches the outdoor I_{SC}. Then the module light *I–V* curve is measured and the V_{OC} at 25°C is derived from this *I–V* curve.

One may ask at this point why the CPV module sun simulator is not used to determine the power output at CSTC? In this case the translation of outdoor *I–V* data to 25°C cell temperature would not be necessary. The answer to this question is the significantly higher uncertainty of the currently available sun simulators compared to outdoor measurement.

Finally, Fig. 10.14 shows the histogram of module efficiency values derived at CSOC after applying the temperature correction to 20°C ambient temperature and at CSTC after translating the CSOC efficiencies to 25°C cell temperature. The rated efficiencies are determined as the average values of the efficiency distribution, ie, 31.7 ± 0.5% for CSOC and 34.3 ± 0.5% for CSTC. The rated power at CSOC is the rated CSOC efficiency times a DNI of 900 W/m^2 and multiplied by the lens aperture of the CPV module, whereas the rated power at CSTC is the rated CSTC efficiency times a DNI of 1000 W/m^2 and multiplied by the lens aperture of the CPV module.

10.4 System performance and energy yield prediction

The objective of this section is threefold: Firstly, we focus on measuring the actual power output of CPV systems and power plants and on recording the prevailing meteorological conditions. Secondly, we quantify the performance of CPV systems by calculating two performance indicators, performance ratio and performance index, which includes comparing the actual measured power output and the expected output from a performance model. Thirdly, we present some frequently used software tools for modelling the prospective power output and the typically involved loss mechanisms. A large part of this section is based on documents of the IEC, namely IEC 61724-1 ('PV System Performance—Part 1: Monitoring') and IEC 62670-2 ('CPV Performance Testing—Part 2: Energy Measurement').

10.4.1 Actual energy output and prevailing meteorological conditions

In a first step, monitoring of the actual performance of a CPV plant requires the installation and maintenance of a data acquisition system. CPV plants are usually equipped with SCADA ('Supervisory Control And Data Acquisition') systems which include data loggers and sensors for the measurement of electrical and meteorological parameters. IEC 61724-1 classifies such measurement set-ups into three categories, depending on whether the sampling rate and the precision of the sensors is considered to be high (Class A), medium (Class B) or low (Class C). These classes may coincide with usage at utility (Class A), commercial (Class B) and residential (Class C) installations. As most CPV plants are built at the utility scale, we may assume that data collection fulfills the Class A criteria, ie, data recording at 1-min intervals, 3% or less uncertainty for the measurement of the DNI, 2% or less uncertainty for DC power and utility-grade AC energy meters. Another relevant document, IEC 62670-2, requires that the measurement system be calibrated and checked for linearity, stability and correctly working integration before data acquisition starts. While electrical meters are installed in protected enclosures and require little maintenance except for occasional checks of the calibration, measurement of the DNI requires significant attention and efforts to ensure high data quality. By their measurement principle, pyrheliometers are exposed to outdoor conditions including dust and rain and require precise alignment and mechanical tracking, regular cleaning of the aperture and regular recalibration. As the uncertainty of the irradiance data often dominates the overall uncertainty of the derived performance indicators, it is strongly recommended to use high-accuracy pyrheliometers and implement a strict cleaning schedule with minimum weekly cleaning of the aperture and documentation of each cleaning event in a logbook.

In a second step, automated and manual quality checks should be applied and questionable data be flagged and filtered, before data is summed to obtain hourly, daily, weekly, monthly and yearly values. IEC 61724-1 lists a number of quality checks, including applying physically reasonable minimum and maximum limits, maximum rates of change and comparing measurements from multiple sensors. Logbooks should be inspected particularly for periods of insufficient pyrheliometer cleaning or delayed recalibration. Periods during which data is omitted from summation or further analysis due to low quality need to be treated with special care, as pointed out in the section on data postprocessing rules of IEC 62670-2. As summed data may be used for the calculation of performance indicators in a later step, it is of high importance to symmetrically exclude periods during a gap that occurred either in the irradiance or power time-series, or in both, in order to avoid misleading results when ratios of electrical and irradiance energy are calculated.

10.4.2 Performance ratio and performance index

Performance ratio and performance index are standardized indicators defined consistently for nonconcentrator and concentrator PVs [22]. Performance ratio, as defined for nonconcentrator systems in IEC 61724-1 and for concentrator systems in IEC 62670-2, is a measure expressed in percent, for the overall effect of losses on the power output of a plant. As shown in Eq. (10.4), it is defined as the ratio of the final AC yield $Y_{f,AC}$ (Eq. 10.5) and the reference yield Y_r (Eq. 10.6). This is equivalent to the product of $(100\% - L_i)$ of i losses. The L_i losses could be shading loss, soiling loss, cell temperature loss, etc. E_{AC} is the AC energy and P_{CSTC} is the plant DC power at CSTC. CSTC stands for 'Concentrator Standard Test Conditions' as per IEC 62670-1, ie, 1000 W/m^2 DNI, 25°C cell temperature and direct normal AM1.5 spectrum. E_{DNI} is DNI energy.

$$PR_{AC} = \frac{Y_{f,AC}}{Y_r} = (100\% - L_0)(100\% - L_1)\ldots(100\% - L_i) \ [\%] \tag{10.4}$$

$$Y_{f,AC} = \frac{E_{AC}}{P_{CSTC}} \quad \left[\frac{\text{kWh}}{\text{kW}} = \text{h}\right] \tag{10.5}$$

$$Y_r = \frac{E_{DNI}}{1\,\text{kW/m}^2} \quad \left[\frac{\text{kWh/m}^2}{\text{kW/m}^2} = \text{h}\right] \tag{10.6}$$

Performance ratio can be regarded as a normalization of the generated AC energy by a 'rough estimate' [23] of the expected energy, namely by name-plate power P_{CSTC} and DNI (ie, E_{DNI}). As loss factors, such as temperature or spectral losses, are not considered in this normalization, performance ratio is below 100% even for a well-functioning plant, which makes it hard to interpret.

When calculating performance index, in contrast, we compare the measured energy output of a power plant to the energy output calculated with a potentially quite complex model. As such models try to accurately consider various loss factors, performance index actually reaches 100% once the plant delivers the amount of energy that is expected from the simulations. In IEC 61724-1, performance index is defined as the ratio of the measured AC energy to the expected AC energy (Eq. 10.7).

$$PI_{AC} = \frac{\text{Measured } E_{AC}}{\text{Expected } E_{AC}} \tag{10.7}$$

As pointed out by Mokri and Cunningham [23], 'deviations from 100% [performance index] can be caused by many factors, including errors or incorrect assumptions during design, poor installation workmanship, equipment failure or degradation, etc.' Diligent analysis may be required to understand the root cause of deviations from 100%, as it may also be the case that the plant works well, but the model may not be precise enough or the meteorological data fed to the model may be compromised.

10.4.3 Typical loss mechanisms and models to estimate the prospective energy output

In Fig. 10.15 we compare the actual (columns) and the modelled (dashed and solid lines) monthly performance ratio of a CPV plant installed in South Africa. The plant is installed at a site with notable temperature differences between summer and winter, as reflected in the monthly DNI-weighted ambient temperature (black crosses) defined in (Eq. 10.8). The modelled performance ratio values plotted as a dashed line stem from performance Model A which does not include typical CPV-specific loss mechanisms. In contrast, Model B (results plotted as dashed line) includes such mechanisms. Obviously, Model B captures the observed seasonality of the performance ratio much better than Model A, so it is worth having a closer look at typical CPV-specific loss mechanisms in the next paragraphs.

$$T_{\text{ambient,DNI-weighted}} = \frac{\sum T_{\text{ambient},i} \cdot \text{DNI}_i}{\sum \text{DNI}_i} \tag{10.8}$$

Temperature dependence of the optics: The efficiency of concentrating optical elements used in CPV modules may show a nonnegligible dependence on the temperature of the optics (T_{optics}), and indirectly on ambient temperature (T_{ambient}). The temperature dependence of the refractive index and different coefficients of thermal expansion of composite materials, such as silicone-on-glass, may contribute to a temperature-dependent internal misalignment of CPV modules, as described by Kurtz et al. [24]. For example, the performance ratio simulated with Model B in Fig. 10.15 includes a simple linear model to approximate the loss due to the temperature dependence of the optics, similar to the model shown in Eqs (10.9), (10.10). Model B thereby captures the lower performance during cold winter months (May–September), while Model A (which does not consider this loss) even predicts higher performance during winter than during summer, as cell temperature losses are reduced with decreasing temperature. Assuming that an arbitrary CPV module would be designed for best

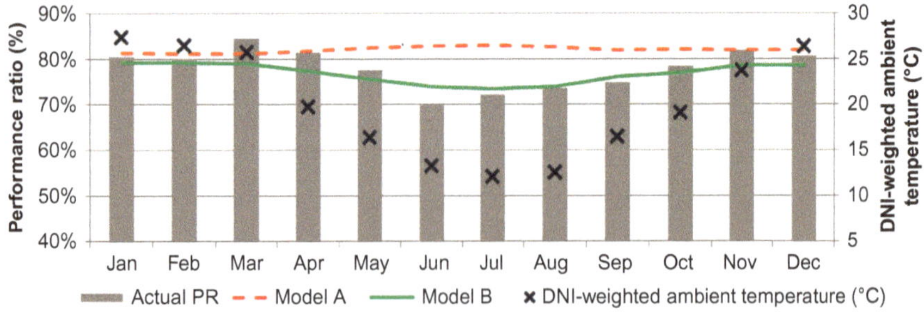

Fig. 10.15 Monthly performance data of a CPV plant installed in South Africa, including actual performance ratio (columns), performance ratio from two different models (*dashed and solid lines*) and DNI-weighted ambient temperature (*crosses*).

internal alignment at a temperature of the optics of 30°C, (Eq. 10.9) could describe a linear model for the loss due to nonoptimal temperature with parameter $\mu = -0.5\%/\mathrm{K}$ for $T_{\mathrm{optics}} < 30°\mathrm{C}$ and $\mu = 0.3\%/\mathrm{K}$ for $T_{\mathrm{optics}} \geq 30°\mathrm{C}$.

$$L_{\mathrm{temp-optics}} = \mu\left(T_{\mathrm{optics}} - 30°\mathrm{C}\right) \tag{10.9}$$

The temperature of the optics could be estimated from ambient temperature and wind velocity with (Eq. 10.10) and parameters like $\alpha = 10°\mathrm{C}$ and $\beta = -1°\mathrm{C}/(\mathrm{m/s})$.

$$T_{\mathrm{optics}} = T_{\mathrm{ambient}} + \alpha + \beta v_{\mathrm{wind}} \tag{10.10}$$

The values given in this example are arbitrary and have to be derived for specific CPV module designs from indoor or outdoor measurements of representative samples, such as described by Faiman et al. [25]. Based on such experimental data it also has to be carefully verified if the simplified Eqs (10.9), (10.10) approximate the observed module behaviour with sufficient accuracy. In a more detailed approach, Steiner et al. [18] used finite element and ray tracing methods to consider the temperature sensitivity of the optics in their 'YieldOpt' model.

Spectrum: Multijunction solar cells are inherently sensitive to the spectrum of the light due to the serial interconnection of several subcells. The subcell which produces the lowest current limits the overall current of the multijunction cell. The spectrum of the light impinging on the solar cells depends on geographical and meteorological parameters such as geometrical AM and haziness of the atmosphere, as well as on the spectral transmission of the optics. As for the temperature sensitivity of the optics, the underlying mechanisms of the spectral loss are complex and need to be modelled with an adequate level of simplification. Strong simplification is applied in the linear AM model of PVsyst [26] and in the parabolic model of Strobach et al. [27]; both do not consider atmospheric parameters. In contrast, Steiner et al. [18] rely on measurements of multifilter rotating shadowband radiometers to obtain atmospheric parameters and modify the clear-sky DNI spectrum accordingly using the SMARTS model.

Cell temperature: Temperature distribution across CPV modules and solar cells is usually nonhomogeneous due to the principle of point-focusing the irradiance. In the technology development phase, heat transfer models may be required to estimate and optimize cell temperature based on conduction, convection and radiation processes. However, linear temperature coefficients, experimentally determined as per IEC 62670-3 or as in the previous chapter, are usually sufficiently precise for modelling the field performance of CPV systems. If measurements of the open-circuit voltage, which serves as a surrogate for cell temperature, are not available, cell temperature may be estimated with heat transfer models from ambient temperature, irradiance and wind velocity, as implemented in PVsyst and NREL SAM [28,29].

Module I–V-characteristics: Electrical models for the voltage–current characteristics of nonconcentrator and concentrator modules may either predict the full I–V curve ('equivalent circuit models') or predict only the characteristic points of the I–V curve, such as I_{mpp}, V_{mpp}, I_{sc}, V_{oc} ('point-value models') [30]. Some examples for equivalent circuit models include the one-diode model implemented in PVsyst [28], the 'five parameter single-diode model' available in the SAM package by NREL [31] and the Spice network model by Steiner et al. [18]. The most widely used point-value model is the Sandia PV Array Performance Model [29], which can also be chosen in the SAM package.

Shading: Modelling the shading loss of CPV systems follows the same principles as for nonconcentrator systems, but may have to consider more complex geometries due to widespread use of two-axis trackers. The solar irradiance that does not reach the CPV modules due to shading cannot be converted to electrical energy and needs to be considered as so-called 'geometrical shading loss', often estimated by applying surface projection and intersection processing on a three-dimensional model of the plant [32]. Depending on the internal wiring of cells and modules in series and parallel strings, the usage of bypass diodes and the shape of the shadows, an additional 'electrical shading loss' needs to be taken into account [33]. It is caused by different current–voltage characteristics of unshaded and partially shaded CPV modules that are connected to the same maximum power point tracker of an inverter. The inverter chooses a common load resistance for all modules, resulting in an electrical shading loss due to nonoptimal power extraction from some modules.

Soiling: Deposition of dirt particles on the surface of CPV modules leads to soiling losses, which may be considered as a constant loss of a few percentage points in a first approximation. More detailed models should consider site characteristics such as soil type and vegetation, frequency of rainfall, wind velocity and direction. Based on a daily soiling rate, a cleaning threshold and a grace period after cleaning, Kimber et al. [34] empirically derived a soiling model. Winter et al. showed that small rain events may significantly reduce soiling loss [35].

Tracker pointing error: The acceptance angle under which incoming sunlight reaches the solar cells inside the CPV modules is limited due to the usage of concentrating optics and cost-efficient designs. The sensitivity of the module power output on misalignment can be characterized using a procedure described in IEC 62670-3, resulting in an 'acceptance angle curve' that relates misalignment in units of degrees to power loss in units of percent. The pointing accuracy of CPV trackers is limited and can be quantified with a method described in IEC 62817, resulting in a frequency distribution of the pointing error in units of degrees. The loss due to tracker pointing error for a specific combination of CPV modules and trackers can be estimated by combining the results of both aforementioned procedures.

Tracker wind stow: Two-axis trackers may only operate in tracking mode up to a maximum wind velocity $v_{\mathrm{wind,max}}$ which depends on the design of the tracker and its foundation. SCADA systems continuously measure wind velocity and, when $v_{\mathrm{wind,max}}$ is exceeded, automatically steer the trackers to a safety position in which the tracked plane is parallel to the ground. If this happens during sunny periods, the resulting loss is 100%, as the modules are no longer aligned with the sun. The loss for longer periods may be estimated by processing a time-series of wind velocity and DNI, replacing the DNI values with 0 during high wind velocity periods and calculating the effectively captured DNI. As wind velocity is a dynamic quantity, it is recommended to use a time-series with a step size of 5 min or less. The time required to move from tracking mode to the safety position and back, as well as hysteresis times of the SCADA systems should be considered in the model.

DC to AC conversion: Inverters used with CPV systems can be modelled in almost all aspects similar to inverters connected to nonconcentrator systems, eg, using the Sandia inverter model [36]. However, there is one point that requires particular attention: As DNI may change much faster than global irradiance, a slowly responding inverter may not be able to follow steep DNI and DC power ramps, leading to so-called 'switch-on loss'. If inverters with short delay times and fast ramping capability are chosen, the switch-on loss becomes negligible. For slowly responding inverters, the switch-on loss can be estimated based on the actual delay times and ramp rates and a DNI time-series with high time resolution (1 min or less).

Parasitic consumption: As CPV power plants usually contain more moving parts than nonconcentrator plants, such as trackers and possibly active ventilation or cooling systems, they usually show a higher loss due to parasitic consumption for powering motors and control boards. Manufacturers often provide typical consumption values on data sheets. Procedures for verifying parasitic consumption data are described in IEC 62670-2.

Frequently used modelling tools:

- The PVsyst software (http://www.pvsyst.com) is not only widely used for nonconcentrator plants, but considers also specific loss factors of CPV by a step-wise linear model for the temperature dependency of the optics and the spectral dependency of the multijunction cells, referred to as 'Utilization Factor' [37]. PVsyst's shading model is rather user friendly and can handle complex geometries of several hundred CPV trackers. The module IV characteristics can either be based on the one-diode model by Duffie and Beckman [38] or Sandia's Array Performance Model [29]. PVsyst offers an extensive graphical user interface, but lacks support for a scripting language. Although the documentation is extensive, some algorithms and details of the PVsyst software remain undisclosed.
- In contrast to PVsyst, the PV_LIB toolbox (https://pvpmc.sandia.gov and https://github.com/pvlib/pvlib-python), initiated by Sandia and extended by various contributors, provides full transparency of its algorithms, as they are disclosed under the Berkeley Software Distribution (BSD) license and can be easily extended with new functionalities. Being a collection of scripts that includes module and inverter models, data handling functions and atmospheric and irradiance models written in Matlab and the Python programming language, this toolbox targets engineers and modellers with programming skills and does not offer a graphical user interface.
- The SAM software (https://sam.nrel.gov) developed by NREL presents a mixture of the concepts used by PVsyst and the PV_LIB toolbox. On the one hand, it presents user friendly graphical guidance through typical simulation tasks (which also include renewable sources other than PVs, and financial modelling) and on the other hand, it includes scripting language support and a software developer kit so that users can extend the functionalities.

10.5 Future trends

CPV technology has made great technological progress during recent years. The highest ever reported efficiencies for PV cell and module of 46% and 38.9%, respectively, were achieved by this technology. Obviously, one can name this technology the 'high end' of all PV technologies. While the technology had progressed extremely fast, the market is still in an infancy status. Only very recently have larger installations of 40-100 MW_p size been installed in the field. The cumulative production volume of CPV (LCPV and HCPV) technology is below 400 MW_p. If compared with flat-plate technology, which is beyond 200 GW, it is obvious that the contribution to the PV market is currently negligible. In other words, marketwise CPV technology is on the same level as flat-plate technology was in the year 1993. However, the costs of the CPV technology are much lower than the cost of flat-plate technology at that time. Therefore, the cost of CPV is promising and the current challenge for CPV technology is to install many more MWs, thus growing the market and benefitting from manufacturing scaling effects. According to the price-experience curves known for other technologies, this will lead to a fast reduction in costs of CPV systems. On the other hand, the performance of CPV still has room for further improvements. Cell efficiencies beyond 50% and module efficiencies of 42% are in reach (see Fig. 10.7). It is important to note that the transfer from R&D results obtained in laboratories into industrial products has turned out to be extremely fast in CPV technology. For example, reported record cell efficiencies from laboratories are commercially available 3 years later.

For the future of CPV technology it is also important to proceed with the development of IEC standards. Standardization has turned out to be very important for market development. Standards help to establish trust in the products of the supplier and pave the way for financing. Currently several IEC standards are under preparation. As detailed in this chapter, some of the standards are very advanced and could be already used, but they are not yet finalized.

Further information on the technology and market of CPV technology can be found in the proceedings of the International Conference on Photovoltaics, which are published in an AIP series. The recent proceedings of CVP-10 till CPV-12 are accessible via open access and can be found here: http://www.cpv-12.org/proceedings.html.

Interested readers are referred to books about CPV technology; see reference list [39,40].

References

[1] H. Helmers, A.W. Bett, J. Parisi, C. Agert, Modeling of concentrating photovoltaic and thermal systems, Prog. Photovolt. Res. Appl. 22 (4) (2014) 427–439, http://dx.doi.org/10.1002/pip.2287.

[2] J.B. Lasich, P.J. Verlinden, A. Lewandowski, D. Edwards, H. Kendall, S. Carter, I. Thomas, P. Wakeman, M. Wright, W. Hertaeg, R. Metzke, M. Daly, M. Santin, A. Neumann, A. Wilson, A. Caprihan, M. Stedwell, World's first demonstration of a 140 kWp Heliostat Concentrator PV (HCPV) system, in: Proceedings of the 34th IEEE Photovoltaic Specialists Conference, 2009, p. 6.

[3] P. Simon, A. Bett, K. Horowitz, S. Kurtz, Current Status of Concentrator Photovoltaic (CPV) Technology, Fraunhofer ISE & NREL, Freiburg, Germany and Golden, USA, 2015. for download: http://www.ise.fraunhofer.de/en/publications/veroeffentlichungen-pdf-dateien-en/studien-und-konzeptpapiere/current-status-of-concentrator-photovoltaic-cpv-technology.pdf.

[4] M.A. Green, K. Emery, Y. Hishikawa, W. Warta, E.D. Dunlop, Solar cell efficiency tables (version 46). Prog. Photovolt. 23 (7) (2015) 805–812, http://dx.doi.org/10.1002/pip.2637.
[5] W. Guter, R. Kern, W. Köstler, T. Kubera, R. Löckenhoff, M. Meusel, M. Shirnow, G. Strobl, III–V multijunction solar cells—new lattice-matched products and development of upright metamorphic 3 J cells, in: Proc. 7th International Conference on Concentrating Photovoltaic Systems, AIP, Las Vegas, USA, 2011, pp. 5–8.
[6] S.P. Philipps, A.W. Bett, III–V multi-junction solar cells, R. Soc. Chem. 4 (2014) 87–117.
[7] M. Meusel, C. Baur, G. Siefer, F. Dimroth, A.W. Bett, W. Warta, Characterization of monolithic III–V multi-junction solar cells—challenges and application, Sol. Energy Mater. Sol. Cells 90 (18–19) (2006) 3268–3275.
[8] S.P. Philipps, W. Guter, E. Welser, J. Schöne, M. Steiner, F. Dimroth, A.W. Bett, Present Status in the Development of III–V Multi-Junction Solar Cells, Springer, Berlin/Heidelberg, Germany, 2012. pp. 1–21.
[9] M.A. Green, K. Emergy, Solar cell efficiency tables (version 02), Prog. Photovolt. Res. Appl. 1 (3) (1993) 225–227.
[10] European Photovoltaic Technology Platform, A Strategic Research Agenda for Photovoltaic Solar Energy Technology, second ed., Publications Office of the European Union, Luxembourg, 2011.
[11] G. Peharz, J.P. Ferrer-Rodríguez, G. Siefer, A.W. Bett, Investigations on the temperature dependence of CPV modules equipped with triple-junction solar cells. Prog. Photovolt. Res. Appl. 90 (1) (2011) 54–60, http://dx.doi.org/10.1002/pip.987.
[12] A.W. Bett, C. Baur, F. Dimroth, H. Lerchenmueller, G. Siefer, G. Willeke, The FLATCON® concentrator PV-technology, in: Proc. 3rd International Conference on Solar Concentrators for the Generation of Electricity or Hydrogen. Scottsdale, Arizona, USA, 2005.
[13] V.D. Rumyantsev, M. Hein, V.M. Andreev, A.W. Bett, F. Dimroth, G. Lange, G. Letay, M.Z. Shvarts, O.V. Sulima, Concentrator array based on GaAs cells and Fresnel lens concentrators, in: Proceedings of the 19th European Photovoltaic Solar Energy Conference, James & James Ltd, Glasgow, UK, 2000, pp. 2312–2315.
[14] G. Peharz, G. Siefer, A.W. Bett, A simple method for quantifying spectral impacts on multi-junction solar cells, Sol. Energy 83 (9) (2009) 1588–1598.
[15] J. Jaus, T. Mißbach, S.P. Philipps, G. Siefer, A.W. Bett, Spectral measurements using component cells: examinations on measurement precision, in: Proc. 26th European Photovoltaic Solar Energy Conference and Exhibition, Hamburg, Germany, 2011, pp. 176–181.
[16] C. Domínguez, I. Antón, G. Sala, S. Askins, Current-matching estimation for multi-junction cells within a CPV module by means of component cells, Prog. Photovolt. Res. Appl. 21 (7) (2013) 1478–1488, http://dx.doi.org/10.1002/pip.2227.
[17] G. Siefer, A.W. Bett, Analysis of temperature coefficients for III–V multi-junction concentrator cells. Prog. Photovolt. Res. Appl. 22 (5) (2014) 515–524, http://dx.doi.org/10.1002/PIP.2285.
[18] M. Steiner, G. Siefer, T. Hornung, G. Peharz, A.W. Bett, YieldOpt, a model to predict the power output and energy yield for concentrating photovoltaic modules, Prog. Photovolt. Res. Appl. 23 (3) (2014) 385–397, http://dx.doi.org/10.1002/pip.2458.
[19] S.P. Philipps, G. Peharz, R. Hoheisel, T. Hornung, N.M. Al-Abbadi, F. Dimroth, A.W. Bett, Energy harvesting efficiency of III–V multi-junction concentrator solar cells under realistic spectral conditions, in: Proceedings of the 6th International Conference on Concentrating Photovoltaic Systems, Freiburg, Germany, 2010, pp. 294–298.
[20] T. Hornung, M. Steiner, P. Nitz, Estimation of the influence of Fresnel lens temperature on energy generation of a concentrator photovoltaic system. Sol. Energy Mater. Sol. Cells 99 (2012) 333–338, http://dx.doi.org/10.1016/j.solmat.2011.12.024.

[21] M. Muller, S. Kurtz, M. Steiner, G. Siefer, Translating outdoor CPV I–V measurements to a CSTC power rating and the associated uncertainty. Prog. Photovolt. Res. Appl. 23 (11) (2015) 1557–1571, http://dx.doi.org/10.1002/pip.2590.
[22] T. Gerstmaier, T. Zech, M. Röttger, C. Braun, A. Gombert, Large-scale and long-term CPV power plant field results. in: AIP Conference Proceedings, 2015, http://dx.doi.org/10.1063/1.4931506. p. 1679: 030002.
[23] J. Mokri, J. Cunningham, PV System Performance Assessment, SunSpec Alliance, San Jose, CA, 2015.
[24] S. Kurtz, M. Muller, D. Jordan, K. Ghosal, B. Fisher, P. Verlinden, A. Hashimoto, D. Riley, Key parameters in determining energy generated by CPV modules, Prog. Photovolt. Res. Appl. 23 (10) (2015) 1250–1259, http://dx.doi.org/10.1002/pip.2544.
[25] D. Faiman, V. Melnichak, D. Bokobza, S. Kabalo, A sun-tracking environmental chamber for the outdoor quantification of CPV modules, in: AIP Conference Proceedings, 2014, p. 106.
[26] T. Gerstmaier, M. Gomez, A. Gombert, A. Mermoud, T. Lejeune, Validation of the PVSyst performance model for the concentrix CPV technology, in: AIP Conference Proceedings, 2011, p. 366.
[27] E. Strobach, D. Faiman, S. Kabalo, D. Bokobza, V. Melnichak, A. Gombert, T. Gerstmaier, M. Röttger, Modeling a grid-connected concentrator photovoltaic system. Prog. Photovolt. Res. Appl. 23 (5) (2015) 582–592, http://dx.doi.org/10.1002/pip.2467.
[28] A. Mermoud, PVsyst Help Section: Array Thermal Losses, PVsyst SA, Satigny, Switzerland, 2015.
[29] D.L. King, W.E. Boyson, J.A. Kratochvil, Photovoltaic Array Performance Model, Sandia National Laboratories (SAND2004-3535), Albuquerque, NM, USA, 2004. p. 43.
[30] P.M. Collaborative, DC Module IV Characteristics, Sandia National Laboratories, Albuquerque, NM, USA, 2015.
[31] P. Gilman, SAM Photovoltaic Model (pvsamv1) Technical Reference, NREL Draft Report, National Renewable Energy Laboratory, Golden, CO, USA, 2014.
[32] H. Capdevila, A. Marola, M. Herrerias Azcué, High resolution shading model and performance simulation of suntracking photovoltaic systems, in: Proceedings of 42nd ASES Annual Conference, Baltimore, Maryland, USA, 2015.
[33] A. Mermoud, PVsyst Help Section: Shading Factor, PVsyst SA, Satigny, Switzerland, 2015.
[34] A. Kimber, L. Mitchell, S. Nogradi, H. Wenger, The effect of soiling on large grid-connected photovoltaic systems in California and the southwest region of the United States, in: 4th World Conference on Photovoltaic Energy Conversion, Waikoloa, Hawaii, USA, 2006.
[35] M. Winter, G. Flynn, J. Foresi, P. Blumenfeld, R. Hoffman, I. Aevy, T. Romero, Reliability testing of high-concentration PV module and soiling issues, in: NREL Photovoltaic Module Reliability Workshop, 2011.
[36] D.L. King, S. Gonzalez, G.M. Galbraith, Performance Model for Grid-Connected Photovoltaic Inverters, Sandia National Laboratories (SAND2007-5036), Albuquerque, NM, USA, 2007.
[37] T. Gerstmaier, S. van Riesen, A. Gombert, A. Mermoud, T. Lejeune, E. Duminil, Software modeling of FLATCON (R) CPV systems, in: Proceedings of the 6th International Conference on Concentrating Photovoltaic Systems, Amer Inst Physics, Melville, 2010, pp. 183–186.
[38] J.A. Duffie, W.A. Beckman, Solar Engineering of Thermal Processes, fourth ed., Wiley, Hoboken, NJ, USA, 2013.
[39] F. Lewis, L.D. Partain, Solar Cells and Their Applications, John Wiley & Sons, New Jersey, Canada, 2010.
[40] A. Luque, V.M. Andreev, Concentrator Photovoltaics, Springer, Verlag, Heidelberg, Germany, 2007.

Large grid-connected photovoltaic power plants

11

Best practices for the design and operation of large photovoltaic power plants

A. Woyte, S. Goy
3E, Brussels, Belgium

11.1 Introduction

How well can a commercial photovoltaic (PV) plant perform today? And what distinguishes a good plant from a bad one?

Recent surveys of the performance of grid-connected PV plants show a large spread in performance ratio (PR). Between 1980 and 2010, the statistical average PR of new PV installations in moderate climates improved from 0.65 to approximately 0.85 [1]. However, also for recent commercial and utility scale plants, a significant spread of PR has been observed. For example, a detailed review of PR for a sample consisting of about 100 plants installed in Germany between 2006 and 2010 shows PR values between approximately 0.7 and 0.9 [2].

In practice, the performance, reliability, service life and safety of a PV plant depend on the quality and reliability of the components used, the plant design, the quality of installation in terms of craftsmanship and diligence, and operation and maintenance. The best-performing PV plants today excel over the medium-tier ones in these aspects.

In this chapter we identify best practices for the design and operation of large PV power plants. In Section 11.2 we review the most important factors that in practice cause a PV plant to work better or worse than another. In Section 11.3 we review the selection and sizing criteria for PV plant design. In Section 11.4 we describe the process of commissioning. Finally, in Section 11.5, we describe good practices for operation and maintenance. The chapter closes with a review of future trends and suggestions for further information.

11.2 The best plants in practice

There is much literature about the operational performance of PV plants, from field test installations to commercial plants. Unfortunately, since these studies are usually conducted in real-life conditions, often parameters cannot easily be controlled. Consequently, many studies are anecdotal and quantitative statements are often

The Performance of Photovoltaic (PV) Systems. http://dx.doi.org/10.1016/B978-1-78242-336-2.00011-2

specific to the plant, climate or period under study. However, qualitatively, the decisive factors for performance, reliability and lifetime of a plant are known.

Whether a PV plant during regular operation performs better or worse than another is mainly dominated by the following factors:

- shading,
- soiling and snow,
- module conversion efficiency as a function of irradiance,
- module conversion efficiency as a function of temperature,
- tolerances, mismatch and maximum power point (MPP) tracking and
- inverter conversion efficiency.

Besides these factors, the overall energy performance also depends on losses due to irregular phenomena like degradation, forced outage due to technical failures, etc. The effect of these phenomena on the overall energy performance can be controlled by a good strategy for operations and maintenance (O&M) along with professional monitoring.

11.2.1 Shading

We can distinguish two types of shading, namely far and near shading. Far and near being relative terms, we may define far shading as shading that leads to larger sections of the PV generator being blocked from direct solar irradiance during certain times. It is mostly caused by landscape, buildings or vegetation relatively far away from the PV plant. For a given site with far shading, the energy yield may depend on tilt and azimuth angle of the PV array. Their effect can be evaluated with a simulation tool.

Near shading is caused by obstacles relatively close to the PV array, for example, trees, fences or poles. Also PV arrays arranged as rows of sheds cause near shading. Near shading causes overproportional losses due to the inhomogeneous irradiance distribution over the modules. Therefore, even small obstacles should be eliminated as much as possible. As a rule of thumb for row-to-row shading, literature recommends designing the PV plant such that no near shading occurs at noon on the winter solstice [3]. The lay-out of the modules, portrait or landscape, is also important here: for crystalline silicon and CiGS modules, row-to-row shading causes significantly higher losses with a portrait lay-out than with a landscape lay-out [4].

11.2.2 Soiling and snow

Soiling and snow on the PV array reduce the light incident on the PV cells, hence causing additional capture losses. Capture losses due to soiling and snow depend on site-specific and installation-specific parameters:

- Site-specific parameters are soil type, rain pattern, wind flow, activities around the site. They are largely uncontrollable.
- Installation-specific parameters are cleaning frequency and tilt angle of the modules. They can be controlled during design and operation.

Soiling losses reported in the literature vary greatly depending on the local conditions and PV power plant installation [5]. Especially in dry and dusty environments they are important. The way cleaning should be considered within the maintenance strategy depends on the long-term weather pattern and its cost versus the energy loss due to soiling. To facilitate the so-called self-cleaning by rain, PV arrays should always be tilted, also in low latitudes. Practical recommendations vary between tilt angles from 5 to 20 degrees. When relying on self-cleaning, maintenance teams during visual inspection should pay particular attention to soil accumulating on the lower side of the module frames, which may shadow the lower solar cells.

11.2.3 Module conversion efficiency as a function of irradiance

The conversion efficiency of PV modules is a function of irradiance, which depends on the technological parameters of the solar cells, mainly their diode ideality factor and series resistance. Especially the decline of efficiency with relatively low and decreasing values of irradiance depends on these parameters and it can vary strongly for different module types. From common modelling practice of large plants in Southern Europe, annual losses due to low irradiance range largely between 2% and 3% of reference yield but, in some cases, are also much higher. The effect of the shunt resistance is relatively less important.

11.2.4 Module conversion efficiency as a function of temperature

Power temperature coefficients of crystalline PV modules today range between 0.4% and 0.5%/K. Since PV modules heat up under sunlight, the difference of module minus ambient temperature is typically positively correlated with the incident irradiance. Consequently, an average 10 K increase in module temperature during daylight means that much more than an additional 4–5% of the reference yield is dissipated as capture loss in the module. Ventilation at the back of the modules helps to keep these losses low.

Most ground-mounted PV arrays today are arranged as free-standing sheds. For them ventilation mainly depends on the microclimate of the site. For roof-mounted systems arranged as sheds, several types of structures and mounting bases are available, allowing for more or less ventilation depending on their particular design. For roof-mounted installations on tilted roofs, sufficient spacing should be allowed between roof and PV arrays. For building-integrated PV, integration concepts allowing convection at the back side of the array will show lower capture losses than those that do not.

11.2.5 Tolerances, mismatch and MPP tracking

In practice PV modules have a predefined production tolerance of typically ±5% of nominal power at standard test conditions (STC) or below. Nowadays, most crystalline silicon manufacturers declare the nominal module power as a minimum value by declaring a negative tolerance of zero. Plants making use of such modules perform better statistically with reference to the contractually installed peak power than power plants with a negative tolerance lower than zero.

Mismatch of PV modules occurs when not all PV modules in a string or not all strings in an array have the same MPP. This happens when the array is illuminated heterogeneously or when the electrical characteristics of the modules vary. For a ±5% tolerance, the mismatch loss for unsorted crystalline silicon modules can reach 0.5% at STC. It can be reduced to 0.1% when the modules are sorted according to their individual MPP currents and modules with the same MPP current are connected in series [6]. Notably, the lower the production tolerance, the lower is the relative benefit of sorting during the construction phase of the plant. Finally, modules may degrade differently over time, which will introduce additional mismatch losses over the years of operation.

11.2.6 Inverter conversion efficiency

By approximation, efficiency of the inverter is a linear factor for the system's energy yield. Consequently, for a good system performance, a high efficiency over the entire power range is essential. Notably, the inverter efficiency also depends on the DC voltage. Particularly for inverters with large MPP windows, efficiency over power does not behave the same over the entire MPP voltage range. For a good system performance, designers should ensure a high efficiency in the expected voltage range.

Finally, the performance and lifetime of an inverter also depend on its operating temperature. Therefore, for large PV plants, sufficient spacing and ventilation help to keep the losses low and reduce the probability of failures. In hot climates, inverters are usually cooled actively. This implies carrying out a thermal study of the shelters in the design phase and respecting the installation guidelines given by the manufacturer in terms of distance to other inverters [7,8].

11.2.7 Operation and maintenance

The best-performing PV plants in practice are usually monitored closely, irregularities are investigated in detail and, if necessary, a maintenance team is sent on-site. Downtimes can be reduced to a minimum by keeping essential spare parts available on site [7,8].

The minimum requirements for proper monitoring of PV power plants are described in IEC 61724 [9]. A recent review of monitoring practices and the required instruments was published by Woyte et al. [10].

Notably, monitoring is more than the regular recording of data. Many commercial PV installations are logged but not followed up closely, or the data are never analysed in detail. Effective monitoring implies at least semiautomated analyses and alerts that enable the maintenance operator to react immediately and resolve the cause for a loss.

11.3 PV plant design

When designing a PV power plant, the main criteria to be considered are the components' technical specifications, their compatibility with each other, their expected lifetime and reliability, and the accompanying warranties. To a large extent quality and safety of components, PV plant design and documentation are governed by international and national standards. For a survey of applicable standards see, for example, DGS [3].

11.3.1 Module selection

The selection of PV modules for large PV plants is based on technical and commercial criteria. Technically, the most important parameters are the module's conversion efficiency as a function of irradiance and temperature, the temperature coefficient and the expected degradation over time. Additional parameters, that can less easily be modelled and controlled, are reflection on the module surface and the spectral sensitivity. When computing the expected energy yield over the amortization period of the project, these parameters are taken into account by means of simplified assumptions. The precision of this long-term yield assessment depends on the quality of these assumptions. Project developers and owners who systematically check these parameters for the modules used, watch them for their operational installations, and take into account this experience during procurement, can achieve higher performances than average.

Commercially, the selection of modules is also based on what is generally referred to as 'bankability', that is, the trust these modules receive from financing institutions. Bankability is based on long-term field experience with the brand and module technology, assessments of the manufacturing process by independent assessors, lab testing of samples and the economic strength and reputation of the manufacturer, which in turn determines the trust in the given product warranties.

Notably, from a techno-economic point of view, the nominal module efficiency at STC does not affect PR as this is based on a normalization per watt peak. This is in line with commercial practices where PV modules are usually accounted for per watt peak. However, from an overall economic point of view, for a PV power plant, surface does matter because costs of land, support structures and maintenance are surface-dependent. Consequently, PV plants built with modules of lower efficiency will tend to have relatively higher costs for their balance-of-system and land.

11.3.2 Inverter selection

For PV inverters, power conversion efficiency over the operating range seems to be the most obvious selection criterion. However, often more practical considerations govern the selection. Particularly the decision whether to use one or few large inverters or a large number of small inverters depends on the particular preference of the system designer. Where large inverters allow for somewhat higher overall efficiencies, smaller inverters may allow for lower costs for operation and maintenance and lower losses in case of individual failures.

The choice of the inverter along with the earthing configuration must be checked carefully in order to avoid corrosion, leakage currents or so-called potential-induced degradation. The latter has first been observed with high-efficiency n-type and back-contact modules, but can also occur with standard modules from p-type crystalline silicon [11]. In case of doubt, the chosen module, inverter and earthing arrangement should be checked for compatibility with the inverter and module manufacturers.

For the sizing of PV inverters versus the PV array, the following main rules apply.

Power

As a rule of thumb for moderate climates, the nominal DC power input of the inverter can range between 80% and 120% of the PV peak power. For this ratio, PV module tolerance and nonstabilized behaviour (for some thin-films modules) should be taken into account. In practice, the designer trades off inverter costs against additional power losses. This trade-off should be evaluated consciously. The evaluation should account for the particular inverter model and the local climate by means of representative meteorological data of high time resolution.

MPP voltage

The MPP voltage of each string for module temperatures between −10°C and +70°C and 1000 W/m^2 should be within the inverter's MPP tracking range.

Maximum PV array voltage

The open circuit voltage of each string at −10°C and 1000 W/m^2 should not exceed the inverter's admissible maximum voltage.

11.3.3 Transformers

Large PV power plants possess their own transformers for stepping up the output voltage of the inverters to the medium or high-voltage grid.

For distribution transformers, loss levels are defined in EN 50464-1 [12] for oil-filled transformers and EN 50541-1 [13] for dry transformers. The loss level is described by two letters in the format L_kL_0, where L_k refers to the on-load losses and L_0 to the no-load losses. L_k can range from A_k to D_k for oil-filled transformers and from A_k to B_k for the dry ones. The no-load losses can range from A_0 to E_0 and A_0 to C_0 for oil-filled and dry transformers, respectively. Starting from A, the losses increase in alphabetical order. In practice, the losses of the selected transformer as indicated by its nominal voltage drop are often limited to 1.5% at maximum, meaning that PV power plants should be equipped at least with C_0B_k transformers.

11.3.4 Support structures

Support structures have to be sized depending on module weight, wind load and the local soil conditions, if ground-mounted. For rooftop systems the roof structure needs to be reinforced in some cases and additional ballast may be required. In any case, a dedicated stability study is required for the final design. This study is usually carried out by an independent civil engineering consultant. It serves for verifying the structural compatibility between the roof or soil conditions, the fixation system and the local wind and snow regime following national construction standards.

11.3.5 Lifetime of balance-of-system components

For a PV power plant, a service life of 20–25 years can be expected with routine corrective maintenance actions and possibly one full replacement of the inverters. For fulfiling this expectancy, outdoor balance-of-system materials made of plastics, like cables, connectors or plastic support structures must be UV resistant and ozone protected. Metal support structures and other metal parts must be noncorrosive, for example, galvanized, or made from aluminium or stainless steel.

11.3.6 Warranties and risk

For the selection of components, warranties are decisive as they serve to mitigate the technical risk of component failures and, hence, unexpected costs, during the service life of the plant. Typical warranties as agreed upon between manufacturer and contractor are listed in the following paragraphs.

Module power warranty

Common practice is to guarantee at least 80% of the nominal STC power after 25 years and at least 90% after 10 years. Best practice is a linearly decreasing power warranty over 25 years or longer.

Module product warranty

A warranty of 10 years is common practice. Best practice is to explicitly cover specific defects (eg, snail trails, hot spots) or conditions (eg, a saline atmosphere) with dedicated warranty clauses.

Inverter product warranty

Common practice is a warranty of 5 years for inverters. Manufacturers usually offer extensions along with a maintenance contract.

As a general good practice, the plant owner should identify the risks not covered by the agreed warranties and those being transferred to him at the end of the different warranty periods. These risks should be covered for the remaining service life of the plant, technically by adequate operation and maintenance and financially by providing adequate reserve accounts.

11.4 Commissioning and acceptance

11.4.1 The commissioning process

Mostly, the developers of a large PV power plant project contract one single party for engineering, procurement and construction of the plant. This party is usually referred to as the EPC contractor. The process of commissioning basically covers the handover of the plant from the EPC contractor to the future owner by verifying the achievement of several essential milestones in technical, financial and legal terms.

Technically, commissioning serves to confirm the completion of works and the proper functioning of the plant. It is the last opportunity for the future owner to verify the proper execution of the works. Legally, at the occasion of commissioning, the full property of the PV plant is transferred from the EPC contractor to the future owner, including all potential risks and responsibilities emerging from this property. At this occasion the warranties on the installation start to run for the legal or contractual warranty period. The warranty period is agreed upon contractually and it typically lasts one to several years. Financially, at this occasion, the future owner can check whether the installation is up to the contractual specifications and whether the performance level as guaranteed in the EPC contract is met. If this is not the case, the owner can claim financial liabilities as agreed upon in the contract.

In practice, the process of commissioning is split into three subsequent acceptance tests as illustrated in Fig. 11.1: acceptance of mechanical completion, provisional acceptance and final acceptance. The steps differ by aim, scope and length. They are described in the following sections.

The commissioning procedures for each of the steps are described in the EPC contract. They typically cover

- content and duration of the different commissioning steps and tests,
- content and detail of the file documenting the plant as it has been built,
- calculation procedures for PR and availability,
- definition of minor and major defects and a contractual period for the removal of minor defects,
- liabilities and
- potential restrictions.

11.4.2 Acceptance of mechanical completion

The mechanical completion check takes place before the system is set into operation. It comprises verifying that the components of the plant are properly installed as listed in the EPC contract and that they meet the contractual requirements and applicable laws and regulations. The mechanical completion check is a thorough verification to make sure the plant is ready for start-up. When no major defects are noted, the

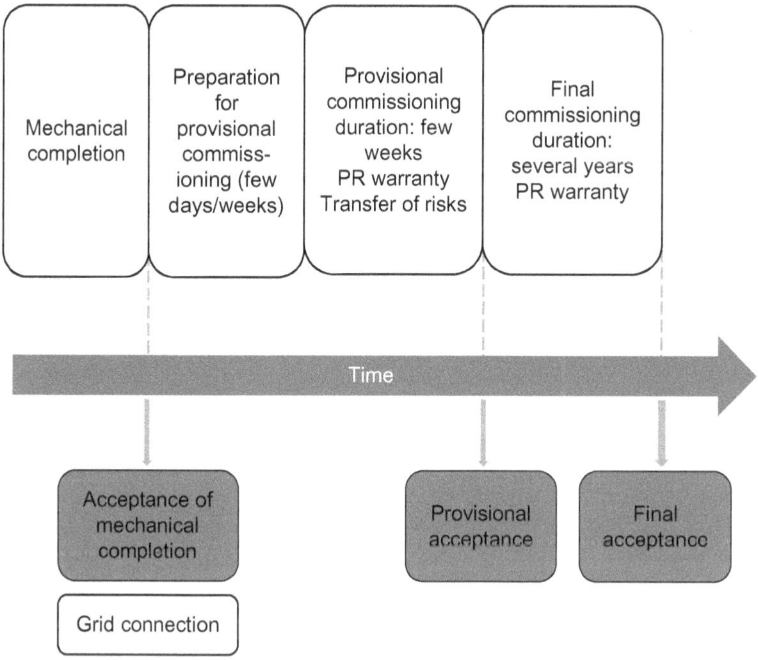

Fig. 11.1 Visualization of the commissioning process.

system is connected to the grid and set into operation. Remaining minor defects are listed on a so-called punch list and should be removed by the contractor before provisional acceptance.

If the acceptance of mechanical completion is delayed at this stage, the EPC contractor incurs delay penalties. The contractual daily or weekly penalties usually cover the lost income due to the delay. Exceptions to the contractor's liability should be listed explicitly in the EPC contract, for example, force majeure. The contractor's liability is usually limited to a certain percentage of the contract value. In exchange the owner then has the right to breach the contract if this limit is exceeded due to the fault of the EPC contractor.

With the acceptance of mechanical completion, the full ownership of the plant is transferred. However, the risk of the ownership still remains with the EPC contractor until provisional acceptance has been achieved.

11.4.3 Provisional acceptance

With the provisional acceptance, also the responsibility for the plant is transferred from the EPC contractor to the owner. Risk is transferred to the owner, as far as it is not covered by the contractual warranties of the EPC contract.

The PV plant normally starts operation after the acceptance of mechanical completion. During the first few days, several tests and trials are performed in order to identify any remaining hidden installation issues. Then the system is monitored for provisional acceptance and the PR and availability of the PV power plant are calculated over a contractually defined period. Ideally this period of provisional commissioning is no shorter than a month; however, in practice as little as 10 consecutive days are sometimes accepted if these days comprise representative irradiance and temperature and a minimal predefined energy yield.

In principle, provisional acceptance requires that three requirements have been met, namely:

- The contractor has provided the owner with a complete executed work file (also called the 'as-built file') describing the installation and following at least the requirements of IEC 62446 [14].
- The plant has been inspected following the requirements of IEC 62446 [14] at least. At this occasion the conformity of the plant with the as-built file has been verified. At this stage also the proper implementation of the punch list from mechanical completion has been verified. Infrared measurements are recommended on components such as modules, inverters and junction boxes.
- The PR for the period of provisional commissioning is above a threshold as guaranteed in the EPC contract. This threshold is usually set significantly below the expected average PR. Not achieving the contractual minimum PR during provisional commissioning causes liabilities for the EPC contractor. In some contracts the contractor also receives a bonus in case the plant performs better than estimated.

Again, the nonconformities noted are classified into minor and major defects. Minor defects do not prevent provisional acceptance for the plant. They are again listed on a punch list and should be corrected by the contractor, typically within 2–4 months. Major defects have to be solved before the plant can be provisionally accepted.

A complete EPC contract contains the following points for performance verification during provisional commissioning:

- The PR definition should be based on the latest version of the standard IEC 61724 [9].
- The PR is to be evaluated for the aggregated PV plant over the provisional commissioning period; good practice is to analyse the PR on the inverter level as well.
- Solar irradiance in the plane of the PV array is measured by at least three pyranometers for a given inclination/orientation; more pyranometers are recommended for PV generators spread over very large areas. The pyranometers must be of First Class or Secondary Standard. Data consistency must be checked regularly and outliers for one pyranometer are corrected by values from the other devices.
- The final system yield is calculated from a calibrated electricity metre, usually the generation metre that also serves as a basis for billing.
- The final system yield is calculated by normalizing electricity output to the peak power of the PV generator; for large plants, the actual peak power values from factory measurements are used rather than the nominal value.
- A minimum data availability is required for the provisional commissioning period, for example, 98%.
- A lower threshold is typically set on the irradiance to exclude daytime data while some inverters are in stand-by mode. Common practice is to set this value around 50 W/m^2.

11.4.4 Final acceptance

The process of final commissioning starts with the provisional acceptance. It runs for several, typically one or two, years, often in line with the contractual warranty on the installation. These warranties commonly include a performance warranty as well as a liability for defects. This liability covers repairs of defective parts of the plant.

At the end of the final commissioning period, the same on-site verification and testing are carried out as for the provisional acceptance. The final acceptance is usually scheduled before the contractual warranty ends.

The guaranteed minimum PR is defined based on the PR guaranteed at provisional acceptance; and then reduced for the expected yearly degradation. As for the provisional acceptance, penalties apply at the end of the final commissioning period if the guaranteed PR could not be achieved. Bonuses may apply as well. EPC contracts specify that underperformance during the warranty period has to be corrected by the contractor. If the EPC and the O&M contractors are different legal entities, their specific contracts with the owner should unambiguously define their respective responsibilities, for example, in terms of module cleaning, replacement of broken parts and supply-of-spare-parts list.

Finally, also for final commissioning a liability limit is explicitly specified. As for provisional acceptance, if this limit is exceeded due to the fault of the EPC contractor, the owner then has the right to breach the contract.

According to most contracts, once the plant has passed final acceptance, the last tranche of the EPC contractor's price is paid. At this moment, responsibility and risk are fully transferred to the owner.

11.5 Operation and maintenance

11.5.1 Operation

After provisional commissioning the system enters operation. For PV power plants, detailed monitoring is good practice. For O&M purposes, monitoring must include data acquisition and recording close to real time with 5–15 min time resolution. The monitoring system provides detailed status information on the installation down to the level of inverters and possibly strings. Ways of monitoring the proper operation of a PV plant in the field have been introduced in Chapter 6. Requirements to be taken into account for operation are the control of the PV plant by the grid operator or by a commercial party that offers the electricity on the power market. Finally, PV plant operators are increasingly required to provide short-term forecasts of the power production for the coming hours or days to the grid operator or for commercialization on power markets. These functions can be integrated with the classical systems for operational monitoring.

During operation, maintenance serves for keeping the plant in a safe, healthy and performant state over its service life. In practice this is done either by the O&M department of the system owner or by a third party, referred to as an O&M contractor.

O&M contracts usually include a performance or availability guarantee. Maintenance can be distinguished between preventive maintenance and corrective maintenance. All maintenance actions need to be properly documented.

The following explanations about maintenance are generally applicable to all kinds of technical installations. Here we largely follow the terminology and classification as described in the French documentation file FD X 60-000 [15].

11.5.2 Preventive maintenance

Preventive maintenance is carried out according to specified criteria in order to reduce the probability of malfunctions. Preventive maintenance can further be split into systematic, conditional and anticipated maintenance. Systematic maintenance is carried out periodically. Conditional maintenance is carried out when specific conditions apply, for example, when a certain temperature would exceed a threshold or after a thunderstorm. Anticipative maintenance is a kind of conditional maintenance based on the assessment of operational data and where the analysis suggests maintenance, repair or replacement.

Standard procedures for preventive maintenance are usually applied to all components of the plant: the PV generator, junction boxes, inverters, transformers, medium-voltage switchgear, auxiliary systems, buildings and security installations. Preventive maintenance requirements for the main components such as inverters or transformers are generally specified by the manufacturers. These requirements are then complemented by requirements from the side of the owner. The most common actions for systematic maintenance are listed here. The maintenance cycles for each of these actions differ, for example, every 6 months, every year, every 2 years or longer.

PV generator: visual inspection of the PV modules, cabling and support structures, tightening of bolts where needed, cleaning irradiance sensors and checking the fixation and orientation of sensors, periodic routine thermography of the generator

Junction boxes: inspection and test of all safety devices (fuses, circuit breakers, surge arrestors), checking and fixing of terminal screws, checking that no water has entered, checking of earth resistance and insulation resistance, replacement of warning labels where needed

Inverters: inspection and test of all safety devices (fuses, circuit breakers, surge arrestors), checking and fixing of terminal screws, checking that no water has entered, checking the operation of ventilation systems, cleaning of filters, heat sinks, etc. from dust

Transformers: visual and aural inspection, checking and fixing of terminal screws, oil level and temperature, integrity of the tank, paint

Medium-voltage switchgear: visual inspection of circuit breakers, disconnectors and isolators, testing of circuit breakers and disconnectors, checking of ventilation, lubrication, locking mechanisms and status indication

Auxiliary systems, buildings, security installations: verifying the proper functioning of the auxiliary electric and data acquisition systems, maintenance and repair of buildings, inspection and cleaning of drainage systems, grass cutting, inspection of fencing and alarm systems

11.5.3 Corrective maintenance

Corrective maintenance is carried out after a malfunction has been detected. It serves for resetting the malfunctioning component into a safe, healthy and performant state.

The adequate corrective maintenance actions are specific to the malfunction that has occurred.

11.5.4 Classification in terms of complexity

Maintenance actions can be classified in terms of their complexity. Complexity usually relates to the required qualification of personnel that perform the different types of maintenance actions. NF X60-000 specifies different maintenance levels according to the complexity of respective maintenance actions.

Maintenance level 1: Simple actions that are applied to easily and safely accessible components, for example, inspection and testing of safety devices.

Maintenance level 2: Actions requiring simple procedures and additional equipment, for example, cleaning of filters that are difficult to access. Such actions can be executed by personnel familiarized with the maintenance procedures for the specific equipment.

Maintenance level 3: Actions requiring complex procedures or application of relatively complex additional maintenance tools, for example, thermography or measurement of earth resistance. Such actions can be executed by qualified technical personnel based on a detailed technical description of the action.

Maintenance level 4: Actions requiring the knowledge of a specific technique or technology or application of specific maintenance tools, for example, diagnosis of inverter malfunction with manufacturer-specific equipment.

Maintenance level 5: Actions requiring know-how that is specific to the applied technology or device, for example, repair of an inverter. Per definition, such actions are executed by manufacturer staff or specialized third parties agreed by the manufacturer.

Based on this classification, it becomes clear that the costs for maintenance of a large PV plant depend on the frequency of preventive and corrective maintenance actions and the maintenance levels of the different actions. To a large extent this is already determined in the design phase.

Example: The decision whether to use a central or many distributed inverters affects the costs for preventive and corrective maintenance. Preventive maintenance for one central inverter usually takes less time than for many small ones. However, in case of an inverter failure, a smaller inverter can easily be replaced by a spare inverter by on-site personnel as a level 2 or level 3 action. On the contrary, when a central inverter fails, it is usually not replaced but repaired by a service technician acting on behalf of the manufacturer as a level 4 or level 5 action.

11.5.5 Logistics for corrective maintenance

The time needed for repair determines the electricity lost due to a malfunction. In order to avoid large losses due to contingencies in the delivery of spare parts, for PV power plants the most important spare parts are kept in stock on or close to the site. In practice, the type and amount of spare parts to be kept in stock differ depending on the particular risk assessment and expectations of the owner. Some general guidelines are listed here, along with exemplary amounts of spare parts to serve as a rule of thumb for PV power plants in the single-digit megawatt range.

PV modules

For PV modules, 0.1–0.5% of the installed peak power should be kept in stock.

Large essential parts

For large essential parts, keeping one spare device of each is advisable as a long-lasting outage will affect a large part of or even the entire plant. This applies mainly to large central inverters and distribution transformers.

As an alternative, for large-scale inverters, keeping stock of all important spare parts for repairing the inverter on site may be acceptable if qualified service staff from the manufacturer side can be on-site immediately and repair the device.

Balance-of-system: Smaller parts

If the plant makes use of a larger number of relatively small inverters, usually a somewhat larger sample is kept in stock, for example, 4–10 pieces, depending on their size and price.

Other components typically kept in stock with exemplary amounts are:

- *Junction boxes*: two pieces of each type
- *High-cost measuring equipment (pyranometers)*: two pieces
- *Low-cost measuring equipment (irradiance sensors, PT100 thermometers, etc.)*: 5–10 pieces
- *Support structure elements*: two standard installation kits
- *Consumables, for example, fuses, connectors, cable*: tens to hundreds of pieces per type

Notably, the precise parts and numbers to be kept in stock are agreed upon between the plant owner and the O&M contractor in a plant-specific contract. In practice they can strongly vary depending on the type and size of the plant and its accessibility for external maintenance personnel and spare parts in case of a contingency.

11.6 Future trends

In design and operation of large-scale PV power plants several high-level trends are visible.

In the project development phase increased attention is currently paid to technical and financial risk. Up to now, mainly the long-term resource uncertainty has been taken into account and some uncertainty on the PV system model. The risk of PV module failure is mitigated by module type testing and factory inspections; however, it has not yet been figured in systematically into the business plans. In future, financial institutions are expected to do this increasingly. From research this will require more statistical evidence on the performance and reliability of PV modules and balance-of-system components in the field.

Also the financial return on investment will become more uncertain in the future. Today most PV power plants receive a guaranteed return per unit of electricity generated over a contractually agreed time. This may be based on a legally fixed feed-in tariff, on a bilaterally negotiated power purchase agreement or some hybrid model as, for example, tender-based power purchase agreements. In future, large PV power plants will increasingly offer under competitive conditions. Business models are needed that are compatible with the time-dependent value of electricity, allowing the owner to earn the investment cost over the service life including a margin.

The operational phase of a PV power plant will gain in importance as compared to the past. Especially for preventive maintenance, methods for on-site characterization of PV modules are increasingly applied. Infrared thermography on the installed plant is increasingly becoming a routine periodic maintenance action either with ground-based equipment or from helicopters or drones. For anticipative maintenance and diagnosis, on-site module testing is applied increasingly by means of mobile test labs that include, for example, solar simulators, current sources for electroluminescence measurements and IV curve tracers for a single module or a string.

In operational monitoring, future trends point towards improved diagnostic functions, typically relying on artificial intelligence methods. Monitoring systems increasingly analyse and interpret the data. Moreover, they develop towards integrated tools for operation and maintenance by integrating options for document management, technical reporting, fault analysis, maintenance scheduling and financial reporting.

Finally, with increasing deployment of PV, constraints originating from power system operations will also affect the operation and overall performance of large PV plants. Ramp-rate control with and without local storage is currently studied in order to mitigate the impact of fast irradiance fluctuations on power system stability. Approached from the generator side, large PV plants could also contribute to power system stability by providing upward or downward reserves. Technically, this is possible; however, particularly the provision of upward reserves will reduce the overall performance of the plant in question [16]. The business case for such operation modes will depend on the incentives available for deviating from the objective of maximum energy yield.

11.7 Further information

For further reading on how to make large-scale PV plants excel in performance and availability, we refer to the activities of the International Energy Agency's Photovoltaic Power Systems Programme (IEA PVPS) and, in particular, its Task 13 Performance and Reliability of Photovoltaic Systems. Latest reports on different PV system aspects are available on the IEA PVPS website [17].

For selection and sizing of components, several good practice guides have been published. The book *Planning and Installing Photovoltaic Systems* [3] has regularly been reedited and it is available in several languages. It may be considered a reference for good practice of planning and installation of PV in general. For the discussion of inverter rating versus losses, several valuable contributions have been published at conferences and in scientific journals, for example, Burger and Rüther [18] and Giesler et al. [19].

For operation and maintenance, the Electric Power Research Institute published a white paper in 2010 addressing O&M challenges for PV and how different types of owners and O&M service providers view and tackle them in their business strategies [20]. The paper concludes with lessons learned and best practices on preventive maintenance, monitoring and warranty. SolarPower Europe published very detailed best practice guidelines for O&M of PV power plants in 2016 [21]. Finally, for very detailed recommendations on best practices for O&M, particularly in terms of safety, we refer to a report of Haney and Burstein [22].

Acknowledgements

The authors thank 3E's PV team for contributions and critical feedback based on their daily consulting practice and, in particular, Alexandre Irissou for final proofreading.

References

[1] A. Woyte, M. Richter, D. Moser, S. Mau, N.H. Reich, U. Jahn, Monitoring of photovoltaic systems: good practices and systematic analysis, in: 28th EU PVSEC, Paris, France, 2013, pp. 3686–3694.

[2] N.H. Reich, B. Mueller, A. Armbruster, W.G.J.H.M. van Sark, K. Kiefer, C. Reise, Performance ratio revisited: Is PR > 90% realistic?. Prog. Photovolt. Res. Appl. 20 (2012) 717–726, http://dx.doi.org/10.1002/pip.1219.

[3] Deutsche Gesellschaft für Sonnenenergie (DGS), Planning and Installing Photovoltaic Systems: A Guide for Installers, Architects and Engineers, Taylor & Francis, 2013. ISBN: 978-1-84971-343-6.

[4] M. Hernández Velasco, F. Fiedler, D. Timm, Comparison of partial shading losses in freefield PV plants with different arrays configurations, in: 28th EU PVSEC, Paris, France, 2013, pp. 4171–4175.

[5] M. Monto, P. Rohit, Impact of dust on solar photovoltaic (PV) performance: research status, challenges and recommendations, Renew. Sustain. Energy Rev. 14 (2010) 3124–3131.

[6] W. Herrmann, S. Kämmer, U. Yusufoglu, Circuit losses in PV arrays caused by electrical mismatch of PV modules—impacts of temperature gradients and a variation of irradiance, in: 28th EU PVSEC, Paris, France, 2013.

[7] S. Goy, C. Tjengdrawira, T. Gheysen, A. Irissou, S. Ortega, C. Morilla Padial, How to mitigate risks in PV projects: lessons learned worldwide in manufacturing, commissioning

and operation of PV power plants, in: 28th EU PVSEC, Paris, France, 2013, pp. 4114–4117.
[8] S. Goy, C. Tjengdrawira, A. Woyte, J. Crotteux, A. Irissou, Assessing design, installation and operational flaws: mitigating risks for PV project financing, in: 26th EU PVSEC, Hamburg, Germany, 2011, pp. 3952–3957.
[9] IEC 61724:1998 Photovoltaic system performance monitoring—guidelines for measurement, data exchange and analysis, ed. 1.0, 1998.
[10] A. Woyte, M. Richter, D. Moser, N. Reich, M. Green, S. Mau, H.G. Beyer, Analytical monitoring of grid-connected photovoltaic systems—good practice for monitoring and performance analysis (No. Report IEA-PVPS T13-03: 2014), IEA PVPS, 2014.
[11] M. Köntges, S. Kurtz, C. Packard, U. Jahn, K.A. Berger, K. Kato, T. Friesen, H. Liu, M. Van Iseghem, Review on failures of photovoltaic modules (No. IEA-PVPS T13-01:2013), IEA PVPS, 2014.
[12] EN 50464-1:2007 Three-phase oil-immersed distribution transformers 50 Hz, from 50 kVA to 2500 kVA with highest voltage for equipment not exceeding 36 kV—Part 1: general requirements, 2007.
[13] EN 50541-1:2011 Three phase dry-type distribution transformers 50 Hz, from 100 kVA to 3150 kVA, with highest voltage for equipment not exceeding 36 kV—Part 1: general requirements, 2011.
[14] IEC 62446-1:2016 Photovoltaic (PV) systems—requirements for testing, documentation and maintenance—part 1: grid connected systems—documentation, commissioning tests and inspection, 2016.
[15] FD X 60-000 Maintenance industrielle—Fonction maintenance, Fascicule de documentation (documentation file), Association Française de Normalisation (AFNOR), 2002.
[16] P. Kreutzkamp, O. Gammoh, K. De Brabandere, M. Rekinger, PV forecasting confidence intervals for reserve planning and system operation, in: Presented at the Solar Integration Workshop, Energynautics, London, UK, 2013.
[17] International Energy Agency—Photovoltaic Power Systems Programme [WWW document], 2014. *www.iea-pvps.org*.
[18] B. Burger, R. Rüther, Inverter sizing of grid-connected photovoltaic systems in the light of local solar resource distribution characteristics and temperature, Sol. Energy 80 (2006) 32–45.
[19] B. Giesler, R. Haselhuhn, T. Komarek, Y. Krüll, S.A. Nisar, M. Zehner, Influence of limitation of the MPP-range from PV inverters, in: Proc. 27th European Photovoltaic Solar Energy Conference, Frankfurt, Germany, 2012, pp. 3770–3774.
[20] N. Enbar, T. Key, Addressing solar photovoltaic operations and maintenance challenges—a survey of current knowledge and practices (an EPRI white paper). EPRI, 2010.
[21] SolarPower Europe, O&M Best Practices Guidelines, Version 1.0, Brussels, Belgium, June 2016.
[22] J. Haney, A. Burstein, PV System Operations and Maintenance Fundamentals, Solar America Board for Codes and Standards, 2013.

Index

Note: Page numbers followed by *f* indicate figures, and *t* indicate tables.

D

FSC
www.fsc.org
MIX
Papier | Fördert gute Waldnutzung
FSC® C083411

Kolibri 360 GmbH
Ferdinand-Jühlke-Straße 7
99095 Erfurt, Deutschland
produktsicherheit@kolibri360.de